Urban Machinery

Inside Technology
edited by Wiebe E. Bijker, W. Bernard Carlson, and Trevor Pinch

For a list of books in the series, see page 341.

Urban Machinery

Inside Modern European Cities

edited by
Mikael Hård and Thomas J. Misa

The MIT Press
Cambridge, Massachusetts
London, England

First MIT Press paperback edition, 2010
© 2008 Massachusetts Institute of Technology

All rights reserved. No part of this book may be reproduced in any form by any electronic or mechanical means (including photocopying, recording, or information storage and retrieval) without permission in writing from the publisher.

For information about special quantity discounts, please email special_sales@mitpress.mit.edu

This book was set in Stone Serif and Stone Sans on 3B2 by Asco Typesetters, Hong Kong. Printed and bound in the United States of America.

Library of Congress Cataloging-in-Publication Data

Urban machinery : inside modern European cities / edited by Mikael Hård, Thomas J. Misa.
   p. cm. — (Inside technology series)
Includes bibliographical references and index.
ISBN 978-0-262-08369-0 (hc. : alk. paper)—978-0-262-51417-0 (pb. : alk. paper)
1. Technology—Europe—History. 2. City and town life—Europe—History.
I. Hård, Mikael. II. Misa, Thomas J.
T26.A1U73   2008
609.4—dc22   2007010333

10 9 8 7 6 5 4 3 2

Contents

Preface   vii

1   Modernizing European Cities: Technological Uniformity and Cultural Distinction   1
Mikael Hård and Thomas J. Misa

I   **Modernism and Mastery**

2   Taming the Rhine: Economic Connection and Urban Competition   23
Cornelis Disco

3   Arenas of Experimentation: Modernizing Istanbul in the Late Ottoman Empire   49
Noyan Dinçkal

4   Appropriating the International Style: Modernism in East and West   71
Thomas J. Misa

II   **Representation and Reform**

5   Spectacles of Sociability: European Cities as Sites of Consumption   99
Paolo Capuzzo

6   Progressive Dreams: The German City in Britain and the United States   121
Mikael Hård and Marcus Stippak

7   Constructing Dutch Streets: A Melting Pot of European Technologies   141
Hans Buiter

III   **Industry and Innovation**

8   Empowering European Cities: Gas and Electricity in the Urban Environment   165
Dieter Schott

9   In the Shadow of the Factory: Steel Towns in Postwar Eastern Europe   187
     Dagmara Jajeśniak-Quast

10   Science in the City: European Traditions and American Models   211
     Martina Heßler

IV   Planning and Power

11   Between Theory and Practice: Planning Socialist Cities in Hungary   233
     Pál Germuska

12   Mediators of Modernity: Planning Experts and the Making of the "Car-Friendly" City in Europe   257
     Per Lundin

13   Greening the City: Urban Environmentalism from Mumford to Malmö   281
     Andrew Jamison

References   299
Contributors   337
Series List   341
Index   343

Preface

This collection of essays investigates the making of European cities and modern urban life. In particular, it brings out the technological dimension of modern European cities. Drawing on wide-ranging examples, it illustrates the intriguing relationships between technological, economic, and social developments in cities and towns across Europe. The book investigates the most important changes in the urban environment across the past century and a half. To cover such a long time span and the whole of Europe is a daunting task. None of the contributors could possibly do so on his or her own.

This book is unique in that it includes studies from northern and southern Europe, West and East—and also in that it discusses the ways in which European cities were viewed from and shaped by interaction with the periphery (for example, Turkey) and the United States. Among our contributors are historians from Germany, the United States, Poland, Sweden, Italy, Hungary, Turkey, and the Netherlands. This is not a typical "conference volume." Since 2000 we have worked together in an international network supported by the European Science Foundation and the U.S. National Science Foundation.

This book is a collective enterprise. It is a joint effort by scholars who study technology, cities, society, culture, and economics from several disciplinary vantage points. Not surprisingly, then, our endeavor has been a lengthy process. The initial task we set ourselves was to examine the state of the art, to get a grasp on the rich body of literature on what we chose to call the history of urban technology. The outcome was two electronic publications, available at ⟨www.histech.nl/tensphase2/Publications/Working/Hard%20Biblio.pdf⟩ and ⟨www.umn.edu/~tmisa/urban/biblios/urban-machine-complete.pdf⟩. We refer readers who are interested in historiographical detail to these publications. Additional city images and teaching resources may be found at ⟨www.umn.edu/~tmisa/urban⟩.

Having taken stock of the existing literature, we set out to make our own contribution. The present volume depended on the generous financial support of two research councils and several other institutions. The European Science Foundation and the National Science Foundation co-funded a series of workshops the group held in

Darmstadt, York, Budapest, and Amsterdam between 2000 and 2004. We would also like to thank the Darmstadt University of Technology and the National Railway Museum in York, England, for their support.

We also need to thank those involved in preparing the book's manuscript. At TU Darmstadt, Iris Ohlrogge efficiently managed a steady flow of papers for the workshops and the book project, and Simon Bihr helped us set up the index. For assistance with editing we are indebted to Stephany Filimon and Ed Todd. Translations were done professionally by Peter McKenna (chapter 9) and Bernard Vowles (chapter 12). At the University of Minnesota, Katie Baumhover Charlet capably took charge of editing and printing the final manuscript, advising with a French translation, and compiling the integrated bibliography. Our long-suffering authors endured untold rounds of questions and revisions.

Last but not least, our thanks to Johan Schot and his colleagues at the Dutch Foundation for the History of Technology at the Eindhoven University of Technology. At the end of the 1990s Johan initiated a Europe-wide network, Tensions of Europe, to which our group belonged from the very beginning. Mikael Hård directed one of nine research themes, "Narratives on the European City." With the present book, we hope that our publications will stimulate further research on the borderlines between the histories of technology, cities, and Europe.

1 Modernizing European Cities: Technological Uniformity and Cultural Distinction

Mikael Hård and Thomas J. Misa

> Between Portugal and Poland, and maybe even further eastward, the national societies of Europe have never been more alike than today.
> Osterhammel (2002, 71)

Citizens, planners, and policymakers confront a world where Poland and Portugal, once separated by politics, economics, and culture, seem to be pushed forward to a common destiny by the unstoppable forces of modernity and global markets. Most of the world in fact appears to march down a similar boulevard. To comprehend why, we often incline to the popular notion that technology and capitalism are the driving forces, with the Internet and the global markets it sanctions being just the latest of the "machines that change the world" (Womack, Jones, and Roos 1990). Industrialization, modernization, globalization, and urbanization allegedly impose homogeneity on society. McDonald's, the archetype of a rigidly standardized consumer product no less than Henry Ford's Model T was in its time, is found in cities all over the globe. Skyscrapers, the preeminent symbol of modern financial capitalism, are going up in once-communist Shanghai. Will all corners of the world, not only the so-called global cities, be dragged into the maelstrom of global capitalism? How much room, if any, do we as citizens have to create alternative paths in this modern world?

Today, Europe is a test case that, if closely examined, can yield valuable insights about meeting the threat of homogenization head-on. Many citizens in Europe view the seemingly unbounded expansion of the European Union with anxiety, concerned that it will unleash standardizing forces that stamp out local practices and overrule traditional ways of life. When a French private company took over parts of the public transport system in Stockholm a few years ago, citizens in the Swedish capital were upset and nervous. Similarly, multinational utilities threaten to take the control of water and energy away from cities and regions. Yet not all the news is ominous. Perceptive observers of urban life can ask why Europe for a long time has been better than other parts of the world at tackling the problems that all large cities face. Despite the social conflicts that plague Parisian suburbs, Europe by and large seems to have

Figure 1.1
Diverse technical systems crammed into compact spaces typify modern cities. A Berlin tourist boat cruises up the Landwehr Canal through a mesh of automobile, pedestrian, rapid-transit, communication, and commercial networks—even a sprig of nature. Photographer: Thomas Misa.

avoided the appalling deterioration and unrestrained sprawl that one finds in too many American city centers and Third World megacities.

In this book we investigate the intersection of Europe, technology, and cities with an eye to these pressing concerns. Our focus is on the social forces, material structures, and cultural practices that created modern European cities in their great diversity. We hope our book gives readers insight into the prospects for Europe, the possibilities of cities, and the future potentials of technology. A history of European cities is filled with cases that might serve as positive models or warning examples for our time.

Cities, with their extreme agglomeration of people and unprecedented density of technology, have been at the core of modern history and modern society. It is no accident that the "founding fathers" of social theory took cities as their paradigm. "This isolation of the individual—...everywhere the fundamental principle of modern society[—]...has been pushed to its furthest limits in London," as Friedrich Engels (1845/1987, 31) put it. Émile Durkheim, Karl Marx, and Max Weber, in various ways, also situated the characteristically modern life in cities. In his essay of 1903, "The Metropolis and Mental Life," Georg Simmel memorably described a new type of human being capable of withstanding the disorienting sensory overload of the modern city. The fast-paced rhythms of life in London, Budapest, Berlin, Vienna, St. Petersburg, and Prague inspired two generations of modernists, and worried many traditionalists (Jazbinsek 2003; Misa, this volume). With good reason, such writers and artists as Charles Dickens, Lewis Mumford, and Fritz Lang drew attention to the inhumane conditions in the city. In an age of unbridled industrialism and rampant urbanization, cities were the cutting edge of a new modern world, the place where people first experienced an environment that was meant to turn them into rational, disciplined, and alienated individuals—in factories, hospitals, offices, department stores, and huge residential areas. Streets that once were filled with the sounds of children playing were turned into traffic arteries, and pedestrians were literally forced onto the newly created "sidewalk" (Buiter, this volume). Cities were also where diverse masses of people congregated for employment and enjoyment. Throughout the twentieth century, Europe's cities were sites and symbols of an often painful history: the arc from the Fascist march on Rome to the internationalist treaty of Rome marks one central strand, while the Berlin airlift and the fall of the Berlin Wall are bookends for the Cold War. Today, the treaties of Rome, Maastricht, and Bologna are recognized signposts for European integration, whereas Brussels and Strasbourg are contested symbols of a new power apparatus.

Jürgen Osterhammel's observation in our epigraph about a homogenized Europe rings true, perhaps even more so for cities than for "national societies" as a whole. Skyscrapers, subways, parking garages, department stores, and suburbs everywhere look more or less the same, while once-pronounced national and regional differences in lifestyles, clothing, and food can be difficult to discern. Walk past the buzzing electronics shops, fast-food storefronts, and glass-enclosed office buildings in the centers of Rotterdam, Paris, or Turin, and it will be difficult to tell just where you are. Globalization reinforces these homogenizing forces. Wiring cities for the global economy, with the requisite business parks, tourist hotels, airport facilities, transport systems, and communications networks, obviously lays down a common material and institutional network. The "global city" is the epitome of these developments (Sassen 1991).

We hardly recognize that the technical infrastructures underpinning these integrating and homogenizing processes are many decades old, in some cases more than a century. Long before the political agreements that founded the European Economic Community (EEC) in 1957, Europe's cities were tied together by border-spanning rivers, railways, and motorways, as well as broadcasting and communications networks. The resulting flows of people, ideas, networks, and consumer goods created a "hidden integration" that prepared the groundwork for formal political integration and whose consequences are still felt today. And if its purely political foundations are no longer secure, the future of Europe may rest on the dynamics of these technological networks, spanning this continent and connecting it to the rest of the world (Misa and Schot 2005).

Cities have long been critical sites for exercising citizenship and configuring state power (Ribhegge 2003; Blockmans 2003). "Stadtluft macht frei" (city air makes one free) proclaimed the city gates of the Hanseatic League of trading cities. Our volume documents the emergence of material, institutional, professional, and political networks, and transnational governance strategies that evolved from the 1850s onward between European cities and across European nation-states. For instance, cities along the Rhine formed international commissions to manage the river for the competing goals of clean water, effective transport, and cheap waste disposal (Disco, this volume). Similarly, continent-wide standards and practices in urban planning, tourism, consumption, and even sustainable development led to informal networks and formal institutions that underpinned and shaped European integration in recent decades.

A crucial question remains whether we have "room for maneuver" in the face of these homogenizing and universalizing pressures. Will these pervasive networks simply make life everywhere the same, or can we facilitate diversity by creatively designing them—or even perhaps by subversively using them? Again, there are crucial lessons to be learned from European cities. Homogenization, in its many forms, has always been contested, and not only in such dramatic protests that regularly accompany the high-profile summits of economic globalizers. In Europe there are many instances when actors—city officials, engineers, planners, even ordinary citizens—have managed to make universalizing forces more appropriate to their domestic traditions and customs (Hommels 2005b; Jajeśniak-Quast, this volume). What have been their successful strategies, and what can we learn from their failed attempts? After all, innumerable local and regional peculiarities have persisted across Europe. Venture beyond Frankfurt's banking district or Stockholm's pedestrian area, and you can easily tell whether you are in Germany or Sweden. Exit the Amsterdam metro, cross over a seventeenth-century canal lovingly reconstructed with steel girders and modern cranes, and see that you of course are not in Greece. Avoid the hamburger bar or pizzeria and instead look for a local café in any smaller European town, and marvel at the menu of regional specialties. Clearly, many cities have understood the potentials of historical and cultural specificities and aggressively market these to an international audience: Kraków, Granada, and Venice come to mind. Perhaps modern society leaves us more opportunities than a menu of doom-and-gloom homogenization and globalization suggests, but if they are to survive, we must actively seek them out. We suggest that a portfolio of European cities, diverse as they are, is the perfect place to start looking.

Figure 1.2
A struggle with water and space has defined Amsterdam from the seventeenth century through today. Canals were even filled in to create space for housing and transport—until their tourist potential proved irresistible. Photographer: Thomas Misa.

The Making of the European City

We take as a point of departure Osterhammel's observation that the nations of Europe are more similar today than they have ever been. Instead of writing an *histoire totale* of European urban modernization, we frame our history with the problem of similarity and difference. Our book examines European cities to consider the menace of creeping homogeneity as well as creative responses and critical reactions. While the international circulation of experts, ideas, systems, and artifacts generated homogeneity on several levels, as we will show, European cities have retained, or at least reconstructed, much of their historically specific character. Our volume asks what this experience can teach us (cf. Bruijn and Norberg-Bohm 2005). It does so by highlighting such fields as urban energy systems, architecture, city planning, traffic engineering, water management, tourism, and consumption. We show that although urban technologies have certainly changed the face of European cities, individual citizens and officials and planners have modified and sometimes even rejected such changes. After World War II in particular, European citizens faced the challenge of reconstructing the old structures or turning their cities into authentic modernist utopias.

Modern urban life is created by and depends on technological systems, urban planning, and an array of artifacts, buildings, networks, and structures. Throughout history, cities everywhere have been massive consumers of energy, producers of wealth, sites of worrisome mortality, and generators of noxious pollution. They have always depended on transport and communication networks to maintain economic dominance and political control over their hinterlands. Europe's cities have many distinct and layered histories. They grew or contracted in the ancient, medieval, Renaissance, baroque, and modern eras, but in each phase the interaction of social and cultural processes with human-made materiality defined how and where people lived, worked, moved about, recreated, and participated in civic culture. In turn, the human-made materialities we understand today as "technology" have largely been urban creations designed with the needs of the urban population and urban commerce in mind. This has been true from yesterday's pioneering gasworks, arcades, department stores, and streetcar lines to today's efforts to create environmentally sustainable residential areas. Throughout the modern period, city officials have invested tremendous effort in keeping chaos at bay and making European cities livable. Weaving together these three strands—Europe, technology, and cities—is our volume's task.

European cities provide us with windows on the role of technology in the making of Europe. Compared to cities on other continents, European cities have been better at keeping their distinct character. No perceptive observer would confuse the characteristically European panorama one sees out the airplane window on approaching Barcelona or Budapest or Berlin with that of Boston, often described as the most European-looking city in the United States. Madrid's suburban sprawl looks nothing like Chicago's. You can trace this distinctiveness on several levels—in the debates on Europe, the historical literature on Europe and its cities, and the ongoing efforts to define the boundaries of Europe.

Today's political and cultural debate about "Europe" is filled with rival conceptions about what it means to be European, what characteristics are common to European countries, and what institutions are best suited to keep Europe together or keep it from falling apart. These conceptions are all loaded with the burden of history. Promoters of an integrated Europe invent or reinvent historical traditions to create a common European identity, while critics point out that surprisingly few historical elements really bind this continent together (Shore 2000). Given the stakes, it is particularly crucial to investigate the historical narratives utilized by actors in the past and present, and as they may be used in the future (Eckstein and Throgmorton 2003).

The literature on "Europe" as a historical construct has exploded in recent years, while urban historians have scrutinized the concept of a "European city" (Schmale 2000; Hassenpflug 2000). Even though technology is typically viewed as a modernizing and integrating force, efforts to understand its role in the project of European integration have just begun (Faulhaber and Tamburini 1991; Zysman and Schwartz 1998; Horrocks et al. 2000). Intriguingly, political scientists and geographers have recently "discovered" cities as a strategic, subnational space for research and analysis. Neil Brenner (2004) and Hubert Heinelt and Daniel Kübler (2005) argue that contemporary European cities are crucial sites where state power is being reconfigured and transformed. Our volume is the collective effort of historians wishing to contribute to this literature by presenting relevant material from the last century and a half. It tries to reveal the *inside of modern European cities*.

We take it as an axiom that the boundaries of "Europe" are continually in flux. The hot wars of the twentieth century ravaged the continent, while the ensuing Cold War brutally cut it in two. Political divisions of the Cold War into East and West can also be understood through segmented or divergent energy and transport networks, industrial production models, and consumption patterns (Strasser et al. 1998). More recently, the European Union expanded to embrace not only Portugal and Poland but also Slovenia and Slovakia, and it may someday stretch to include Turkey. Clearly there is no fixed political definition of Europe. For centuries Europe has also had clear links with its overseas colonies and with North America. The circulation of municipal engineers, urban planners, and city managers in fact spanned the whole globe. Colonial cities became laboratories of modernity in which engineers and planners often had much more freedom of action than back home (Arnold 2005), while the incomplete efforts to assimilate the colonies and colonial peoples into European society has left a volatile legacy in cities such as Marseille, London, and Rotterdam.

The recent attempts to bring Turkey closer to Europe, while appearing unprecedented, are in fact more than a century old. When Ottoman Empire officials and leading citizens initiated wide-ranging urban renewal and sanitary reforms in the late nineteenth century, they explicitly wanted to "modernize" Istanbul following the metropolises Berlin, London, and Paris. Germany and Britain offered technological models for fresh running water, while France suggested public health measures (Dinçkal, this volume). The Ottoman elites' promise that new technologies would bring about a "modern" way of life was repeated in many languages and many places. The compelling if ill-defined notion of modernity led many cities to install hygienic water systems during the second half of the nineteenth century, as well as to impose automobile-friendly traffic schemes after 1950 (see Lundin, this volume). While individual technological projects were often contested because of their high social and cultural costs, the overall project of modernity was seldom questioned (Rohkrämer 1999).

Cities are also windows on the making of the modern world. Modernists in literature and art gave cities personality and a voice. The Italian futurists reveled in "the frenetic life of our great cities," painted kinetic images such as *The City Also Rises*, and wrote urban homages such as "Il canto della città di Mannheim." As painter-theorist Piet

Mondrian phrased it, "The genuinely Modern artist sees the metropolis as Abstract living converted into form; it is nearer to him than nature, and is more likely to stir in him the sense of beauty" (Mondrian quoted in Banham 1960, 152). Poet Walt Merin even wrote staccato-sounding verses in the 1920s to represent the mechanical character of modern, urban life—an aspect of what we term, for short, *urban machinery*. And, not least, modernist planners and architects had their own pronounced impact on cities (Misa, this volume).

Cities were everywhere the primary object of the modernizers' dreams. Indeed, cities deployed select technologies to secure a "modern" reputation. Paris, already with a well-developed suburban rail and electrified tram network, built the underground Métropolitain railway as a showcase project for the 1900 world's fair. Known simply as the Métro, it was archetypically modern and characteristically French: fast and clean, rational and technically advanced, its entrances designed by leading art nouveau figures. Like the contemporaneous buildings designed by the Austrian architect Josef Hoffmann, the Métro was conceived as a "piece of total art" (*Gesamtkunstwerk*). Other European capitals, most famously Moscow, followed its lead. Interestingly, *Webster's* defines *metro* as "a European subway."

The coming of subways obviously restructured patterns of life and work, and this volume tries to understand the relation between a number of urban technologies and urban life. The city is a purpose-built environment where technologies to some extent shape human behavior and affect human well-being. Streets and bicycle lanes, residential areas and shopping malls, the barriers that mark off where you can and cannot walk—these urban artifacts are *dispositifs* that create what Michel Foucault (1975/1977) called a "spatial order." This order is also upheld by a number of institutions and organizations. We maintain that a proper understanding of cities requires acknowledging the profound interdependencies of technological systems with the multiple levels of urban life—*everyday-practical*, *institutional*, and *discursive*. Over the past century and a half, city dwellers were effectively habituated to use water closets, gas or electric stoves, streetcars and automobiles, highways and sidewalks. These technologies were once new, but they have now become so common and pervasive in everyday life that they mostly inhabit only the background of our consciousness. They are now "second nature," an artificial structure that appears entirely natural. Daily life in cities completely depends on the technological systems that provide water and energy, remove sewage and trash, deliver information, and transport us between homes and workplaces. Unless they fail, we hardly notice them (Edwards 2003).

These technologies did not fall from the sky. Their funding and maintenance depended upon such institutions as local utility boards, city governments, regional energy providers, and, more recently, multinational corporations. Urban technological systems are creations of finance, regulation, and the prevailing political powers. In the nineteenth century, private companies erected energy, water, and even public trans-

Figure 1.3
The "spatial order" of cities simplified urban activities and disciplined urban residents. Here along the famous Andrássy boulevard in Budapest, there are defined spaces for shopping, walking, biking, and driving automobiles. Photographer: Thomas Misa.

port networks for profit, but fairly soon municipalities expanded their authority beyond overseeing and regulating private business. Under the banner of "municipal socialism," city authorities founded public utilities to guarantee that services were provided equitably and reliably (Hård and Stippak, this volume). The institution of municipal ownership deserves a place alongside insurance and health care schemes as a fundamental building block of the twentieth-century European welfare state.

The third, discursive level needed to properly understand modern urban technologies becomes clear if we recall that few of them were anything more than hazy dreams in the mid-1800s and that none of them was available to the common person. Even the "need" for ample supplies of running water and correspondingly adequate sewage systems—taken for granted in European cities today—was by no means self-evident.

City officials, private entrepreneurs, and visionary promoters all mounted campaigns to persuade urban residents to adopt these technological luxuries as "necessities." The introduction of urban technologies was always contested and accompanied by rhetorical work. Changes in discourse—the pervasive "ways of thinking" that structure what we know and in what terms we can know, involving language, argument, images, and symbols—are just as important as changes in institutions and daily practices.

Circulation and Appropriation

Circulation is a paradigm of modern urban life. In *Horse-Drawn Cabs and Omnibuses in Paris*, Nicholas Papayanis (1996) documents that the "circulation" of goods and people pervaded Paris as early as the eighteenth century. Within the city, there was the movement of vehicles and people, books and information, capital and labor, resources and waste. The modern nation-state took form when cities were interconnected and the urban flows became national flows (Vleuten and Kaijser 2005). Urban circulation and the exercise of state power were at issue in Paris in the wake of the 1848 street barricades, when Napoleon III sanctioned the plans of Georges Haussmann for reconstructing the old medieval city by forcibly building a matrix of boulevards through it. "Napoleon and Haussmann envisioned the new roads as arteries in an urban circulatory system," in the words of Marshall Berman (1982, 150–151):

The new construction wrecked hundreds of buildings, displaced uncounted thousands of people, destroyed whole neighborhoods that had lived for centuries. But it opened up the whole of the city, for the first time in its history, to all its inhabitants. Now, at last, it was possible to move not only within neighborhoods, but through them. Now, after centuries of life as a cluster of isolated cells, Paris was becoming a unified physical and human space.

In time, such grand boulevards with their homogenized and unified space would typify modern cities across the continent and around the world (see figures 3.4, 6.2, 7.3).

In many diverse fields, the interurban circulation of people, artifacts, and knowledge resulted in similar solutions appearing throughout Europe. British cities were models for water supply systems on the continent; a British company even built Berlin's first centralized water supply system. Spanish cities often chose French technologies. German models were especially strong in electrification and transportation. Europe's 220-volt electricity standard originated in Berlin and, in time, spread across the continent and eventually around much of the world (Schott and Buiter, this volume).

The successful spread of the Paris Métro and the Berlin 220-volt electricity standard illustrate how circulation brings about uniformity. Our chapters demonstrate how the concept of circulation can account for the manifest similarities across Europe in energy, transport, and road systems, urban planning concepts, traffic engineering, and a host of technical standards. *Circulation was an engine of homogenization.* Circulation was

Figure 1.4
Modern boulevards "opened up" the medieval city to the circulation of people, goods, and ideas. Here the classic homogenized, open modern street in Berlin (with a view into Berlin's distinctive *Mietskasernen*, or "tenement barracks"; see chapter 6). Photographer: Thomas Misa.

made possible by and through formal and informal study tours, international congresses and societies, and the emergence of professional journals in such fields as engineering, planning, public health, and architecture. It was not only progressive social reformers who constituted the "Atlantic crossings" documented by Daniel Rodgers (1998); countless engineers, architects, medical doctors, city planners, and civil servants traveled between North America and Europe looking for pioneering models, sympathetic colleagues, and evidence to influence developments back home. After World War II many European experts spent time in the United States on Fulbright grants, and after returning home they often actively "Americanized" Europe's roads, factories, and airports (Lundin, this volume). The Marshall Plan also contributed to this process (Kipping and Bjarnar 1998; Zeitlin and Herrigel 2000).

Circulation, then, understood as a dynamic constellation of institutions and forces, helps us to comprehend Osterhammel's observation that "the national societies of Europe have never been more alike than today." There is, however, another side of the coin. When technologies are introduced into a new setting, they are often substantially modified and even given new meanings. When water closets were introduced into cities on the European continent, they were considered to be British, a meaning they did not have in the United Kingdom; the English "WC" still persists in many languages. To function, technologies have to be domesticated into routines of daily life, incorporated into existing institutional arrangements, and assimilated into prevailing cognitive and linguistic structures (Hård and Jamison 2005). In short, actors must *appropriate* them.

Indeed, significant modifications in technologies and practices are the common outcome of local domestication processes. Apparently universal "modernistic" images, ideologies, and practices were in circulation during the first four decades of the twentieth century, but as Hård and Jamison (1998) show, countries in Europe appropriated them in distinct and culturally specific ways. Postwar reconstruction was molded by international funding and international models, which sometimes amounted to the same thing, but also by institutions and laws that differed from country to country. Just think about the different visions that guided the postwar cities in West Germany and East Germany: car-friendliness in the West, monumentality in the East. The contributors to this volume adopt the concept of appropriation as a conceptual tool for understanding the manifold *differences* that exist among European cities, as well as between them and other cities of the world.

The task of this volume is to investigate the place of modern technology in the making and reshaping of European cities. To do so, we emphasize the international *circulation* of technological knowledge, artifacts, services, and people. In parallel we highlight the *appropriation* processes that have modified the more or less globally available technologies. The two concepts enable us to understand the tension between homogenization and differentiation. Attention to circulation gives a historically meaningful account of technological convergence: namely, that technologies, to a great extent, appear to be "the same" everywhere you look—at least on the surface of things. And likewise, appropriation helps reformulate the social shaping of technology in historically grounded ways when looking inside individual cites. These concepts frame the stories that we tell.

Technological Systems and the Urban Matrix

A deeper exchange between technology studies and urban studies has been a long time in the making and yet still seems overdue (Hommels 2005a, 2005b).[1] Areas of overlapping concern include transport, water, waste, energy, communication, and other infrastructures and systems. Other compelling topics are urban planning and public health, architecture and housing, sites of production and consumption, and spaces of entertainment and sociability. The social histories of sanitation, urban medicine, and public health are also promising research sites (see, e.g., Luckin 1986; Reid 1991; Hamlin

Modernizing European Cities

1990, 1998). Richard Evans's model study *Death in Hamburg* (1987), for example, examined the relationship between public health and urban reform. Urban environmental history has positively boomed in recent years (Melosi 1993; Stine and Tarr 1998; Bernhardt 2001; Rose 2004), while historians of science have also developed an increasing interest in urban sites and spaces (cf. *Osiris*, vol. 18 [2003] on "Science and the City," and Heßler, this volume). In gender studies the city appears as a complex field, with diverse emancipatory and oppressive potentials (Wilson 1991; Green 1997; Bijker and Bijsterfeld 2000; Capuzzo, this volume). Even intellectual historians such as Andrew Lees (1985) anatomized the debates accompanying the spread of urban structures and urban lifestyles.

Figure 1.5
Sanitary and hygienic reformers at the turn of the century cleaned up city streets by deploying diverse technologies. One of Amsterdam's numerous outdoor street urinals, still keeping the canals clean. Photographer: Thomas Misa.

Recently urban historians have opened the "black box" of urban technologies. The movement began with special theme issues of the *Journal of Urban History* (in 1979, 1987, 2004) as well as selected articles in *Technology and Culture*.[2] Joseph Konvitz, Mark Rose, and Joel Tarr (1990) reviewed several different approaches to understanding technology and the city, emphasizing how public and private choices shaped urban and technological developments. Since the founding in 1989 of the European Association of Urban Historians and its biennial international conference, urban history—often foregrounding technological matters—has also flourished in Europe (Funck and Reif 2003). *Technology and the Rise of the Networked City in Europe and America*, edited by Tarr and Gabriel Dupuy (1988), focused on case studies of transport, water, waste, energy, and communication systems and how these systems shaped modern cities. As Julie Johnson-McGrath (1997) does in her admirable survey of the twofold "shape and shaping of urban technology," we conceptualize urban developments as complex, multifaceted processes that cannot be reduced to unidirectional impacts or constructions.

Many historians of technology have treated urban infrastructures, even though they have seldom problematized the city in its own right. Their interest in comprehending the evolving form of technological systems and networks led them naturally to focus on sites characterized by high concentrations of population, challenging spatial problems, and intellectual and financial stimuli for innovation—that is to say, cities. The histories of electricity, transit, water, sewage, roads, and building construction are to a large extent the histories of urban technologies. In his wide-ranging history of electrification, Thomas Hughes (1983) included detailed comparative analysis of Berlin, Chicago, and London. Some historians picked up Hughes's concept of "large technological system" and applied it to the development of city-based gasworks, electric power grids, and other energy networks (e.g., Schott 1999c), while others took up the idea that the diffusion of technological structures is dependent on local or regional particularities, leading to the unfolding of various "technological styles." Historians of technology have energetically debunked the "impact" that these systems supposedly had on cities, spotlighting instead the myriad social and cultural influences that affected the *urban machinery* in different settings. For example, instead of stressing the impact of telegraphs on cities, technology historian Paul Israel (1992) traced the influence of urban information markets on the telegraph industry.

Urban historians and historians of technology thus have a lot in common. The early works of Joel Tarr (1979) and Martin Melosi (1980, 1981) examined urban environmental problems that later found favor in the history of technology. Carl Condit's (1977, 1980–1981) comprehensive works on railroads and harbors in urban settings were warmly received. Tarr and Dupuy's edited volume, noted above, was unique in that it treated extensively both European and American developments. Tarr's award-winning *Search for the Ultimate Sink* (1996) reached a wide audience, and Melosi's *Sani-*

tary City (2000) was even awarded a prize by the Society for the History of Technology (SHOT). Another sign of intellectual convergence was the Open University Press's four edited volumes on the history of "Cities & Technology" in Europe and the United States (Goodman 1999; Goodman and Chant 1999; Roberts 1999; Roberts and Steadman 1999).

This volume follows in the traditions sketched out above, with an emphasis on exploring the cultural or linguistic "turn" in both urban and technological history (Gilfoyle 1998; Ladd 1997; Norton 2007). We write contextually rich histories that combine the technical with the social, the economic with the cultural, and the discursive with the spatial. We show how social, political, economic and cultural factors shape the dynamics of technological developments, and we try to find an adequate conceptual language for dealing with the technological shaping of cities. We demonstrate that modern European cities and their technological systems evolved in distinct ways. European cities were surprisingly open to ideas about urban modernization, and governments and citizens were often willing to adopt new technologies and transform them into public utilities. "In what was effectively a pan-European discourse," as Driver and Gilbert (1999, 9) write, "national models were implicitly and explicitly defined in relation to other national models, in a spirit of competition as much as emulation." The circulation of ideas through publications, conferences, exhibitions, personal visits, and multiple networks created "a kind of European market in urban ideas, strategies and models."

To understand the phenomena of uniformity and distinction, we adopt a perspective that highlights conflict and negotiation, power and control, inclusion and exclusion. From the very beginning the first urban gasworks and electric power plants were objects of and sometime subjects in political fights, economic deliberations, social differentiation, and cultural reinterpretations (Gugerli 1996). The contributors to this book foreground the interactions between humans and their surroundings and bring out the "co-construction" of technological structures and urban life. In doing so, we go "inside" the city and try to uncover its hidden technological structures and reveal its character as second nature. Instead of relying on older terms like "impact" and "influence," which have many troubling mechanical associations and do not truly capture the fluidity of history, we focus on processes of circulation and appropriation.

Our book has been inspired by cultural studies and cultural history. Accordingly, we treat seriously such topics as meaning and discourse, representation and perception. Machines and structures such as shopping arcades, department stores, high-rises, streetcars, and street lights do more than move people around and keep shoppers out of the rain or dark; they also embody values, carry meanings, and transport norms (Reid 1991; Nye 1997; Brooks 1997). Even such unlikely structures as water and sewage systems were icons of modernity a century ago. It is important, however, not to forget that the symbolic meanings of institutions and buildings are never stable. At first, contemporary critics denounced the Eiffel Tower (1889) as ugly, useless, monstrous, a latter-day Tower of Babel, a "disgraceful skeleton," a "gigantic factory chimney," "the disgrace of Paris." Only in the early twentieth century did modernists fully embrace it as a positive symbol of modernity and French grandeur (Thompson 2000). Similarly, for decades political controversy dogged the sewer system installed in Berlin in the second half of the nineteenth century. While some observers regarded it as progressive and environmentally sound, others criticized it for being expensive and unhealthy. *Urban technologies are often highly contested.*

We also stress the social side of urban technology. The Haussmann clearances of central Paris pleased the officials charged with keeping order and the middle classes who could enjoy the newly opened urban circulation patterns, whereas they had dramatically negative consequences for poor tenants and small shopkeepers. The transition to mass commercialism that is characteristic of modern European cities began more than a century ago (Capuzzo, this volume). And while reformers focusing on sanitation and public housing might have had benevolent intentions, the reforms they instituted often increased the gap between the haves and have-nots. Urban sociotechnical structures frequently contribute to the separation, even segregation, of social groups and to the control and streamlining of behavior (Ben-Joseph 2005). *Urban engineering is always social engineering.*

Topics and Themes

Our chapters show that modern European cities took form amid an international flow of people and ideas. In his chapter, Hans Buiter maps the international character of urban engineering, paying close attention to how foreign knowledge about water, sewage, traffic, and street design was appropriated in the Netherlands and transformed into local practices. Dutch engineers traveled to Germany, Hungary, and Britain to identify technological solutions that would fit local Dutch conditions—not to find the "one best way" of making streets. Dutch streets embodied this selectively international background. Similarly, Mikael Hård and Marcus Stippak show how American reformers during the Progressive Era looked for inspiration to German and British cities. Urban engineers, reformers, and city officials kept current with transatlantic developments by scrutinizing an international literature, attending conferences, taking study trips, and inviting colleagues from abroad to make reciprocal visits. Multicentered flows of people and ideas also are prominent in Thomas Misa's chapter on modernism in Europe. He maintains that modernism in architecture and planning took form in a specifically European context. The version of modernism that focused on workers' housing, rational site development, and functionalist urban planning in the 1920s and 1930s responded directly to the severe problems of European cities, but did not resonate in the United States. After his proposals repeatedly fell on deaf ears during a much-anticipated 1935 visit to the United States, Le Corbusier despairingly labeled America "the land of the timid" (M. Bacon 2001).

Modernism's faith in technological solutions contributed to uniformity across the European continent and beyond. Especially after the Second World War, modernist planners and architects stamped a certain look on residential quarters, shopping districts, and office buildings nearly everywhere. The champions of this ideology, once claiming the label of a universalistic International Style, too often generated oppressively homogeneous results. As with many machines that allegedly "change the world," however, modernism claimed a victory it only partially deserved. Misa traces the fate of modernist concepts in the Netherlands and Czechoslovakia, where rival schools lent modernism distinct inflections.

Homogeneity thus seems to be contingent and precarious, especially in Europe where there are strong local traditions and resilient local institutions. Schemes of top-down control are frequently partial and piecemeal. Le Corbusier's comprehensive plans for reconstructing the capital cities of Paris and Moscow went nowhere—while his planning ideals were most fully embodied in Chandigarh, Brasília, and public housing complexes in the United States and United Kingdom. Dagmara Jajeśniak-Quast shows that the Soviet model of Magnitogorsk—itself derived from Gary, Indiana—simply could not serve as a practical blueprint. Steel cities in East Germany, Poland, and Czechoslovakia took form, a bit like Buiter's streets and Misa's modernism, in an interplay of international circulation and local appropriation. Pál Germuska also analyzes the uneven character of top-down city planning in eleven Hungarian cities. Local geography and even local political traditions significantly modified the Soviet-inspired models. Despite the extreme centralizing forces, Germuska also discerns a multilayered appropriation process.

Opposition to large schemes was particularly vocal in the decades after 1968. For politicians and urban experts the change came rather suddenly. For a generation, they had reconfigured many European cities in the name of progress, modernity, and mobility, but now such efforts faced serious criticism. In Stockholm, postwar urban planners reshaped part of the city explicitly to fit the demands of the automobile. As Per Lundin shows in his chapter, their role model was the American city with its ring roads, oversize parking garages, and automobile-dependent suburbs, all of which promoted American values and an American way of life. After having embraced the American-style car-friendly city in the 1950s, many Europeans began to realize that the automobile threatened the European urban fabric. As the old saying goes, "They got what they wanted, but lost what they had."

Active and effective criticism of misguided urban planning schemes and architectural adventures followed Europe's encounter with the modernist city. While too many New Yorkers passively accepted the infamous Cross Bronx Expressway, literally paving the way for devastating urban blight, residents of Utrecht successfully mobilized against a similar scheme to create a car-friendly city that would have paved over their city's canals. As Anique Hommels (2005b) makes clear, the Dutch urban system tenaciously resists changes imposed from outside. When the citizens of Munich recently banned the erection of skyscrapers, they did so in a conscious effort to preserve the historically defined character of their city. Meanwhile, citizens in Malmö, Sweden, adopted a much more ambivalent attitude to modernist Santiago Calatrava's design for a skyscraper. In his chapter, Andrew Jamison brings out the tensions in this project. Skyscrapers are signature icons of modernist architecture, and Calatrava's building is, of course, meant to show that Malmö has transcended its industrial past and become a thoroughly modern city. In fact, the project was originally to be an environmental exemplar, but Jamison argues that a sustainable skyscraper is a sad contradiction in terms.

Urban technologies often constitute a rich iconography. Calatrava's skyscraper in some way stands for Malmö, and in Garching, north of Munich, the town's central symbol is an atomic research reactor. Martina Heßler tells the story of this "atomic egg" and shows how it became emblematic for this village-turned-research-center. Her chapter highlights the synergies between research politics and urban planning ideals. When local politicians in the 1950s were asked to turn farmland and meadows into a high-tech R&D area, they readily accepted, and only a decade later the egg-shaped reactor had become an element in the official town crest. Ironically, however, the research center itself was hardly integrated in the rest of the village, and the knowledge workers remained strangers. The modernist planning principle of functional separation was just as inappropriate in Garching as elsewhere on the globe. Once the research center was built, it also proved hard to change Garching. When the modernist model of separated urban functions had lost its dominance in the 1980s and Silicon Valley had become a model for politicians and planners of science the world over, local politicians instead tried to turn Garching into a Bavarian version of Palo Alto.

Garching's policy has to be seen as an attempt to attract scarce resources in a knowledge economy. Due to Europe's distinctive geographic density, competition between cities has had a long history. No other continent has so many sizable cities, each with substantial legal independence, crammed into such a compact space. Malmö politicians launched their modernizing campaign expressly to compete with Copenhagen, situated just across the Öresund strait. Competition between cities along the river Rhine also has a long history. In his chapter, Cornelis Disco discusses how French, German, and Swiss cities in the river basin have competed (and sometimes cooperated) over the centuries. He shows that the Rhine as a modern artery of transport connecting Switzerland with the North Sea is a creation of humans and not a fact of nature. Owing to heavy investments in various hydraulic engineering projects, the circulation of goods and people increased between such cities as Mannheim, Strasbourg, and Basel, but so did competition between them. Disco shows that the creation of the Rhine as an integrated international space was hardly a smooth process; there were many political and jurisdictional battles (not to mention several wars). Yet, again, Europeans developed intercity and international institutions to manage these conflicts.

The emergence of the Rhine as a continental artery for goods went hand in hand with its discovery as a prime area for tourism. The first steamships brought tourists to the famous Loreley cliff in the 1820s. As Paolo Capuzzo shows, travelers began to turn certain parts of Europe into touristic sites. For northern Europeans Rome was already an established goal, but soon select other cities joined the required itinerary for a Grand Tour. Tourism also created such entirely new cities as Brighton on the English coast as well as Cannes and Nice on the Mediterranean. Applying Guy Debord's intriguing notion of "spectacle," Capuzzo shows how these seaside resorts became places of consumption in their own right. The construction of railroad—and later, automobile and airplane—connections was necessary. In order to attract visitors, the municipal authorities consciously built modern urban infrastructures while marketing their city as a place for relaxation.

Marketing may serve as an inroad to the ways in which urban technologies have been given meaning, and how such meanings have enabled actors to create what Pierre Bourdieu (1979/1984) once called "distinction." Mary Blume's book on the Côte d'Azur, subtitled *Inventing the French Riviera* (1992), shows the emergence of tourist cities as an inventive act that involves the design of technological structures as well as the cognitive definition or redefinition of towns and places. This cultural perspective can be found in all our chapters; tourism is hardly unique. For instance, Dieter Schott emphasizes how the emerging electricity industry used the 1891 Frankfurt International Electricity Exhibition to bring its products to the attention of city officials and urban residents. Here we also find elements of "spectacle": electrification during the early decades focused on up-market shops, restaurants, and theaters in the fashionable districts.

In their search for relaxation, Europe's holiday makers fled the urban machinery they might have found restrictive or even overwhelming at home. Increasingly, however, they found themselves in tourist areas that were as thoroughly modernized and mechanized as the cities they sought to escape. With time, the wish to escape turned into a desire to relax while having standards of living like those back home. In this way the pervasive circulation of people around Europe—tourists, engineers, city officials, planners, and experts—contributed to the establishment of homogeneity and common standards across the continent.

We can now understand why "the national societies of Europe have never been more alike than today." The creation of urban homogeneity in Europe is an outcome of wishes and visions on the part of the population at large as well as the result of the hegemonic ambitions of modernism. We hope that our book will help citizens, policymakers, and professionals gain insight into how Europe for the most part avoided the "universal urban homogenization" believed by some to be the inescapable destiny of cities worldwide (e.g., Goodman and Chant 1999, 353). In the twenty-first century we will need to reinvent institutions and mechanisms in order to keep our cities livable.

Notes

1. Citations to the sizable literature in urban studies and technology studies can be found in Hård (2001) and Hård and Misa (2003).

2. The special theme issues of the *Journal of Urban History* included "The City and Technology" 5 (May 1979); "The City and Technology" 14 (November 1987); "Technology, Politics, and the Structuring of the City" 30 (July 2004).

I Modernism and Mastery

2 Taming the Rhine: Economic Connection and Urban Competition

Cornelis Disco

Cities have an affinity for rivers. Easy access to fresh water, food, and energy, as well as facile transportation, waste disposal, and defensibility are some obvious reasons. These advantages also explain why river settlements had a better-than-average chance of developing into sizable towns and even cities. Especially in Europe it is hard to find a city of any size that is not located either on the sea or very near a substantial river, while "dry" villages number in the thousands.

As the river's proximity fostered a settlement's growth, it also *shaped* it. The river was first and foremost a prominent feature of the local geography and economy. Waterfronts, harbors, and an opposite shore were decisive features of the morphologies of all river settlements. Moreover, the riverside location stimulated the development of a variety of riverine trades like fishing, haulage services, and shipbuilding. These in turn encouraged secondary enterprises like fish markets, credit banks, warehouses, and ship's chandlers. The river added extra dynamism to the local economy and fostered a culture with better than average tolerance for heterogeneous practices and persons.

But not all rivers are equal. Some rivers are central arteries of large valley systems reaching many hundreds of kilometers inland from the sea. Among European rivers, the Rhine and the Danube are certainly preeminent, though others like the Elbe, the Weser, and the Volga are important national and regional streams. Settlements located on such grand thoroughfares experienced not only profound local effects from the river's proximity but also cosmopolitan effects due to the relative accessibility (by water, roads, and ultimately railroads along the valley) of the other centers of population along the river—and, in time, even the world's seaports via downstream sea harbors. On such rivers, each of the riverine settlements became linked to other settlements along the river. Urban development was no longer limited to interactions with the local geography or with local economic and political relations. Instead, it also became embedded in a cosmopolitan network of cities-in-the-making.

The quality and intensity of cosmopolitan riverine interactions depended in some measure on the physical condition of the river, and hence there tended to be an incentive to "improve" the river as a watercourse (flood control, reclamation, pollution control) and as an artery of commerce. Such improvements often required the intercession of riparian states and encouraged specifically riverine modes of diplomacy and in some cases even transnational institutions for riverine governance to regulate matters like shipping and sewage disposal.

In the case of the Rhine, this transnational order was originally framed by the powers gathered at the Congress of Vienna in 1815, where they established a Central Commission for Rhine Navigation as a permanent organ of the riparian states. The Commission's aim was to establish and enforce agreements about the technical and legal details of Rhine navigation.[1] The Act of Mainz (1831), which abolished staple rights and guild privileges on the Rhine, was the Central Commission's first major diplomatic triumph. More than a century later, widespread concern about the sorry state of the Rhine's waters due to urban and industrial pollution led to the formation of another "central" commission, the International Commission for the Protection of the Rhine against Pollution (1950). Here again, Rhine cities did not participate directly but were represented (along with other interests) by their national governments.[2]

In sum, large river basins were seedbeds not only of a kind of protoglobalization, but also of a protoglobal institutional order. In these geographically extended linear networks, river cities functioned as mutually interdependent nodes of commerce, communication, politics, and culture. By "improving" the river, engineers enabled a higher degree of circulation of goods, passengers, and ideas among cities.

The Rhine: "Route" and "Line of Rupture"

This chapter examines a specific instance of the cosmopolitan shaping of the development of river cities, namely the struggle for the "head of navigation" on the Rhine. The "head of navigation," to put it briefly, refers to that harbor city upstream of which "standard" river shipping is impossible or so onerous as to be unprofitable. In general this difficulty is a consequence of the inhospitable physical condition of the river upstream of the navigational terminus: swift currents, insufficient depth, and unpredictable shoals. The city that is head of navigation is a privileged entrepôt at which goods coming from downstream are unloaded and transferred to other modes of transportation for shipping further upstream—and vice versa. This heightened commercial activity stimulates an influx of wealth and population and creates extraordinary opportunities for derivative physical, industrial, and financial development. On the Rhine, the status of head of navigation moved successively upstream from Mainz after 1800, to Mannheim, to Strasbourg, and finally to Basel around 1960. Each of these cities was profoundly affected—albeit in different ways and in different measure—by its tenure as head of navigation. In the process the river too was transformed from an unruly natural stream into a domesticated commercial and industrial conduit (Cioc 2002).

In unraveling this complex history, it is important to appreciate the dualistic topology of rivers. Longitudinally they are drains, conduits, arteries, routes connecting cities and river states *through* a geopolitical space. In a transversal sense, they are ruptures, barriers, borders, and sometimes armed lines of defense *between* states and cities. Thus, Albert Demangeon and Lucien Febvre introduced the concepts of *route* and *frontière naturelle* in a chapter on "Les thèmes du Rhin" (Demangeon and Febvre, 1935). Cities along the Rhine have been suspended in two orthogonal force fields, one depending on their relative position *along* the Rhine, the upstream-downstream dimension, and the other depending on their position *across* the Rhine—which bank they happen to be on and what and who is on the other side (and what means there are for crossing, or preventing crossing).

The location of the head of navigation may at first sight appear to be defined simply by the most upstream harbor city that is not beyond the pale of profitable navigation. For the Rhine, however, this is certainly a simplification, at least for the epoch of state formation that set in after the seventeenth century and which resulted in the incorporation of formerly autonomous city-states into the political fabric of new riparian states. Cities increasingly became strategic pawns in political and economic projects pursued by the states to which they had become subject. Insofar as the river functioned as a *frontière naturelle*, former hinterlands on the opposite shore might even become hostile territory, thus severing the city from its "natural" political and commercial environment. This new logic of transversal competition and hostility cut across the simpler upstream-downstream logic governing the head of navigation. To put it simply, a city's claim to the status of head of navigation by virtue of its longitudinal position along the river could be severely frustrated because of political tensions with the opposite shore—quite aside from inevitable upstream and downstream conflicts. This was especially so where becoming head of navigation first required improvements to stretches of river forming a *frontière naturelle* between politically antagonistic states.

The Rhine valley occupies a crucial position within the European peninsula. In the north-south direction, it is part of a transalpine conduit between the Low Countries and the Po valley. In the east-west direction, it is a geographical rupture and sometime political border between the Franco-Iberian salient of the European peninsula and its Germano-Slavic base to the east. So, from prehistoric times, the Rhine valley both facilitated the circulation of people, goods, and culture in a north-south direction and impeded it in an east-west direction. But, just as the north-south journey was hardly devoid of obstacles, neither was the east-west barrier impermeable. At certain points, such as the "Burgundian Gate" between Strasbourg and Basel, access to the valley through the hilly regions to either side was less arduous. These points of access fostered the construction of ferries and bridges that further facilitated crossing (Livet 2003, 105). These natural features (and the bridging projects they encouraged) gave rise to a stabilized network of east-west trade routes that intersected with the north-south routes along the valley. The points where trade routes converged near the river crossings were especially favorable to the development of centers of population. Demangeon and Febvre (1935, 285) argue that this explains the locations of the most ancient and illustrious of the Rhine cities: Basel, Strasbourg, Mainz, and Cologne.

During Roman times and in the Middle Ages, the Rhine and its cities were oriented both economically and culturally toward the south: to southern France via the Burgundian Gate and to Rome and northern Italy through passes in the Alps. However, in the course of the nineteenth century, three developments shifted the center of gravity to the north. First, there was the blossoming of mass tourism in the Rhine gorge through the Taunus Mountains between Cologne and Mainz. Novel means of mass transport like the steamship and the train lured large annual contingents of (especially British) tourists to the "romantic Rhine" in pursuit of those picturesque riverscapes and ruined castles that had already been made famous by an earlier generation of elite romantic voyagers (Cepl-Kaufmann and Johanning 2003). Second, the increased demand for coal to fuel Germany's industrial revolution (in particular as a raw material for iron smelting for the new Bessemer steelmaking process) stimulated intensified exploitation of the vast Rhine-straddling coalfields in Westphalia. The Rhine, whose transport capacity was steadily increased by river engineering and new shipping technologies, soon became transformed into an artery for distributing the Ruhr's "black gold" throughout a good deal of Germany, the Netherlands, and Switzerland. Third, this "carboniferous Rhine" (Cioc 2002) with its new coal-based industrial and chemical complexes, provided an important impetus for the further development of North Sea harbor cities, especially Rotterdam and Antwerp, hence linking Rhine port cities ever more closely to worldwide maritime commerce.

The Rhine's development into a conduit of culture, empire, and commerce was abetted by its unusually stable annual discharge cycle. Most rivers exhibit great seasonal variation. Alpine rivers tend to be high in late spring and summer when the snow and glaciers are melting and to be low in fall and winter when all is frozen. Pluvial, or rain-fed, rivers tend to be low in summer and high in fall and winter when it rains most frequently. The Rhine is a felicitous mix of the two types: the main river and its Swiss tributaries are essentially alpine, and the numerous large downstream tributaries are pluvial. The upshot is that, as far upstream as Mannheim, more than 600 kilometers from the North Sea, the river exhibits an unusually regular flow throughout the annual cycle, a boon for shipping, harbor construction, flood control projects, hydroelectricity production, and water supply. These natural advantages encouraged engineering projects to further improve the navigability of the river and thereby established the Rhine as a major force in the development of its riparian cities.

From Mainz to Mannheim

Mainz's tenure as head of navigation, spanning a period from the Middle Ages to the 1840s, was rooted in medieval privileges. Like several other of the Rhine's left-bank cities, Mainz was founded by the Romans as one of the chain of *castella*, or fortresses, built to defend the border of their empire against right-bank Germanic tribes. Mainz's location at the northern end of the upper Rhine's fertile floodplain, its commercially advantageous position opposite the mouth of the Main, and its proximity to the Rheingau wine-growing region all conspired to make the settlement around this *castellum* an important medieval ecclesiastical center and marketplace. The basis of its status as head of Rhine navigation was, however, the *Stapelrecht* and the *Umschlagrecht* granted by princely decree in 1317. In its unadulterated form, the right of *Stapel* stipulated that riverborne merchants had to unload their cargoes and to offer them for sale to city residents at the public markets for a certain number of days. Only after this period could the unsold wares be offered for sale elsewhere. The *Umschlagrecht*, or right of transshipment, further stipulated that the remnant goods could be transported further only in ships owned and skippered by members of the Mainz shippers' guild. Practically speaking, the rights of *Stapel* and *Umschlag* meant that upstream ships had to terminate their voyages at Mainz. From the point of view of skippers bound upstream from the Netherlands or Cologne, Mainz was thus the head of navigation. By the eighteenth century, of the seven or eight Rhine cities that had enjoyed the *Stapelrecht* and *Umschlagrecht* in the Middle Ages, only three still exercised them: Mainz, Cologne (205 km downstream of Mainz), and Strasbourg (195 km upstream). By that time the formal stipulations were typically being waived in lieu of a meticulously calculated duty, i.e. no actual unloading or transshipment of cargo took place. Nonetheless, something like a head of navigation still existed in the form of delays, bureaucratic regulation, and increased costs. Upstream cities remained at a competitive disadvantage because of the increased cost of goods upstream of a city exercising its *Stapel* rights.

Despite these burdens, commercial life did exist upstream of Mainz. Strasbourg, another former Roman garrison town with a lively medieval shipping tradition, enjoyed the same navigational privileges as Mainz, and, even further upstream, Basel's shippers regularly plied the Rhine down to Strasbourg. In view of this upstream mercantile activity, Mainz's status as head of navigation is obviously ambiguous. Was it really a more decisive head of navigation than Cologne or Strasbourg? In relation to Rhine seaports like Amsterdam and Rotterdam, Cologne also functioned as a head of navigation. And Strasbourg was formally the terminus for shipping emanating from all downstream ports, so it too could lay claim to the title of head of navigation.

In fact, given the technologies of navigation prevailing in the eighteenth century, one can only conclude that *no* city possessed a unique status as definitive head of navigation. A more sensible picture might be that of four successive Rhines, stretched end to end from the North Sea harbors to Basel, each with its own "mouth," its own head of navigation, and its own locally adapted shipping fleet. First, there was the stretch from Amsterdam or Dordrecht to Cologne, then from Cologne to Mainz, from Mainz to Strasbourg, and finally from Strasbourg to Basel. Although the resulting fragmentation of Rhine navigation in this period is usually condemned as a throwback to medieval particularism, the system seems to have been well adapted to local circumstances. From a geotechnological standpoint, it was "appropriate technology." That, at least, is what the *Stapel* cities Cologne, Mainz, and Strasbourg argued in response to the Treaty of Paris (May 1814), which came down firmly in favor of free trade on the Rhine and hence directly threatened their privileges. With a view to the upcoming Congress of Vienna in 1815, charged with working out the details of the new post-Napoleonic European order, the three cities asserted that, far from obstructing the Rhine trade, their tripartite *Stapel* and *Umschlag* system was actually a practical boon to the Rhine trade. They argued that the physical condition of the Rhine required different kinds of boats for different stretches as well as highly specific local knowledge in order to safely navigate the many shoals and rapids. Moreover, the transshipment demanded by the *Umschlag* rights improved efficiency and decreased freight prices, inasmuch as each stretch of the river could be navigated with the largest boat possible. Traffic on the deep and roomy lower stretches would not be uneconomically confined in the small ships that would have been necessary had a single vessel been required for the entire trip upstream. Finally, the *Stapel* cities argued that the strict bureaucratic control over ships and cargoes which they exercised in order to levy their fees protected merchants against fraud and thievery (Chamberlain 1923). The upstart city of Frankfurt am Main, outside this charmed medieval circle, quite naturally challenged these arguments and championed the new liberal economic order under construction at the Congress of Vienna.

Although the *Stapel* cities were obviously arguing from self-interest, they clearly had a point. Due to the limited scope of river engineering in the eighteenth century, the river had to be taken as it was. The extremely arduous propulsion technologies of sail and haulage prior to the age of steam placed a premium on efficient and locally adapted ships (Cepl-Kaufmann and Johanning 2003). And in fact the fragmentation inherent in the *Stapel* system also created a fine-tuned organization of riverine navigation based on local vessels and local pilotage. The *Stapel* and *Umschlag* system—despite its inefficiencies and patent unfairness—represented a workable organization of riverine trade in a period before the advent of large-scale river engineering and steam propulsion. One might question whether the location of the *Stapel* cities corresponded precisely to the discontinuities in navigational requirements, and yet the system did succeed in supporting a regular Rhine trade from the sea to Basel under arduous and difficult navigational conditions.

Notwithstanding its technical advantages, the exclusionary *Stapel* system increasingly came under fire from new competitor cities and the powerful new nation-states that emerged in the wake of the French Revolution. These modernizing forces converged in the Treaty of Paris and at the Congress of Vienna. The Congress established a Central Commission for the Navigation of the Rhine, located at Mainz, with a mandate to hammer out the navigational details of a "free Rhine." A treaty was delayed for many years by a lengthy conflict between the Netherlands and Prussia concerning whether the commission's jurisdiction should also apply to the Dutch tidal Rhine. Finally, in 1831 the states represented in the Central Commission—France, the Netherlands, and the post-Napoleonic German principalities of Prussia, Baden, Hesse-Darmstadt, Bavaria, and Nassau—signed the Act of Mainz, which limited tolls all along the river and, more particularly, abolished the ancient rights of *Stapel* and *Umschlag* (Eysinga 1935). It is probably not coincidental that this timing corresponded with the advent of steam-powered vessels on the Rhine.

With the Act of Mainz, navigation on the Rhine became dictated by economic and technological considerations rather than political privilege. Cities formerly excluded from the Rhine trade by medieval statutes could now compete on an equal footing. New riparian states could now bend their ambitions and energies to ameliorating the river and—in league with their Rhine cities—specifically to improving its navigability. In this fluid situation it was anyone's guess where the effective head of navigation of mainline Rhine shipping would ultimately come to rest. The only certainty was that it would be well upstream of Mainz.

The city of Mannheim was a leading contender. This erstwhile military outpost, established in 1606 by the Bavarian palatine prince elector at the confluence of the Neckar and the Rhine, had early attracted a sizable civilian population. By 1669 it also boasted a floating bridge across the Rhine capable of bearing the weight of 100 mounted soldiers. The bridge was guarded by a redoubt on the left bank called the Rheinschanze, the site of the future city of Ludwigshafen. In 1716 the palatine court settled at Mannheim, where it remained for six decades. This move inaugurated an era of prosperity and urban expansion. The city's social and cultural life revolved around the court, which was an important center of Enlightenment culture. But unfortunately the city's economy also was built on the quicksand of court dependency and not, as might be expected, on the city's advantageous position at the confluence of two navigable rivers. However, Mannheim's isolation from the Rhine trade was only in part due to the seductions of court life; during this period Cologne, Mainz, and Strasbourg still held the Rhine in thrall to their ancient navigational monopolies. Disaster ensued when in 1778 the Palatine court relocated to Munich, taking Mannheim's prosperity with it. Lacking an alternative, the local bourgeoisie was thrown back onto the city's geographical advantages: its position at the confluence of the Rhine and Neckar and its central location in the agriculturally prosperous Palatinate.

Hopeful visions of the city as a future Rhine harbor notwithstanding, Mannheim's population dropped from a high of 25,000 in 1778 to 18,800 by 1802.

Although the French Revolution ultimately inspired the liberalization of Rhine navigation and hence provided Mannheim with a new future, in the short term it also brought French rule to Mannheim's opposite shore. This was another setback for Mannheim. It was now sundered by an international boundary from much of its "natural" hinterland, the left-bank portion of the former Palatinate. On the right bank, Napoleon forged the new Grand Duchy of Baden, into which Mannheim was incorporated. The Congress of Vienna conserved the spirit of Napoleon's handiwork by recognizing Baden and dividing the left-bank Palatinate between Hesse and the Kingdom of Bavaria. With Hesse just downstream and Prussia encroaching on the left bank up to Bingen, Mannheim had clearly lost a strategic position at the center of a prosperous hinterland. At this critical juncture, the sudden appearance of steamships on the Rhine breathed new life into Mannheim's navigational ambitions.

In 1816 the British paddle-wheeled steamer *Defiance* docked at Cologne after a trip upriver from Rotterdam. "A rather large ship without a mast, sails and rudder approached with uncommon speed," the *Kölnische Zeitung* reported (Cepl-Kaufmann and Johanning 2003, 137). Steamships' "uncommon speed," especially against the current, launched an entirely new era in Rhine navigation. Initially, steam was regarded only as a more dependable means of propulsion for passenger transport, and it soon enough provided the impetus for the development of mass tourism on the river. Yet freight shipping was not far behind. In 1830 a steam-powered tugboat towed a train of fully loaded freight barges up to Altrip, just upstream of Mannheim, and proved that Mannheim could, in principle, become the head of navigation of a new, steam-powered regime for Rhine navigation. In a word, thanks to steam power, the Rhine suddenly appeared to be a much more navigable river with its upper reaches tantalizingly close to the sea. With the liberalization of the Rhine trade confirmed by the Act of Mainz in 1831, it seemed only a matter of time before Mannheim would assume Mainz's venerable place in relation to Cologne and the North Sea seaports.

But how was Mannheim to realize its promise? Where were the harbors, the cargo-handling equipment, the warehouses, and the freight connections to the hinterland? By 1830 Mannheim had nothing to show in the way of harbor facilities but a short riverside quay and a crane. The city itself lacked the funds to build harbor facilities on its own. But the role of the river as *frontière* galvanized matters. The political crowbar turned out to be the competitive threat presented by developments at the Rheinschanze (now under Bavarian jurisdiction) just across the river. Since 1808 the Bavarian government had leased the Rheinschanze to an innkeeper, who also installed a crane and a number of jetties on the Rhine shore—in short, producing a rudimentary port. Mannheim's alert chamber of commerce saw this as a threat and an opportunity.

Ludwig Basserman, one of the chamber's leading members and also a member of Baden's Chamber of Deputies, explicitly referred to the Rheinschanze "port" when he petitioned the Baden government in 1825 to allow Mannheim to install its crane and to develop its Rhine quay. Within three years, having acquired a broader view of its riverine interests, the Baden government granted Mannheim the status of "free port." Work started on an *Inlandhafen* (domestic harbor) and an *Auslandhafen* (foreign harbor) in the Giessen, an old Rhine branch just north of the city. The festive opening of the new harbors in 1840 was attended by the Grand Duke of Baden himself, a gesture underscoring Baden's interest in Mannheim and its expectation of sharing in the new prosperity. In that same year, mirabile dictu, Mannheim also opened a main railway station to the southeast of the old city, thus directly connecting the harbors with a huge hinterland encompassing both Frankfurt and Basel. In 1840 Mannheim had only a passenger connection, but rail freight connections with the Giessen harbors were built in the next few years.

Mannheim's new harbors were completed just in time for the city to capture a handsome share of the new coal trade that was emerging on the Rhine as a result of the development of bituminous coal mining in the Ruhr. Mid-nineteenth-century industrialization centered on the mining and transport of coal both for the increasingly ubiquitous steam engine and as a raw material in iron smelting, steelmaking, and the coal tar chemicals industry. The industrialization of the Rhine valley, which took off in the 1840s, was in the first place predicated on cheap and massive transportation of coal. Although railways sometimes provided stiff competition, river shipping managed to remain competitive (at least for bulk goods) by dint of continual improvements in navigational and cargo-handling technologies and incessant "improvements" to the river itself. Rhine harbors—thanks to the secondary distribution system provided by the railroads—became key entrepôts in the rapidly increasing circulation of coal and other commodities that were the lifeblood of an industrial society, including the bulk cereals required to feed rapidly growing urban populations. In 1839 the Ruhr produced more than one million tons of (anthracite) coal; by 1853, more than two million tons were brought to the surface, including bituminous coal for blast furnaces and steel mills. The total figure would rise to some 115 million tons of coal in 1913, of which 3.5 million tons were unloaded in the harbor of Mannheim (Cioc 2002).

Mannheim's ambitious civic elite sought to fulfill what it perceived as a new destiny for their city as head of navigation on the Rhine: a river port for bulk commodities destined for southern Germany and Switzerland. To achieve this, they pressured Mannheim's municipal administration and the Grand Duchy of Baden to transform the immediate environs of the old baroque city into a huge machine for the handling, storage, and processing of bulk cargo. The Giessen harbors of 1840 were the first incursions into the Mühlau, the roughly triangular area to the north of the city at the confluence of the Rhine and Neckar (see figure 2.1). This harbor area was soon connected to the new railway station at the opposite south side of the city by a semicircular rail line built on the free space left after the city's baroque battlements had been torn down in 1818.

The expansion into this formerly swampy area was facilitated by a major civil engineering triumph in those days: the upper Rhine "ameliorations" carried out by the Oberdirektion des Wasser- und Straßenbaues (Directorate of Waterway and Road Construction) of the Grand Duchy of Baden under its chief engineer, Johan Gottfried Tulla. Launched in 1817, the ambitious Tulla project aimed to improve the course of the Rhine between Basel and Worms (just downstream of Mannheim) by reducing braided portions of the river to a single channel, removing islands, and cutting off the larger meanders. The massive effort took nearly six decades to complete. Tulla's project lowered the water level all along the river and greatly reduced the risk of flooding. For Mannheim, the project conveniently supplied new real estate for the expansion of its port. A major local feature was the cutting off of a large meander just downstream of the city opposite the Mühlau by a new straight channel called the Friesenheimer Durchstich. This channel, like all of Tulla's reconstruction works, was dug by hundreds of laborers wielding little but picks and shovels. The Durchstich was originally some 100 meters wide and 4.5 kilometers long, but by 1870 the current of the river itself had widened it to the standard Rhine width of 300 meters and it had become the main channel. The old meander, now a backwater, was put to good use by Mannheim after 1906 as the site of an impressive new industrial harbor, the *Industriehafen*. Tulla's project was not everywhere perceived as a boon to civilization and progress, though; many Rhine villages were far from sanguine about the outcomes, and the project ran into stiff resistance in several locations (Kalweit 1993; Kieser 2003). Grand ducal troops had to be called in to protect the workers, and Tulla himself often went armed. After its completion in 1876, Tulla's project also turned out to have a number of deleterious effects on navigation, to which we shall return below.

As Ruhr coal sought new markets, it was in Prussia's interest as suzerain of the Ruhr and major riparian power on the Rhine to facilitate Rhine navigation as much as possible. Improved Rhine navigation would make steady increases in the scale and reliability of shipping possible and make Ruhr coal competitive even at great distances. Starting in 1851, the Prussian Rhine River Engineering Administration, under the engineer Eduard Nobiling, began work to improve the navigability of the middle and lower Rhine between Mainz and the Dutch border (Cioc 2002). This project improved access to Mannheim and strengthened its position further, especially relative to potential upstream competitors like Karlsruhe, Strasbourg, and Basel. The Act of Mannheim, enacted by the Rhine riparian states in 1868, clinched the matter. This agreement did away with the last vestiges of protectionism and enabled the Rhine trade—and hence Mannheim's harbor—to compete on a stronger footing with the railways.

Figure 2.1

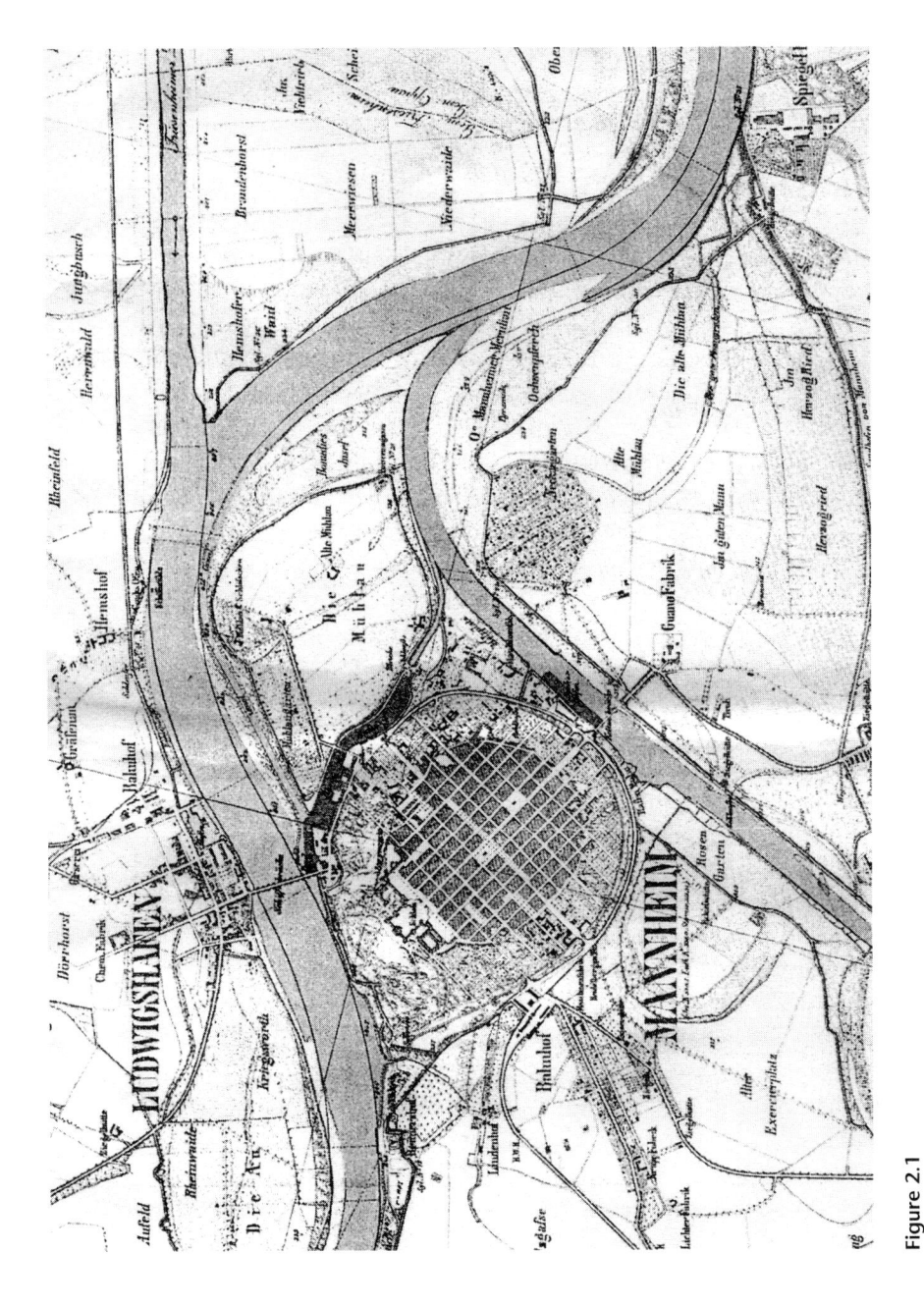

The river city of Mannheim in 1856. The gridded baroque city is wedged between the Rhine and Neckar, and encircled on the east side by a ring railroad. The city's first harbors (dark gray) were built in the triangular Mühlau region where the Rhine and Neckar merge. The straight cut of the Friesenheimer Durchstich is at the upper right. (North is to the right.) Source: Baer 1928, plate 3.

From Mannheim to Strasbourg

Strasbourg, literally "city of roads," was a former Roman *castellum* town on the site of earlier prehistoric settlements. It owed its medieval preeminence to its position on the upper Rhine, at the crossroads of major axes of circulation in medieval Europe. Up to the seventeenth century, it had been a "free city" of the Holy Roman Empire. Strasbourg emerged from the ruinous wars of that troubled century subject to French rule. In the course of the following century, the city was able to recover some of its medieval wealth and glory, again exploiting its "crossroads" position. In addition, the city's merchants profited from a special status outside the French Tariff Union, which allowed them to buy and sell freely without having to pay French customs, for example on agricultural produce from the German right bank (via the Kehl bridge) or on goods shipped via the Rhine, mainly colonial products from Amsterdam. Thanks to their *Umschlagrecht*, Strasbourg's shippers, using the then current techniques of sail and haulage by horses, enjoyed a monopoly of trade on the river upstream from Mainz. At Strasbourg they would unload for further transshipment in yet smaller ships via the river Ill or for transport by road into France, Switzerland, or Italy. With its Rhine-based commercial orientation to Germany and Switzerland, rather than to France, the city became the dominant center of population on the upper Rhine, counting 50,000 souls by 1789 (Göhner and Brumder 1935).

But by then, great political transformations were already altering Strasbourg's newfound relation to the Rhine. In Alsace, at least, the French Revolution and its aftermath fundamentally changed the Rhine from an economical *route* into a political *frontière naturelle*. In its vigorous pursuit of national unity, the Revolutionary Directorate lost no time in firming up the borders, which included abolishing Strasbourg's special customs privileges. From a Parisian perspective, Strasbourg was to be a tributary and military outpost of the French state, and not an international entrepôt oriented principally to Germany and Switzerland. Although Strasbourg's trade flourished briefly under the Republic, thanks to the city's superior accessibility (as surrogate for Napoleon's Channel ports, which were suffering a British blockade) and its role as a staging area for the ill-fated eastern campaigns, by 1815 this too was all past glory.

Desiring to maintain the city's position as a hub of trade, Strasbourg's chamber of commerce had no choice but to integrate itself more effectively into the French canal system. Two new canals were proposed to the French government: one connecting Strasbourg with Lyon and Marseille via the Saône and Rhône rivers and another connecting Strasbourg with the river basins of the industrialized French north via Nancy. The chamber of commerce thus sought to "unite Marseille, the entrepôt of the commodities of the East, with Strasbourg, to which accrues the merchandise of the North," and to contest the "transit from Holland and the northern portions of Germany to Switzerland and Italy" by means of road traffic along Baden's right bank (Looveren

1948, 187). The Canal du Rhône au Rhin was opened for traffic in 1833. A decade later the canal was extended around and to the south of Strasbourg, thus finally creating the long-desired through route between the north and the Mediterranean seas. Unfortunately, both the poor condition of the Rhine up to Strasbourg and the initially very modest dimensions of the canal (even by the standards of the time) precluded a significant role for this new transportation link.

The canal to the French northwest, the Canal du Marne au Rhin, was intended to cheapen the transport of "colonial commodities" then transported to Strasbourg either overland from French Channel ports or up the Rhine via Dutch North Sea ports. The chamber of commerce estimated that transport over the Rhine would take about 50 days longer than by means of the new canal. This seems an extremely optimistic estimate for the canal, considering that there were 212 locks between Paris and Strasbourg (Looveren 1948, 197–198). The comparison likely says more about the impossible state of the Rhine and the arduousness of upstream navigation before steam power than about the efficiency of the proposed waterway. By 1853 the canal had reached Strasbourg, and in 1858 it carried 239,000 tons of freight, twice the amount then being transported on the Rhône-Rhine canal between Strasbourg and Mulhouse. However, the bulk of the freight on the Marne-Rhine canal was not colonial commodities but coal from the Saar mining region in Lorraine. Strasbourg's merchants had apparently foreseen this development and had agitated for a branch canal from the Marne-Rhine canal to the Saar coalfields. By 1869—on the eve of the German takeover of Alsace—coal transports to Strasbourg on the Canal des Houillères de la Sarre amounted to 623,000 metric tons, and expectations were high. Strasbourg's chamber of commerce wrote: "the consignments by way of the two canals are such as to console us for those which have abandoned the Rhine" (Looveren 1948, 198).

Strasbourg's retirement from the Rhine trade was not for lack of trying. Some of Strasbourg's merchants had made a serious effort to retain a hold on Rhine shipping by adopting the new technology of steam. From 1832 on, a regular steamship line carried passengers from Strasbourg to Rotterdam and even directly to London. And in 1849 a direct freight line to Rotterdam was established employing steam tugs and barges. Two small steamers also maintained a regular connection with upstream Basel. Improvements in harbor facilities were carried out in 1837 and 1849 to accommodate steamers and to connect the new canals with each other and the existing harbor infrastructure. These initiatives were, however, frustrated by the poor navigational state of the Rhine. In the 1840s the coming of the railways marked the beginning of the end. By means of trade agreements and strategic pricing, the railroads tried to force traffic off the river and into their boxcars. The Baden Railway, for example, made a deal with the Mainz shipping companies by which the latter agreed to unload their cargoes at Mannheim, leaving the Baden Railway to transport the goods to southern Germany and Basel, thereby circumventing Strasbourg. In 1854 the eastern branch of the French railways delivered the coup de grâce by reducing its tariffs so that shipping via the port of Strasbourg was no longer competitive. Rhine traffic at Strasbourg literally disappeared after 1868.

Straßburg as Capital of a German Elsass: 1871–1918

Strasbourg's effort to reorient itself away from the Rhine and toward the French canals and railways had begun to pay off, but in 1871 the city's hard-won commercial renaissance was abruptly terminated by the German annexation of Alsace-Lorraine. The Germans conceived of Strasbourg not as a commercial city but as the beachhead of German imperial rule in the new left-bank territories. The occupying forces revived Strasbourg's old role as a garrison city, and before long it was again bristling with new barracks, fortifications, troops, and armaments. Parallel to this military presence, Strasbourg became the political and administrative center of the new *Reichsland* of Elsass. German officers and bureaucrats became the city's new elite class. A German-language university and extensive medical clinics were established. Strasbourg's traditional commercial elite found themselves sundered from France, their painfully achieved web of infrastructural connections with the French hinterland next to useless because of the new customs border at the crest of the Vosges mountains. Their only option now was to try to reestablish a presence in the east and on the Rhine. Alas, the river, while it was no longer a *frontière*, was not yet much of a *route*, either.

Reestablishing relations with the old right-bank German hinterland in fact proved much easier than reclaiming a place on the Rhine. A network of narrow-gauge railways soon connected Strasbourg via the Kehl railway bridge (built in 1860) with the major towns on the opposite shore. In 1896 only 2 percent of the traffic on the railways of Alsace-Lorraine (Elsass-Lothringen) was destined for France, as compared to 47 percent heading to the rest of Germany. As for the Rhine, Strasbourg now faced three closely related problems: the sorry navigational state of the river, the virtual absence of port facilities, and Mannheim's consolidation of its position as head of navigation.

Tulla's program of "ameliorations" to the Rhine, finally completed in 1876, had left behind a mixed legacy. Although the upper Rhine had been straightened, the faster current and the consequently increased sediment transport created several unexpected problems. The faster current made upriver transport slower and more costly. Where the river flowed fastest—for example, just below the rapids at Istein near Basel—the strong current scoured the river bedding to a depth of some seven meters. This erosion made the rapids an even greater obstacle. Worse yet, the scoured-out sediment gave rise to new shoals and sandbars at numerous downstream sites where the current was slower, particularly between Strasbourg and Mannheim. So while initially the Tulla project appeared to improve navigational conditions on the upper Rhine, it actually resulted

Figure 2.2
Strasbourg with its back still to the Rhine in 1883. The walled city is connected to the Rhine only by a small diversionary stream (the Petit Rhin), and the considerable space between the city and the Rhine is occupied by fortifications, forests, and a hippodrome. Source: Göhner and Brumder 1935, plate X.

in closing down this segment of the river entirely, at least for the high-volume steam-powered transport that had become the norm on the Rhine downstream of Mannheim.

The condition of Strasbourg's Rhine harbor facilities corresponded entirely to the poor navigability of the river. As figure 2.2 shows, Strasbourg was not actually a Rhine city but had been founded on the river Ill (lower left corner). This meant that a Rhine port did not come naturally, but had to be created in the considerable space between the city and the Rhine; on the other hand, once the need arose, there would be more than enough space, unlike the situation at Mainz or Mannheim. Until 1900, however, the future site of Strasbourg's impressive Rhine harbors was still occupied with fortifications, woods, and a hippodrome—eloquent testimony to the irrelevance of the Rhine for Strasbourg in this period. Up to 1880, in fact, Strasbourg's port facilities were all still within the city walls. In that year, construction started on a connecting canal between the Rhine and the Ill along the city's eastern and southern perimeter. A modest harbor with a railway quay was constructed on the southern portion of the canal (Göhner and Brumder 1935, 25).

In their strivings to rejoin the Rhine navigation, there was one last hope for Strasbourg's merchants to clutch at: enlisting the help of Prussia, the dominant power in the new Reich. Unlike Baden, the Berlin government was not necessarily wedded to the idea of Mannheim as head of Rhine navigation. For downstream powers like Prussia it made economic sense for the head of navigation to be as far upstream as possible, and since Strasbourg had now become an imperial German city, it was preferable to potential competitors like Karlsruhe (not far upstream) or the Swiss city of Basel.

Strasbourg's chamber of commerce moved quickly. Only a year after the German takeover of Alsace in 1871, it petitioned the chancellor of the Reich to create a waterway to Strasbourg suitable for heavy barges. In view of the technical difficulties of ameliorating the Rhine itself, the chamber of commerce proposed digging a capacious lateral canal along the left bank between Strasbourg and Ludwigshafen.

Opposition from Mannheim and its twin city Ludwigshafen compounded the geographical and hydraulic obstacles. From the French point of view of the matter, the source of German resistance lay in the refusal of Mannheim-Ludwigshafen to cede their hard-won position as head of navigation. The twin cities were supported by the railways, and especially the Baden State Railways, which derived considerable revenue from the redistribution of bulk goods unloaded at Mannheim, both to the German hinterland as well as upstream along the Rhine to Switzerland. "The history of the Rhine regulation from Mannheim to Strasbourg is a dark chapter of German particularism," concurred one German commentator in 1925 (Metz 1927, 202n).

Despite Strasbourg's quick action in 1872, ten years passed before detailed plans for the lateral canal were submitted to the Reich government in Berlin. Mannheim's and Baden's strong lobby in Berlin delayed decision making on the project another eight years, at which point it was decided that a lateral canal would be too expensive and that the option of improving the riverbed itself should be explored. This required tripartite negotiations among the German states of Elsass, Bavaria, and Baden about the technical details and the division of costs. It took three years just to get these talks under way and until 1901 before an agreement was reached. In the event, though, Baden refused to sign and it took a further five years before the project was finally begun.

However protracted, the negotiations over the improvement of the upper Rhine between Mannheim and Strasbourg were fateful for both cities. Mannheim understood that it would sooner or later lose its position as head of navigation, while Strasbourg hoped to recover its dominant pre-1789 position on the upper Rhine, based on transshipment of goods overland or by canal. By 1913, on the eve of the First World War, the second round of improvements to the Rhine up to Strasbourg was completed—this time aimed specifically at improving navigability—and the city once again became the head of navigation.

Meanwhile, Strasbourg's new German masters had been preparing the city itself for her new role, a transformation that coincided with industrialization, strong population growth, and the "networking" of the urban space with public utilities and transport systems. The trigger for the morphological transformation was the construction of a new railway line between Strasbourg and Kehl, routed more southerly than the existing one and built on a dike seven to eight meters high (Göhner and Brumder 1935). Military authorities considered this dike an adequate replacement for the existing southern and eastern ramparts and, with these torn down, the path was cleared for urban expansion southward, as well as for building new modern Rhine harbors on the so-called *Sporeninsel* between the "small" and the "big" Rhine to the east of the city. There was even room left for enlarging the hippodrome, though it had to be moved southward.

Mannheim Bets on a New Horse

Strasbourg's looming status as head of navigation had understandably stimulated some hard thinking in Mannheim. In June 1891 Mannheim's chamber of commerce suggested to the Baden government at Karlsruhe that, in view of the scarcity of industrial sites adjacent to navigable water, it would be prudent to turn a section of the abandoned meander created by Tulla's Friesenheimer Durchstich into an industrial zone. But Baden refused to finance the venture. Mannheim's new mayor, Otto Beck, whom Dieter Schott (1997b, 153) describes as an "energetic and lucidly analytical personality, committed to strategic and long-term thinking," took up the challenge. In early 1895, Mannheim and Baden came to an agreement about the organization and financing of the *Industriehafen* (industrial harbor). The city was to pay the bulk of the projected costs, about six million marks, amounting to about one and a half times the annual municipal budget.

Mannheim's campaign for a new industrial harbor not only marked a shift of local power from a commercial middle class to an industrial one (complete with an industrial proletariat and the inevitable "social question"), but was also, as Schott (1997b, 153) phrases it, the "axis of a planned, long-term concept for structural change" in the city. Becoming an industrial city became the "highest goal" of Mannheim's self-conscious urban redevelopment.

This municipal reconstruction commenced with purchasing and annexing the new island of Friesenheim, on the edge of which the *Industriehafen* was to be built. Though cut off by Tulla's Friesenheimer Durchstich from the left-bank Bavarian Palatinate, the new island had remained under Bavarian jurisdiction. By 1897 the town of Käfertal was also joined to the city, which brought important right-bank tracts of land under Mannheim's purview as well as a sizable forest in which the municipal waterworks could expand its facilities—goaded by the projected industrial demand for clean water. Finally, in 1899 the privately built and owned harbor facilities at the upstream town of Neckarau were bought up and the town itself annexed. As a consequence of this territorial expansion, Mannheim had become an immensely elongated urban agglomeration stretching some 15 kilometers along the Rhine's right bank. The city's elongated topography created new problems for networked infrastructures like gas, electricity, and urban transport, which were only exacerbated by the impressive leap in population between 1890 and 1900, from 80,000 to 140,000. The extra costs involved in providing services to the far-flung extremities lay "like an eternal mortgage on the shoulders of the municipal administration...and the individual citizens" (Hook 1962, 12).

The period during which Mannheim's new *Industriehafen* was conceived and built coincided with the wave of electrification then sweeping through cities all over the continent. The expected development of the industrial harbor determined the choice for both the type of system (AC polyphase) and the location of the power plant. With the wisdom of hindsight, we can say that Mannheim, thanks to the *Industriehafen*, opted for an extremely progressive electrical plant, shaped not by the usual requirements of lighting or the electrical tram, but by the needs of the industrial harbor, primarily for powering large polyphase electric motors for cranes, elevators, and factories. Although Mannheim's public tram system ran on direct current, it was considered more efficient to produce polyphase alternating current in a large central plant and then to change it into direct current for use by the tram, than to continue to generate direct current in a separate power station. Locating the plant in the new *Industriehafen* would also save on fuel costs (coal) and provide plentiful supplies of cooling water and room for future expansion.

Strasbourg Becomes Head of Navigation

In 1918, with the end of the First World War, Strasbourg and the portions of Alsace-Lorraine under German jurisdiction since 1871 reverted to France. The Rhine again became a frontier between nations, but due to stipulations in the Versailles treaty designed to facilitate the Alsatian transition back to French rule, it was a rather low-key border. Under German rule, Strasbourg had prospered economically. The restoration of its prerevolutionary commercial links with the opposite shore had handsomely compensated the loss of its French hinterland. And, most importantly, under German rule and with German engineering, the Rhine had become a navigable *route* again up to Strasbourg, and of such dimensions that the city was now head of large-scale navigation and effective port of entry for Switzerland. It was Strasbourg's good fortune that Paris realized the possibilities and saw the city and its harbor as the beachhead of a new French economic policy on the international Rhine. Mannheim's historian Sigmund Schott (1929, 34) was moved to exclaim enviously: "Where one could characterize Mannheim as almost the stepchild of German trade policy, Strasbourg is most certainly the darling of the French."

By a law of 26 April 1924, Strasbourg's harbor was promoted to the status of "Autonomous Port," emancipating it from direct rule (and financial constraints) by the municipal government. A project for the further expansion and improvement of the port was to be financed almost completely by the French state. The French government granted the port a favorable customs regime and put it on a par with the large maritime harbors. A second set of measures sought to neutralize competition on the Rhine, especially by German parties. This task was made easier by France's position as plaintiff in the reparations arrangements included in the Treaty of Versailles. Using this political leverage, France commandeered German vessels to build up its own fleet of Rhine steamers and barges to service Strasbourg, arguing that Strasbourg's original fleet had likewise been appropriated by German companies during the 1871 to 1918 occupation. The bulk of the vessels were tendered by Mannheim shipping companies.

Another potential threat was posed by the German city of Kehl, Strasbourg's German neighbor across the river. It was feared that the Germans would seek to develop Kehl as preferred head of navigation for Ruhr coal and other bulk commodities, in direct competition with Strasbourg. Articles were included in the Treaty of Versailles to neutralize this danger: the ports of Kehl and Strasbourg were to be managed by a single administrative body for at least seven years after the armistice. This arrangement effectively placed the harbor of Kehl under Strasbourg's jurisdiction and allowed Strasbourg to expand its port facilities at leisure, meanwhile utilizing Kehl's facilities to its own advantage.

In pursuing this expansion, as argued above, Strasbourg's distance from the Rhine did not force it to develop the expensive elongated morphology of other Rhine cities like Mannheim, Rotterdam, and Mainz. Nonetheless, the city was a major railway hub and a point of intersection for several canals. Connecting these systems to the new Rhine harbor entangled the city in a dense web of transport conduits: canals, railways, and in time highways. Strasbourg continued to face a formidable challenge in interweaving these various conduits with each other as well as in managing urban expansion through such fragmented terrain.

Notwithstanding the continuing investments in the harbor left behind by the Germans, Strasbourg remained vulnerable to the Rhine's vagaries and to Mannheim's enormous momentum as an established port—not to mention Basel's growing insistence on free access to the Rhine. Although the amelioration works carried out by the German Reich between 1906 and 1913 had supposedly made Strasbourg's port accessible "year-round," there were regular periods of low water during which the channel's depth did not allow passage for fully loaded barges. Since Mannheim continued to be accessible for the largest barges year-round, the only recourse for Strasbourg's shippers during periods of low water was to unload part of their cargo at Mannheim, in order to "lighten" their ships. Hence, according to an Alsatian author, Strasbourg

remained tributary to Mannheim whose assistance it had to solicit in order to work the "lightening" demanded by low water. The operations of unloading were onerous, the cargoes resulting from the discharges were rarely reloaded onto other barges; they were reloaded onto wagons or left at Mannheim to be sold on site. It was therefore easy to justify the prejudice that at Mannheim the regularization of the Rhine had the effect of allowing a portion of the traffic to pass on to Strasbourg. (Kronberg 1924, 54)

The figures support this conclusion, although it is hardly likely that the discrepancy was due solely to the "imperfect" state of the regulated channel. In spite of Strasbourg's postwar renaissance as a port and Mannheim's emphasis on industrialization, Mannheim managed to remain by far the largest port on the upper Rhine. In 1914 Mannheim alone handled some 6.7 million tons of freight against 1.5 million for Strasbourg. While these figures had become almost equal by 1928, at 5.8 million and 5.4 million respectively, by 1938 Mannheim was processing 13.1 million tons (or 14.8 percent of the entire tonnage shipped through all Rhine harbors, save Rotterdam) against Strasbourg's mere 4.1 million tons. Of course, by 1938 new political and economic developments, first and foremost the rise of Germany's Ruhr-centered military-industrial complex under the Nazi regime, had advantaged Mannheim (as an important manufacturing center) while it had disadvantaged Strasbourg.

The rise of Nazism, with its threat of another European conflagration, disadvantaged Strasbourg in another way as well. As France's easternmost outpost, Alsace became incorporated into a national defensive strategy wherein it served the role of *glacis*, or first—and in some measure expendable—line of defense. Military strategy came to have a commanding voice in the planning of infrastructure and the location of new industrial plants in and around Strasbourg. A project to locate a refinery in Strasbourg's harbor, for example, was vetoed by the Ministry of Defense. Work on the Grand Canal d'Alsace, a big hydroelectric and navigational project between Basel and Strasbourg, was put on hold. A proposal to bar new industrial plants in a zone 30 kilometers from the border was even passed into law. The politics of the *glacis*, imposed by international tension, chased industry away (Traband 1966). Strasbourg would have to await the outcome of a new war—after which, for a change, it would not be subject to new masters—to lay the groundwork for a new niche in the French nation and a new position on the ever more European Rhine.

Taming the Rhine: Economic Connection and Urban Competition

From Strasbourg to Basel

On June 2, 2004, the Swiss city of Basel celebrated the hundredth anniversary of the mooring at the city quay of the steam tugboat *Knipscheer IX*. The twin-screw tug, fitted with a 350-horsepower engine and towing a fully laden barge with 300 tons of coke for the city's gasworks, had taken only nine days to make the 614-kilometer trip from Duisburg-Ruhrort. Its arrival at Basel prefigured a new upstream extension of Rhine navigation. The test voyage had been masterminded by a young Basel engineer, Rudolf Gelpke, with the financial backing of Basel's gasworks director. Gelpke's demonstration of the technical and economic feasibility of navigating the open Rhine upstream of Strasbourg to Basel propelled Basel into the age of steam navigation and inaugurated a spate of harbor building adapted to bulk transport by steam tug and barge. In the local mythology, the *Knipscheer IX*'s arrival marked the beginning of Basel's renaissance as a twentieth-century Rhine port.

Figure 2.3
The Dutch-built, steam-powered propeller tugboat *Knipscheer IX* moored at Basel in June 1904 after its nine-day trip upriver from Ruhrort. Moored behind the tugboat is the freight barge *Christina*, laden with 300 tons of Ruhr coal for Basel's gasworks. Source: Verkehrsdrehscheibe Basel ⟨www.verkehrsdrehscheibe.ch/transport/geschichte/g_schiff.html⟩ (accessed November 2006).

But in actual fact, the *Knipscheer IX* was not the first steamship to arrive in Basel. Aside from the successful passage of the steam tug *Justitia I* in 1903, organized by Gelpke as proof of principle for the *Knipscheer IX*'s journey, there had been earlier ships. In 1832, only 16 years after the first appearance of steam on the lower Rhine, a paddle-wheeled passenger steamer called *Stadt Frankfurt* had also made the journey upriver to the Swiss port. But even though the possibility had thus been established, the unregulated Upper Rhine with its numerous braids and islands remained a daunting obstacle for steamships between Strasbourg and Basel. When the river was high, the fierce current easily overpowered the modest steam engines of the period; when it was low, shoals and snags were everywhere. As a result, steam navigation to Basel was possible on average only 60 days per year. In 1844 the French Compagnie de l'Est completed a railway line between Strasbourg and Basel that quickly ran the unreliable steamship lines out of business. Their patchy and unpredictable service simply could not compete with rail transportation that functioned irrespective of the Rhine's water level. By 1850, steam navigation to Basel had ceased.

What had changed between 1850 and 1904? For one, Tulla's upper-Rhine regulation had eliminated numerous braids, meanders, and islands and given the Rhine a single channel from Basel to Worms. *Grosso modo* this had made the river deeper—with the exception of local shoaling—but it had also, on average, made the current stronger. This change suggested a niche for more powerful tugboats, inevitably of greater draught. With the increase in power and efficiency of steam engines in the course of the nineteenth century, and the adoption of propellers instead of paddle wheels, which allowed for more streamlined and maneuverable hulls, the journey to Basel became technically feasible on the basis of hydrodynamic efficiency and brute power alone. And indeed, the 350-horsepower engine that drove the *Knipscheer IX*'s twin propellers nearly doubled the average of 200 horsepower per tugboat reported for the lower Rhine even as late as 1900. Moreover, the *Knipscheer IX* was towing a barge of no more than 300 tons burden, which meant that each horsepower had to tow only 1.16 tons. The largest (paddle wheel) tugboats on the lower Rhine generated about 1,800 horsepower and towed six or seven barges totaling 7,200 tons; hence, they towed at least 4 tons per horsepower. Even though its economics must have been highly sensitive to coal prices and engine efficiency, the voyage to Basel could now be practically accomplished under a wider range of river levels than in 1832–1850 and hence, on average, more days per year.

By 1902 it was also clear to Gelpke that the German rulers of the upper Rhine were seriously committed to a new round of Rhine regulation between Mannheim and Strasbourg. With the potential head of navigation at Strasbourg, soon the only question would be how to subdue the remaining 130 kilometers to Basel. Unfortunately for Basel, this stretch of the river was now controlled fully by Germany. Gelpke and his supporters somehow needed to convince the Germans that more improvements on the *Oberrhein* were also in their interest. However, German interests increasingly aligned with Strasbourg's (and Mannheim's) against giving Basel direct access to Rhine shipping.

There was, of course, also the old Canal du Rhône au Rhin, the northern portion of which between Strasbourg and the southern Alsatian city of Mulhouse was effectively a lateral canal to the Rhine, with a branch opening near Basel. The great advantage of a lateral canal is that, lacking a significant current and having a fixed channel depth, it is open to navigation year-round (barring ice or maintenance). Moreover, canal navigation demands little in the way of motive power. Moderate-size steam engines, even haulage by horses or electric locomotives, are sufficient. Hence, a lateral canal like the Canal du Rhône au Rhin could have been a lifesaver for Basel. But canals achieve their tranquil waters at a price: the locks necessary to take the canal over gradients in the terrain slow down navigation and set a hard limit to the size of ships. And under the constraints of nineteenth-century lock and weir technology, with its fall-per-weir limit of three meters, weirs and locks multiplied at a rapid rate. To bridge the 129 meters difference in altitude between Strasbourg and the Rhine at Basel, the 122-kilometer-long section of the Canal du Rhône au Rhin required 48 locks—on average, a lock every 2.5 kilometers. Most locks were still only 34.5 meters long, making them unsuitable for the Freycinet *péniche* (barge) of 38.5 meters length, the new French standard established in 1879. Worse yet, the final Rhine lock near Basel was only 30 meters long and was to remain so until 1925, a sterling example of the navigational obstacles cultivated at many waterway border crossings. In sum, both the navigational delays and the modest size of ships that could navigate the canal gave it limited relevance to Basel's future as a Rhine port. Against this background, the arrival of the *Knipscheer IX* at Basel in 1904 was big news indeed.

All the same, Gelpke's daring demonstration might never have borne economic fruit—in spite of a vigorous spate of harbor building at Basel—had not the First World War again transformed politics on the Upper Rhine. After 1918 when Alsace-Lorraine was returned to France and the Rhine became an international border again, German policy ceased to be a factor in shaping the Rhine above Karlsruhe. Not that France was any more favorable to Basel's shipping interests than Germany had been, but the former nation's vigorous pursuit of hydropower development on the newly "French" Rhine quite inadvertently promised to transform the difficult stretch from Strasbourg to Basel into a nautical paradise. By terms of the Treaty of Versailles, France had been granted a monopoly on the use of the Upper Rhine for hydropower production. Energized by this unique historical opportunity, it set about to construct a huge lateral canal—bigger in cross section than either the Suez or Panama canals—from Basel to Strasbourg. This Grand Canal d'Alsace would divert some 80 percent of the Rhine's flow and, by means of a series of eight dams and hydropower plants, produce an immense amount of energy. At the same time, due to locks flanking the dams, it would offer a grand waterway of uniform and stable dimensions. Despite the current (up to 1.2 meters per second) required for the primary hydropower function, the canal was touted by the French as an immense improvement. In 1932 the first of the power plants on the Grand Canal d'Alsace came on line at Kembs about 20 kilometers downstream of Basel. Besides supplying France with more cheap electricity than it could use, it allowed ships bound to and from Basel to bypass the rapids at Istein and thus removed the most formidable obstacle for Basel's accession to the Rhine trade.

Ultimately it took until 1970 before the Grand Canal was extended to Strasbourg—in a strongly modified form, due to German concerns about lowered groundwater levels on the Rhine's right bank. Ships traveling from Strasbourg to Basel now had to negotiate eight locks, certainly a hindrance, but still a far cry from the 48 locks on the old Canal du Rhône au Rhin; and for most days of the year, and certainly in an upstream direction, the canal was an improvement on the open river. Moreover, the new locks were dimensioned to accommodate the largest Rhine vessels. Over the course of a drawn-out construction project lasting nearly 50 years, Basel gradually achieved the status of head of navigation on the Rhine, a distinction it retains to this day.

Conclusion

This tale of four Rhine cities has highlighted how their shared river shaped each city's urban development. My central argument has been, in general terms, that it is impossible to understand the contingent process of urban development and the way utilities and transportation infrastructures take shape without taking account of the various contexts—geophysical, technological, economic, political—in which municipal governments and agencies are embedded.

I have taken what is perhaps the exceptional case of cities on the Rhine to underscore the importance of technological and geographical factors for framing these contexts. There is no doubt that cities like Mainz, Mannheim, Strasbourg, and Basel developed through positioning themselves on the one hand vis-à-vis the Rhine as a major feature of the local geography—sometimes as a political border and inevitably as an unpredictable but more or less malleable body of water—and in a cosmopolitan sense vis-à-vis other Rhine cities as nodes in the Rhine's navigational network. This positioning was embedded in overall municipal strategies that functioned as master plans for expansion, even when, as in the case of Strasbourg, the political landscape changed radically and repeatedly. In his account of the "networking" of cities in the Rhine basin, Dieter Schott (1999c, 736) even suggests that the close proximity of the river, with its spatial and structural challenges, forced river cities to be more astute about their strategies than "dry" cities like Darmstadt: "Networking did not produce its greatest effect in the pioneer case of Darmstadt, where the direct-current system drew narrow limits around its strategic potential. It showed itself to much greater effect in places where—as in Mannheim and Mainz—the municipal administrations, facing grave developmental problems, were challenged to develop more perspicacious strategies, within which subsequently the energy and local transport technologies could be integrated." The "grave developmental problems" in the case of Mainz were the result of its cramped situation on the river, whereas for Mannheim the primary challenge was building up and maintaining a leading position in the Rhine's commercial and industrial networks, despite the imminent threat of losing its status as head of navigation.

At the same time, my narrative shows that cities can be important agents, directly or indirectly, in modifying their geographical and technological environments. The political and commercial elites of these four cities collectively reshaped the Rhine and simultaneously transformed technologies pertinent to riverine navigation and inland harbors. They acted not in isolation, but in the context of dense networks of technological learning and exchange. In this account, the forceful circulation of ideas and practices among Rhine harbor cities has been somewhat overshadowed by an emphasis on competition and rivalries in a contentious political landscape. Nonetheless, mutual learning based on technological exchange and cooperation was no less responsible for the wholesale transformation, within the span of two centuries, of both the cities and the river with which their fates were inextricably intertwined.

Notes

1. The original participants in the Central Commission were the riparian states Prussia, France, the Netherlands, Baden, Bavaria, Hesse-Darmstadt, and Nassau (Eysinga 1935).

2. Charter members of the 1950 commission were the Netherlands, France, Germany, Switzerland, and Luxembourg, with the European Community signing on as contracting party in 1976 (Cioc 2002).

3 Arenas of Experimentation: Modernizing Istanbul in the Late Ottoman Empire

Noyan Dinçkal

In the fifteen years between 1908 and 1923 [Istanbul] completely lost its old identity. The Young Turk Revolution, three major wars, a whole series of fires large and small, financial crises, the dissolution of the Empire, and finally in 1923 our complete acceptance of a civilization whose doorstep we had been occupying, scratching our heads for a hundred years, completely effaced [Istanbul's] old identity.
Tanpınar (1969), quoted in Duben and Behar (1991, 27)[1]

Istanbul has grappled with the question of urban identity for centuries. Here, the city as a self-devouring and yet perennially resurrecting monster is old news: Fernand Braudel (1972, 1: 348), for example, described sixteenth-century Istanbul as "an urban Monster, a composite metropolis." Looking for remains of late classical architecture in the sixteenth century and seeing only feverish Ottoman building activities, Pierre Gilles (1988, 216–233) complained that the city was destroying its Byzantine past. Not without a reason, historians such as Doğan Kuban (1996, 439–441) and Semavi Eyice (1973) lament that rapid growth in the past five decades has created an uncontrollable megacity that devours what little remains of Ottoman Istanbul.

It would be rash to write off these complaints as simple outbursts of nostalgia, to dismiss the attempts to preserve the remaining Ottoman constructions—like the currently popular fountains—as doomed efforts to save the city from change. Still, contemporary developments prompt many legitimate questions about the city's past. In 1996 the World City Istanbul exhibition tried to show the city's history from its beginning to the present (Batur 1996). Various local periodicals, magazines, and books have conducted an intense debate on the integration of the city into the international division of labor, as well as the contradictory effects on its character and identity. The debates circle around a set of urban symbols. Like the first flush of modernity 150 years ago, transportation networks and buildings have played an important part, with the difference that buildings now are no longer made of stone—the building material of early modernity—but of glass and reinforced concrete.

The mid-nineteenth century marks the beginning of Istanbul's radical transformation. Framed by the Westernizing reforms known as the Tanzimat (1839–1876), this period of far-reaching modernization dramatically affected the Ottoman capital, with its many symbolic and representative functions. This development is commonly labeled "Westernization" (Haneda and Miura 1994; Eldem 1999; Berkes 1998). Though certainly problematic, this term does highlight a time when the "West" became increasingly dominant and when the so-called sick man of Europe was born. This modernization led to a redefinition of Istanbul, even the emergence of an entirely new city. "Istanbul underwent profound changes in the nineteenth century," according to Kemal Karpat (1974, 267): "Its social organisation, government, population, and even physical appearance changed so radically as to make it appear a new city at the end of the century." And Zeynep Çelik (1993) titles her historical portrait of the city *The Remaking of Istanbul*.

Modernization of cities led to profound changes in the structure and functions of urban space, as several contributors to this volume show. In the decades of the Tanzimat, Istanbul built modern technical networks such as gas, electricity, telegraph lines, water supply, and sewage systems. This chapter argues that these technical systems were symbols of modernity, and not merely pragmatic adjustments to challenges such as population growth. These urban infrastructures were "arenas of modernity" where the search for a new urban and imperial conception was staged. Using the case of Istanbul, the Ottoman Empire capital on the southeastern periphery of Europe, this chapter explores the role of urban technology in the process of Westernization. First, it presents a general overview of Istanbul's urban modernization during the Tanzimat period. Then, it discusses the direct impact of the West by focusing on the regulation of urban streets and the creation of the district of Pera as a laboratory of modernity. Finally, the chapter considers how actors in the Ottoman Empire, especially in Istanbul, appropriated modern urban technologies, and how they interpreted and placed them in a new context. It focuses on the late Ottoman Empire, roughly from the mid-nineteenth century to the years after the Young Turk Revolution, often referred to as the Second Constitutional Period (1908–1918).

Urban Reform and the Place of Technology

From an Ottoman or Turkish perspective, the reforms known as modernization or Westernization were strongly conditioned by European phenomena and European actors (Berkes 1998; Ortaylı and Akıllıoğlu 1986). At least from the mid-nineteenth century, the city of Istanbul tried to follow what domestic actors understood to be the European way. Modeling their efforts on the French *préfecture de la ville*, the Ottoman government established an urban administration for Istanbul (Şehremaneti) in 1855. This administrative change was provoked in part by the complaints of European settlers living in the city, whose numbers had considerably increased during and after the Crimean War (1853–1856). A Commission for the Order of the City (*Intizam-ı Şehir Komisyonu*) began its work in May 1855. Its founding document explicitly stated that, if Istanbul wanted to compete with the capitals of leading European countries, then it urgently needed to be embellished, cleaned, and regulated. Its roads had to be illuminated and extended; and the city's building types, methods, and materials had to be improved and regulated. According to the commission's subsequent reports, setting up standardized routines and creating a secure legal framework for urban developments were the best way to reach these goals. Paris was taken as the leading model (Ortaylı 1974, 95–111; Çelik 1993, 42–48). Here was the Europe-centered blueprint for becoming a "modern" city.

The commission's report showed immediate results, initiating a phase of general urban transformation that implemented several modern urban technologies. The rapid changes in Istanbul depended, however, on a complex of various factors. The commission's insistence on standardized decrees, rules, and laws cannot be explained by any abstract preference for formalism. Rather, it was an attempt to increase the control and authority over the capital city by the Sublime Porte (*Bab-i Ali*, the administrative center and government of the Ottoman Empire). In other words, the efforts to modernize Istanbul were an integral element of a policy that aimed at centralizing the Ottoman realm (Yerasimos 1992).

Istanbul had never been easy to control. Yet urban conditions changed substantially as the city's population grew to previously inconceivable dimensions. After lost wars and imposed territorial losses disrupted their homes, Muslim refugees streamed into Istanbul from the Balkans and the Caucasus in the second half of the nineteenth century. By 1900 the population exceeded one million (Karpat 1985, 13). The rapid demographic and social changes touched off several difficult conflicts, and the situation in fact grew worse because the Tanzimat reforms created a power vacuum (Keyder et al. 1993). The surge of immigrants, most of them impoverished refugees, dramatically altered the city's social structure. Especially in the poorer quarters of the city like Kasımpaşa, dire sanitary conditions led to epidemics of cholera and typhoid (Dinçkal 2004, 160–175). Devastating fires broke out regularly in Istanbul's densely built inner city, since the majority of the residential buildings were traditional wooden houses (Tekeli 1994, 11–12).

Ostensibly to improve safety and security, local authorities straightened and extended streets and lanes. The lanes and alleys in particular were difficult to access in case of fires and not easy to control at any time. Removing the tangle of traditional lanes, city officials imposed modern right-angled streets and thoroughfares. "Presently it is as difficult to maintain order in Istanbul as it is to protect a forest that has never seen an axe," the Commission for Road Improvement reported in 1866: "If streets were built whose one end could be seen when standing at the other end, security could be provided by half the security forces we need now" (quoted in Ergin 1914–1922, 5: 990).

For the Sublime Porte, the European model of urban modernization appeared attractive, since it allowed the government to claim increased control over urban space (Yerasimos 1999, 6). Such a drive for control on the part of modernizers was hardly unique to Istanbul (see Schott and Lundin, this volume), but it is only one side of the story. In a world that was increasingly influenced by European powers and Western culture, the strategy of the Ottoman Empire was to preserve the empire's integrity by adopting Western science, technology, and organizational structures. In this respect, the Ottoman Empire could be seen as an example of so-called defensive or reactive modernization. That is, the ruling elite initiated social reforms in an attempt to accommodate the external and internal pressures to adapt the empire to the demands of modern times. Consequently, Istanbul's administration chose urban technologies already familiar in Western European cities. Appropriating these modern technologies was more than a mere response to current deficiencies or problems; it was also a symbolic act. Urban reforms and urban technologies thus had functions and meanings beyond their institutional or everyday, practical purposes.

Ottoman reformers believed that modern urban technologies—such as central water supply, sewers, and streetcars—would raise the general level of culture, which was defined and redefined in a continual and contested process. Prestigious urban modernization projects, carried out in the spirit of science and technology, became an indicator for the city's progressiveness and for the country's political, economic, and cultural power. It is not at all surprising that Istanbul, the Ottoman Empire's capital, was especially influenced by the Paris of Georges Eugène Haussmann. In his description of how Ottoman and Turkish novelists and poets constructed, comprehended, and even mystified the "other space," Enis Batur (2000, 285) concludes that Paris was the center and the symbol of Europe: "Is Paris just a bridge to a different world, to the West? No, it is the transition to a different era as well, to a new calendar ... a stepping stone to modern civilization." By adapting Western technologies and customs and providing modern urban services, the "progress of civilization" should be brought to the Ottoman capital, at least to its wealthy quarters. This stance also had an ideological motive: the modernizers claimed that the Ottoman Empire was not an old, declining empire at all but had become a modern state (Faroqhi 2000, 254).

The Transformation of Istanbul

The capital's remodeling unfolded in several phases. Up to the 1870s, Istanbul's modernizers focused on laying out and building streets in the wake of several devastating blazes, as well as on drafting and enforcing new building regulations. After 1870, the focus shifted to terminals, docks, water supply, and the planning and construction of large, prestigious buildings. New residential quarters for the rapidly growing population went up throughout these decades. With the increasing division of labor and the differentiation of social strata, the city's residential pattern also became socially and spatially segregated. For instance, the emergence of "middle-class" areas led to a boom in the construction of terraced houses by the end of the nineteenth century (Göçek 1996; Kırsan and Çağdaş 1998). Furthermore, the central business districts, especially Galata, were reconstructed to meet the requirements of an expanding foreign trade. The outcome was a modern urban network of streets, sewage, docks, central water supply, and local transportation. As we will see, this was not accomplished without significant cultural strain.

Members of the local elite took up a prominent role in modernizing the city. Foreign trade companies also pressured the municipal administration to replace the old docks, which were no longer suitable for larger, modern ships. These demands fit well with the interests of wealthy local residents, who wished to develop seaside roads and pedestrian areas into prestigious places where—to paraphrase Paolo Capuzzo's contribution to this volume—they could follow strategies of spectacle and sociability. Even though a new quay had been built in Karaköy in 1849, it took another 30 years to effect a complete remodeling of the whole shore area. Finally, a modern space of urban circulation emerged when a strip of 20 meters was cleared between the quay walls and the buildings to make space for the brisk traffic on the seaside road and promenade. These reforms helped give Istanbul the "unified physical and human space" keenly sought by urban modernizers everywhere (Hård and Misa, this volume).

Regular streetcar service tied the Karaköy quay effectively to nearby districts. In 1869, a city decree authorized four horse-driven lines in Istanbul, although not all of them were actually opened. The streetcars ran from dusk until midnight, with timetables in four languages. The city's first subway line joined the transportation network in 1875, when the *tünel* was opened between Karaköy and Pera (today Beyoğlu). The *tünel* was the world's third subway, after those in London (1863) and New York (1868). The line climbed more than 60 meters at a steep gradient of 15 percent, and carried an average of 40,000 passengers per day. The engineer who constructed it also proposed a 4-kilometer subway line from Kumkapı to Eminönü and then from Karaköy on along the shoreline up to Ortaköy, but these lines were never built (Çelik 1993, 90–103).

The importance of the urban rail system should not be exaggerated. Ships and boats have remained, even through today, one of Istanbul's most important means of transportation. In 1844 the three main boat lines in the city employed 19,000 rowboats; the first steamship arrived in 1850. Şirket-i Hayriye, the first Ottoman steamship company, was founded in 1851 and was soon granted a monopoly on the valuable inner-city routes (Tekeli 1994, 3–49). By 1888 seven different steamship lines served passengers at half-hour intervals and were subject to strict municipal regulation.

Figure 3.1
Regulated seaside road in Büyükdere and the Bosporus, between 1880 and 1893. Photographer: Abdullah Frères. Abdul Hamid II Collection, Library of Congress (LC-USZ62-38083).

Due to the disastrous economic situation of the Ottoman Empire and the lack of technical know-how, the establishment and extension of urban technologies required the acquisition of loans and credits, and private companies typically managed the enterprises. The Ottoman state had to declare bankruptcy in 1875, and in 1881 the international Public Debt Administration, directed by European financiers, stepped in. The private companies involved in urban engineering projects were mainly foreign firms that tried to explore new markets, exploiting the wave of urban modernization and the common euphoria for technology (Tekeli and Ilkin 1996). For the national and municipal authorities, especially the Ministry for Public Works, the question was whether they should erect and operate the infrastructure facilities themselves or instead grant concessions to private companies, and if so, under what conditions. In Istanbul, as in many other European cities, the urban infrastructures that promised sufficient profits (gas, water, and streetcars) were built and operated by private companies, while the city more or less took charge of unprofitable services such as sewage.

Regulating the Streets

For several decades the municipal efforts to modernize Istanbul focused on street regulation. Although the city's main purpose was to eliminate dead ends and to straighten and extend the streets, it had multiple, interrelated motives. As has already been mentioned, the rapid growth of the population and the need to enhance order and security were important factors. However, street regulation was necessary as well for implementing such urban technologies as transport and sewage. Street regulation also conspicuously affected urban order in that it entirely reconstructed the border between public and private urban space. Finally, in many areas of the city, the laying out of new, wide streets was a means to exhibit the city's power and authority.

Street regulation was a pronounced area of conflict between municipal modernizers and representatives of the traditional structures of the city. Istanbul exhibited a functional division between residential and business areas, and between private and public spaces. Officially the city was divided into modern administrative districts, but in practice it was divided into *mahalles*, traditional residential quarters in which almost all of Istanbul's inhabitants lived. For a long time the *mahalle*s remained the basic building-block of the urban fabric. Comprising a true maze of lanes and streets, a *mahalle* was not really an open public space, since its inhabitants controlled the coming and going of people. Prior to modernization, Istanbul, like pre-Haussmann Paris, had only a few large publicly accessible thoroughfares.

Up to the second half of the nineteenth century, the inhabitants themselves, not the city administration, controlled the use of the streets. When streets were to be narrowed by the building of shops, the inhabitants took charge, initiating and managing projects like the building of new cesspits. In the traditional Ottoman system, essential municipal services were provided by the inhabitants of the *mahalles*, under supervision of the urban administration. Urban administrative duties were the responsibility of the *kadis* (religious judges). Once the urban modernization projects of the Tanzimat period began, the *kadis* were stripped of their administrative power and lost many of their traditional powers as representatives of the local urban communities. Street regulation projects became arenas of power struggle between the traditional and modern administrative structures. In many cases, the city's modernizing administration refrained from pushing through a radical and disruptive solution, since they did not want to risk open confrontation with the inhabitants (Yerasimos 1999, 11–12; Çelik 1993, 55–57).

The tangled relations between street regulations, changes in the existing building structures, and the implementation of modern infrastructure were by no means peculiar to Istanbul; in this volume Hans Buiter explores this same relationship in the case of Dutch cities, for example. Haussmann, the driving force behind the modernizing of Paris in the middle of the nineteenth century, faced similar challenges. Whereas street regulation was also a centerpiece of urban modernization in Paris, there the problems were solved by destroying the poorer quarters and their narrow lanes. Given the importance of Paris as a model for Istanbul, it is remarkable that Istanbul's regulation measures and changes in its building structure were by comparison quite flexible and far-sighted. Instead of radically destroying the traditional structures, Istanbul's municipal administration only gradually increased its influence on urban development and little by little diminished the importance of private initiatives in the quarters. Although the residential quarters changed quite radically as a result of the street regulations, their social composition showed remarkable continuities. In stark contrast with Haussmann's Paris, Istanbul's low-income groups were not displaced to the periphery of the city (Çelik 1993, 79–80).

Fires, Transportation, and *l'Embellissement*

In Istanbul, as elsewhere, many aspects of urban transformation—the remodeling of the streets, various building codes, the required building materials, and the erection of central water supply works—were related to the dramatic impact of nightmarish fires on the mentality of inhabitants (cf. Schott, this volume). Due to the lack of water for firefighting, the use of wood as the predominant building material, and the high density of construction, even small fires could turn into catastrophic blazes that reduced entire quarters of the city to charred rubble. The narrow lanes impeded the fire brigades, so that they often arrived too late to be of any help. In the 1860s alone, four major fires destroyed more than 3,200 buildings. Municipal administrators argued that the opening-up of dead-end streets and the straightening of others were needed to protect the city from devastating fires. The authorities even went so far as to ban the construction of wooden houses. In spite of a variety of building codes, however, not all of these aims were fully reached. Especially the poorer house owners could not afford stone or brick, and so the municipal administration contented itself with enforcing the building codes that prescribed the use of these materials in the major, fashionable or showcase streets. In other areas it was usually acceptable to set up fireproof walls that could serve as barriers to the wider spread of fires (Tekeli 1994, 24–38; Çelik 1993, 52–55).

Fires and street regulation were mutually dependent in a certain aspect. Ironically, the catastrophic fires themselves allowed the city government to straighten and extend the streets when the burned-out areas were reconstructed. Zeynep Çelik (1993, 49) notes that these spaces became "arenas of experimentation" for Western-inspired models of urban planning. A quarter with a certain number of burned-down houses was simply declared "undeveloped land" and then divided into new building lots. Furthermore, after 1882, owners could be compelled to give up parts of their lots for street extensions. In this way Istanbul's "old city" was dramatically altered, and the old web

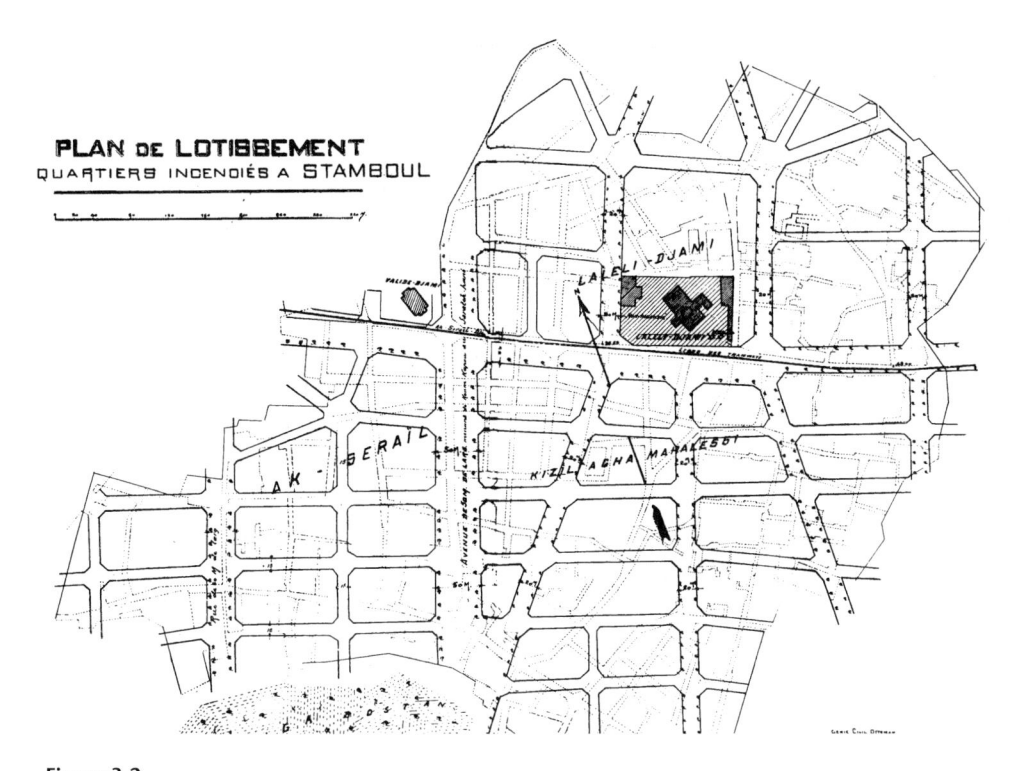

Figure 3.2
Street regulation in the course of rebuilding burned-out areas between the Valide mosque and the Laleli mosque in the quarter of Aksaray in 1912. The rebuilding included an extension of the already existing system of sewers and the building of new sewers. In the background, faint dotted lines indicate the former street system, which consisted of both the classic structure of winding lanes and alleys and geometrically planned thoroughfares that had been built after fires in the nineteenth century. Source: "La reconstruction de Stamboul," *Le Génie Civil Ottoman* 2, no. 1 (1912): 5.

of streets and lanes almost completely disappeared. After a major blaze in 1856, the new streets in the central quarter of Aksaray were classified hierarchically in relation to their breadth, even though residents condemned the measure as contemptuous of local heritage.

The reconstruction of the burned-down areas also allowed the laying out of new sewers. The planning process that followed the fires of the 1860s, which destroyed much of Istanbul's old city, shows the close relation between the installation of sewers and the widening of streets. After this huge fire, generously proportioned avenues were built, one running from east to west between the Place of Beyazıt and the Hagia Sophia and one from north to south, from Divanyolu to the Golden Horn. The adjacent streets were straightened and extended, and sewers were laid underneath the newly built streets (Dinçkal 2004, 155).

The sewer grid was extended even further after the 1908 revolution. As part of their modernization strategy, the Young Turks tried to reorganize the municipal administration and to restructure urban services, concentrating in particular on Istanbul. When Ethem Halil Bey, Istanbul's mayor at the time, had to find a director for the city's technical department, he offered the position to André Joseph Auric, then chief engineer of the city of Lyon. Auric accepted the position for three years, bringing his French staff with him. Among other things, Auric developed plans and projects for the reconstruction and regulation of the burned areas. His reconstruction was far more than a simple response to an emergency; it was part of a program that went under the label *l'embellissement*. As Auric (1911, 1) himself explained: "boulevards are nothing but big streets, yet of greater dimension, wider, breezier, brighter, sunnier, and in particular bordered by trees, which lends them a more dignified and majestic appearance, something characteristic for big cities." Broad streets and generously laid out boulevards followed the contemporary urban aesthetics and added to the expressive or showcase character of the capital. By contrast, the dark maze of streets and lanes in Istanbul's old quarters were regarded as shameful symbols of an antiquated and unhealthy past. In fact, sanitary concerns played an important role in the redevelopment of previously densely populated areas. While the boulevards had symbolic functions, they also provided inhabitants with sunlight and fresh air—a policy that included extending the green areas of the city.

Not surprisingly, traffic considerations and social status also entered into the debate on street reform. Deliveries of food, wood, and other goods needed for the rapidly expanding city clogged the traditional street network. Nevertheless, the inhabitants of the affected quarters resented the modern thoroughfares as an invasion of their privacy, because they regarded the municipal (and sometimes national) street regulation as an invasion of a space that they considered a nonpublic area. After all, streets do more than merely move people around the city. According to Suraiya Faroqhi (2000), the city's streets were widened and straightened to handle larger carts for cultural as well as practical reasons. The wealthy residents' use of oversize carts and carriages that required broader streets indicated their higher social status and were preferred for reasons of self-representation. An important Ottoman novel called *The Carriage Love* portrays heavily decorated coaches carrying secluded groups of women. Such coaches were also popular among European photographers and their clients (Evin 1983; Faroqhi 2000, 254–255). The importance of carriages for the better-off inhabitants of Istanbul is illustrated in the diaries kept by a wealthy bureaucrat, Said Bey, between 1901 and 1909. Bey vividly describes how his social decline started when he was forced to sell his carriage (cited in Dumont and Georgeon 1985; Exertzoglou 2003).

In a somewhat darker vein, Palmira Brummett describes the contested nature of the modern street in her work on the satirical press in the Ottoman Empire. Popular cartoons condemned the economic marginalization of the Ottomans through their exploitation by European business circles and the disastrous effects on traditional lifestyles and occupations. The satirists were not at all enthusiastic about modern technologies, and they repeatedly made readers aware of the physical hazards that came with modern streets and traffic: "citizens were shown killed or injured by cartoon streetcars and cars that bore down on unsuspecting pedestrians accustomed to more traditional modes of transport" (Brummett 2000, 296). Clearly, some of Istanbul's residents sensed that modern, Western-style streets also carried into their midst the most troubling aspects of "a civilization whose doorstep [they] had been occupying...for a hundred years."

Pera: Laboratory of Modernity

Pera, the sixth district of Istanbul, took on a pioneering role in the city's modernization. Modeled explicitly on the *sixième arrondissement* of Paris, Pera was literally and figuratively a "laboratory of modernity." New models of communal administration and new urban technologies and services were introduced and tested in this quarter. Modernizers appropriated these new technologies to fit local circumstances and intended to apply them to the whole city (Kuban 1988; Yerasimos 1999; Rosenthal 1980b). Due in large measure to the government's embrace of modernity, efforts to modernize the district took on extraordinary importance. Through the modernization campaigns, Pera, as well as Beşiktaş on the shore of the Bosporus, evolved a new kind of social and ethnic segregation. People were moved around, and not always in directions they preferred. When the court abandoned the old Topkapı palace and moved to the new Dolmabahşe palace in 1856, the shore area toward Beşiktaş gained enormously in value. The move symbolized the ruling dynasty's determination to modernize; it also shifted Istanbul's old, inner-city center to the thriving areas north of the Golden Horn, the center of which was Pera (Rosenthal 1980a; Reimer 1995).

In demographic terms, "modern" Istanbul was something of an urban island and hardly representative of the whole city. The census of 1885 showed that in the sixth district (Pera), 47 percent of the inhabitants were foreigners, 32 percent were non-Muslim Ottomans, and only 21 percent were Muslim Ottomans. In its share of foreigners, the fourth district (Beşiktaş and the settlements on the European side of the Bosporus) ranked second, with 10 percent foreigners and 43 percent Muslims. Clearly, European settlers made up an important share of the population of both Pera and Beşiktaş. These two districts contrasted sharply with the structure of the old-town districts on the southern side of the Golden Horn. There, Muslims made up 55 percent of the population, and foreigners a mere 1.5 percent (Shaw 1979, 268; Eldem 1992). Because of its social structure, in combination with its economic dynamic, Pera seemed an especially suitable target for modernization.

Sharp differences in ethnicity, social structure, and everyday culture characterized Istanbul's urban scene. The specific structure of the sixth district affected the cityscape and the urban culture decisively. Most of the banks and joint-stock companies had their offices in Pera, new ministries were built in European style in Pera, Istanbul's first city hall was erected in Pera, and the European embassies were situated in Pera. On the whole, a recognizably modern type of urban life developed in this "laboratory." The Westernization of habits and the changes in the urban culture were thus far more pronounced and clearly visible in the sixth district than in any other part of Istanbul. Pera's cityscape displayed all the elements of the modern, *alafranga* (*alla franca*) way of life: public squares, hotels, restaurants, cafés, shops, office buildings, apartments, and theaters. Men and women dressed differently there from the inhabitants of Istanbul's other districts. The distinct iconography gave this district a different, definitely more European appearance (Mardin 1974, 419–426; Çelik 1993, 80–81). One traveler stated, "one could think that one was in a second or third-rate Italian city ... though annexed to Istanbul this city [Pera] is as different from it as it would be from Peking or Calcutta" (Ubicini 1855, 443).

Some of the European foreigners living in Pera were able to give expert advice on questions of modern municipal administration. In 1857 the Commission for the Order of the City initiated administrative reforms in hopes of strengthening the powers of the different local urban areas. The commission divided the city into 14 districts, and explicitly labeled the sixth district as a "laboratory" for reform: "Since to begin all things in the above-mentioned districts [except Pera] would be sophistry and unworthy, and since the Sixth District contains much valuable real estate and many fine buildings, and since the majority of those owning property or residing there have seen such things in other countries and understand their value, the reform program will be inaugurated in the Sixth District" (quoted in Rosenthal 1980a, 233). The members of the first council were wealthy citizens of the district. Curiously enough, foreign consultants who moved there even had the right to vote, as long as they had financial means of at least 500,000 kuruş. The term "consultant" was meant to diffuse the resentment of those who were prone to regard as unacceptable the involvement of foreigners in city administration.

The new administration began its practical work by extending and resurfacing the main streets of Karaköy and Pera. New laws allowed the city to confiscate private property for such public purposes as the extension of streets. Gas lighting was installed along the sixth district's main avenue, Grand Rue de Pera / Cadde-i Kebir (today Istiklal Caddesi), and a refuse collection system was developed. However, a prestigious business complex called the Karaköy Han turned out to be a bad investment. When it was

Figure 3.3
View of the Galata Bridge, between 1880 and 1893. Photographer: Abdullah Frères. Abdul Hamid II Collection, Library of Congress (LC-USZ62-7948).

finished, tenants for the modern shops and offices proved almost impossible to find. This project underlined the considerable discrepancy between the aspirations of the city council and the actual needs of the population. In order to win the population's consent, the administration slowly learned that it could no longer focus only on the "embellishment" of a few prestigious buildings and that it had to offer extensive improvements in urban services.

In the fall of 1864, one of the most important restructuring projects in Pera began with the destruction of its old city walls. After nearly thirteen million cubic feet (370,000 cubic meters) of stone and debris had been removed, the area was perfectly suited for the construction of modern Western-style avenues and boulevards. A prestigious building for the municipal administration was erected at the corner of a newly designed square. Both the square and the building were greatly influenced by the architecture in Paris; in fact, though considerably smaller, the square distinctly resembled the Place de l'Étoile. But unlike the Parisian model, the streets that led from the square had no organic connection to the rest of the city; two of them actually ended abruptly after only a few meters. To some extent, this piecemeal development was typical for the era. Instead of a complete structure that integrated the parts of the city, the result—even in Pera—was a patchwork that had only formal similarities to the European models (Çelik 1993, 70).

Infrastructure and Dual Urban Development

"Dual urban development" implies a specific type of unequal urban modernization that was prevalent during the peak of colonialism in the nineteenth and the beginning of the twentieth century. It is a useful term that describes the building of certain new, modern quarters in European style, while allowing the old parts of the city to remain more or less neglected (Abû-Lughod 1980). While such development was quite blatant in cities under direct colonial rule, it was less obvious, though still present, in Istanbul. Istanbul's modernization was not based exclusively on foreign impulses, nor was it characterized by strict segregation; instead, it was marked by a partial integration of "islands of modernity" into the fabric of the city. Nevertheless, the modernization of Istanbul clearly showed certain characteristics of a dual urban typology. The case of water supply illustrates this point.

The first concession agreement with the French Compagnie des Eaux de Constantinople in 1874 prescribed neither the extent of the water system nor the number of households to be connected to the water pipes. There were, however, instructions detailing which specific districts of the city were to be provided with tap water. Not surprisingly, when the first central waterworks was put into operation in 1885, the "modern" districts of Pera and Beşiktaş were the first to be served.

The economic, cultural, and political influence of these districts' population gave them a specifically European character. Increasingly, a modern lifestyle required individual hygiene and the domestic use of water. Already before 1885 the district's demand for water was considerable, and it seemed that the old Ottoman water mains could not possibly meet the long-term demands. The Europeans who lived in the sixth district were accustomed to having tap water on demand in Europe and regarded domestic water supply as a necessity. This specific population structure was, of course, advantageous for the Compagnie des Eaux de Constantinople. Pera promised to be an extraordinary lucrative market because of the potentially high consumption of water, not to mention additional and expensive sanitary articles like bathtubs.

Not surprisingly, beliefs about the financial strength of potential clients played an important part in the water company's preference for Pera and Beşiktaş. This can be confirmed by the scheme for pipe laying and distribution designed by Paul Boutan, the engineer in charge, two years before the works was put into operation. The list of premises designated for water was a virtual roster of the European and administrative

Figure 3.4
A typical street in Pera (Beyoğlu), between 1880 and 1893. Photographer: Abdullah Frères. Abdul Hamid II Collection, Library of Congress (LC-USZ62-78325).

buildings of the city: the embassies of France, Britain, Italy, Spain, and Russia; the hotel complexes Bycance, Pétala, Belle Vue, Universe, and L'Apierre; the shopping arcades Hazzopoulo and Oriental; the German and Italian hospitals; the churches of St. Constantin, St. Paraskeri, and St. Nikita; and finally, the municipal administration building at Rue Kabristan, a mosque, and the palace of Ali Paşa.

This blatantly preferential water supply policy lost its importance as time passed. In the revised contract between the French consortium and the Ministry for Public Works in 1887, the supply area was extended to cover the whole "European" side of Istanbul. The Compagnie des Eaux de Scutari et Kadi-Keui, the city's second water company, supplied Istanbul's "Asian" side from 1893 on. Similar to the "European" side, the supply area of this company was extended to the periphery of the "Asian" side in 1914. By the first decades of the twentieth century, the entire city had been provided with water mains (Dinçkal 2004, 82–121).

But even if access to water was made less discriminatory over time, water still had a pronounced impact on the city's social structure. Daniel Headrick (1988, 147) traces how evolving standards of sanitation led to residential segregation in countries affected by imperialism: "[A]mong Europeans and those few who could afford to emulate them, a growing concern for hygiene provided the motive for segregation, while new technologies of water supply, sewage disposal, and building materials provided the means. And since these technologies were costly, they were reserved for the better-off." This statement certainly applies to the initial stages of Istanbul's water supply. The company's supply strategy overlapped with the well-articulated demands of the supply area's wealthy inhabitants. In the initial period, the focus lay mainly on the central water supply of Pera, that is, on the European citizens and the Ottoman upper classes. The supply of modern technology in the wealthier districts was given priority over a fair or at least a wider distribution of water, thus buttressing the tendencies toward segregation and distinction that Headrick observes for colonial countries.

Circulation of Knowledge—Appropriation of Technology

How did technical standards and norms, along with scientific and technological knowledge, circulate to Istanbul, through which channels, and by which means? Basically, one can say that circulation roughly followed the same pattern as in the case of other European cities (Buiter, this volume). Private companies played a prominent role. Lacking homegrown experience and trained technical staff, Istanbul's authorities assigned the building of urban infrastructures mainly to companies based in Europe. These companies were also a means for securing the immense investments necessary (Dinçkal and Mohajeri 2002). Foreign joint-stock companies built modern waterworks in almost all the big cities of the Ottoman Empire. A concession for Beirut was granted to the British Compagnie des Eaux de Beyrouth, while two Belgian companies built the waterworks of Salonika and Izmir. In Istanbul, water concessions were granted to the German-owned Compagnie des Eaux de Scutari et Kadi-Keui and the French Compagnie des Eaux de Constantinople, a subsidiary company of the well-known Compagnie Générale des Eaux pour l'Étrangère. In addition to Paris, this company operated in such cities as Porto, Venice, Naples, and Trieste.

One early consequence of this model of water concessions was a flow of European contractors and suppliers to Istanbul. A French engineer directed the waterworks in Istanbul, overseeing local construction by such companies as Righetti, Abdulrahmen Agha, Vassilaki Johannides and Bariola, and Mavrogordato. European hardware largely constituted Istanbul's water system. The water mains and pumping plant were provided by La Compagnie de Fives-Lille (Nord), R. Laidlaw and Sons, and Thomas Edington and Sons (Glasgow). The household supply points were installed by the French company Dalmas, which also sold sanitary facilities in Istanbul. We know that approximately 900 workers (about 75 percent of them Ottoman) were employed in the construction of the waterworks. Similar conclusions can probably be drawn concerning the workforce of the Compagnie des Eaux de Scutari et Kadi-Keui. For its waterworks on the "Asian" side of Istanbul, this consortium turned to the Swiss engineering company Gruner AG of Basel. These companies were the principal means for circulating water technology from various European locations to the city of Istanbul (Dinçkal 2004, 82–85, 91).

The extension of the urban infrastructure and the resulting entry of engineers into municipal service created a new foreign-dominated technological elite in Istanbul. To an even larger extent, the city authorities relied on the experience and knowledge of foreign experts in urban planning. André Joseph Auric, the director of the city's technical department, has already been mentioned. A list of the French, British, Italian, and German engineers, architects, and urban planners who worked in Istanbul would include people such as Helmut von Moltke, Joseph Antoine Bouvard, and Luigi Storari, as well as Raimondo D'Aronco and Marius Michel (Çelik 1993, 53–55, 74–76; Tekeli 1994, 40–41). Many of them were hired by the municipal administration and the Ministry for Public Works in order to provide state-of-the-art knowledge—a crucial factor when making decisions about infrastructures that might persist for many decades. Many expert committees were appointed whose members had to sort out the advantages and disadvantages of existing technologies and make suggestions for new systems. Such committees are important connecting links in circulating knowledge on the individual and institutional levels (cf. Lundin, this volume).

An exemplary figure in this context of pervasive circulation was Eugène Henri Gavand, a French engineer who lived and worked in Istanbul for much of his life. He was a remarkable man in many ways. In the late 1860s, Gavand worked for the first communal administration of Pera and served on a committee created by Istanbul's mayor Server Efendi. Its mission was to suggest how to overcome the recurring water shortages. Other members of the committee included Leval, chief engineer of the sixth district; Tridon, an engineer working for the Ottoman administration; and Barborini, an architect and urban planner; as well as a number of domestic Ottoman experts. Within its framework, Gavand proposed a plan in 1869 for the central water supply. However, when he applied for a concession, the Ministry for Public Works rejected his proposal because of the project's financially insecure situation (Dinçkal 2004, 73, 78–79). Clearly disappointed, Gavand nonetheless went on to become the driving force behind Istanbul's subway, a project that was eventually completed with British capital after the French administration had denied financial support.

These developments led the authorities to become dependent on foreign expertise and foreign capital. With hopes of lessening this reliance on foreign technical experts, the Ottoman government created a new technical education system that sent numerous students on state-funded scholarships to Western Europe. In this way, direct student experiences complemented the traditional Ottoman ambassadors' reports from the capital cities (such as the reports of Mustafa Reşid Paşa, to name a popular example). Students were sent mainly to Paris, Vienna, and London to study and acquire technological knowledge (Yerasimos 1999, 3–5; Gül and Lamb 2004). Subsequently, these cities became the implicit models for Ottoman urban infrastructures. According to Ekmeleddin Ihsanoğlu (2000, 35–36), already in 1839 36 students from the Ottoman military and engineering schools were studying in these cities. About 50 students went to Paris between 1848 and 1856, and 61 between 1856 and 1864. A total of 93 students studied in Paris during 1864 and 1876, 42 of whom went to study science, and 51 to receive apprenticeship training in various technical vocations. Additionally, in 1857 the Mekteb-i Osmanî, an Ottoman school, was opened—conveniently enough, in Paris—with the aim of educating students from the Ottoman Empire in science and the arts.

As Hård and Stippak show in their chapter in this volume, images and representations of a city draw partly on local traditions, but are also based on contrasts and comparisons at varied levels of experience. The same holds true for technology, which is also acquired and interpreted in different ways and is subject to significant changes and mutations. Two examples illustrate this point: the streetcar and the water supply. There are several ways to look at Istanbul's streetcar lines. Emrence (2001, 9) presents a picture of the chaos he calls "Istanbulization": the juxtaposition of the city's hectic pace with the extremely slow pace of the streetcars pulled by horses on the city's hilly streets. Brummett (2000, 300) stresses the development and extension of a net of broad thoroughfares that eventually became the main streetcar lines. These interpretations may be complemented by yet a third perspective that underscores that the new urban technologies were often subject to multiple appropriation processes. The Ottoman streetcar asserted traditional culture in a modern conveyance: it was divided, as was the Ottoman house, into men's and women's sections, *selâmlık* and *harem*. The same was true for the Istanbul *tünel*: the subway cars had two (fare-based) classes, and each class had separate compartments for male and female travelers (Çelik 1993, 97). This assertion of culture in the division of space is a neat metaphor for the Ottoman appropriation of foreign technologies. Although these technologies were certainly of foreign origin, in the process of circulation they were culturally conditioned and transformed to fit into local cultural patterns and preferences.

A distinct cultural adaptation and technical transformation also took place in the case of water supply. Private companies erected Istanbul's waterworks (Terkos and Elmalı) primarily to supply the European citizens and the Ottoman upper classes, as discussed above. Ottoman authorities, however, especially the Ministry for Public Works, pressured both foreign companies to erect public fountains at certain locations, supplied with water from the modern waterworks. As with the centuries-old system of classical fountains, these fountains were to provide water at no cost to the users. In adopting modern technologies and hygiene standards, the Ministry of Public Works secured the trappings of modern life that accompanied a central water supply. At the same time, it delivered water to the majority of the population that lived in strained circumstances—much to the regret of the water companies, since they had to supply this water for free.

Hybrid Cultures and the Quest for Adventure

Through processes of circulation and appropriation, urban technology became a vital part of Istanbul's culture and identity. Residents experienced urban technology through its visible expressions in architecture and the urban landscape, as well as through the changes, both welcome and unwelcome, it brought to daily life. Urban technology was an important symbol of a modern urban identity. Modernity was inseparably linked to new technology, which itself was also seen as directly related to the "West." The differential access of Istanbul's citizens to urban technologies also generated social inequality within the city, and the reliance on European urban technologies symbolized the economic and political dependency of the Ottoman Empire, the "sick man of Europe." For all these reasons, the transformation of Istanbul was a fundamental break that transformed the functional logic of the urban society.

Of course, this kind of urban transformation could be dismissed as yet another aspect of foreign dominance and one-sided modernization. Many scholarly studies premised on a strict distinction between different cultural entities describe modernization as the impact of different (foreign) values on (domestic) behavioral and consumption patterns. In this perspective, modern urban technologies are the result of the diffusion of the above-mentioned cultural patterns, with foreigners and Ottoman minorities being the agents of change (Kazgan and Önal 1999, 24–25). Such a perspective, however, entails the severe liability of relying on static cultural concepts and the use of such overly simple binary oppositions as Europe/Non-Europe, Occident/Orient, or Christianity/Islam. It is highly problematic to believe that installation of central water supply, sewerage, streetcars, and other modern urban technologies are, in simple terms, met by a traditional order, characterized as Islamic or oriental, that hinders innovation.

This often-repeated perspective neglects the clearly articulated interest of the Ottoman elites in introducing new technologies. Even the neo-absolutist regime of Abdul-Hamid II (1876–1909), with its strongly religious grounding, exhibited an active interest in technical innovations, especially when they seemed suitable for enhancing the power or prestige of the regime (Davison 1973). Religious belief and progressive thinking were not at war; indeed, Islamic religion was brought in line with the knowledge of modern science in order to overcome the economic, technological, and social deficiencies of traditional society (Shaw and Shaw 1977, II: 221–223). Although the Ottoman Empire was in many ways influenced by and dependent on European powers—especially, as we have seen, in urban technologies—it was never colonized politically. For the most part, the Ottoman upper classes kept the process of urban modernization under their own control.

It is a simple fact that modernization and urban renewal were strongly favored by the Ottoman elites, who eagerly adapted Western technology and habits. As students or attachés, they had become habituated to modern urban life in Western Europe. Especially during the Young Turk period in the early twentieth century, wealthy Muslim families took up residence in the newly erected apartment buildings in the district of Pera and the new neighborhoods of Nişantaşı and Şişli because of their access to such newly introduced municipal services as street lighting or a central water supply. All the same, there is yet another aspect to consider. The modernizing of Istanbul clearly created new borders of social inequality and new areas of conflict. But did it also create new areas of social interaction and communication? Emrence (2001, 10–13) suggests that streetcars in Istanbul helped women to gain a new presence in public urban space (see the striking parallel with London's West End discussed by Capuzzo elsewhere in this volume). Based on the study of memoirs, Emrence finds that streetcars became spaces of real or imagined adventure because the passengers traveled in the same streetcar, even though they were in different sections or separated by a curtain according to gender and class. Yakup Bektaş argues a similar point in his study on technology and cultural diversity in the Ottoman Empire. For Bektaş (2000; 2003, 145–146), the telegraph office represented a new unique space for intercultural experimentation. He writes that in a telegraph office of the 1870s one could "find an Armenian director, a Turkish inspector, a Greek mechanic, and a British or French operator." Simplistic models of coercive "cultural imperialism" cannot account for such meaningful subtleties and consequential social innovations, even (perhaps especially) if they took place in the symbolic shadow of the West.

Finally, can it be said that the Ottoman elite's increasing interest in Europe and its modern urban technologies was in any way linked to the political and economic crises that seem to have unsettled the Ottoman Empire? Was the adaptation of modern technology and science merely a strained effort to keep up with the "superior" West (Lewis 1968, 41–43)? However plausible this interpretation may be, it is also unduly biased because the adaptation of modern urban technologies was not confined to any necessary imitation of Western patterns. A more satisfactory account recognizes that the cosmopolitan Ottoman elite paid eager attention to Western artifacts due to their intellectual curiosity and their willingness to experience novelty (Faroqhi 2000, 240). When setting up urban laboratories of modernity, members of the elite also began to experiment with themselves.

Note

1. Quotations not originally in English have been translated by the author.

4 Appropriating the International Style: Modernism in East and West

Thomas J. Misa

Modern Paris is a strange paradise in Jacques Tati's *Playtime* (1967). The airport, office blocks, high-rise apartments, and restaurants are well-ordered and spotlessly clean. Everything is brand-new and everyone is in their proper place, except of course Tati's genially bumbling Mr. Hulot, an Everyman lost in modern life. The technified environment does take some getting used to. Cars, trucks, and buses clog the streets. An Orwellian surveillance system keeps Hulot out of protected spaces. Huge sheets of plate glass form the walls, doors, and floors. The residents are so habituated to this glass menagerie that the restaurant's doorman continues to pocket tips for "opening" the floor-to-ceiling glass door even after it has been shattered and long since swept away.

Historic Paris slips into view only when a reflected image of the Eiffel Tower or Arc de Triomphe momentarily appears in a real glass door swinging open, but not even the visiting tourists in the film pay the slightest attention. They have come to Paris to see an international exposition on modern conveniences for the home and office. "Wait until you see how modern it is—and they even have American stuff!" one exclaims. On display are such wonders as stylish women's eyeglasses with tilting lenses, Greek-themed waste cans, and battery-lighted floor mops. A German salesman proudly shows off modern office doors that really do slam noiselessly "in golden silence." It is a world where modern technology is firmly, if absurdly, in the saddle.

Playtime gains force as a critique of modern life precisely because the modern city it depicts is anyplace and everywhere. Tati's Paris could easily be anywhere in Europe; the travelers arriving from Frankfurt show no sign whatsoever of culture shock. The tourism posters display the same runaway modernism in Mexico and Hawaii, too. This critique calls out for historical understanding. How and why did cities in the twentieth century develop such striking similarities in their outward forms and internal structures? How have they created and structured "modern society," with its highly choreographed ways of life? Can we frame historical studies that account for the obvious similarities among modern cities around the world, and yet remain attentive to the manifold differences that persist between them?

Figure 4.1
Modernism anyplace and everywhere: Antwerp's modernist district, across the river from its touristic old city, with tall residential buildings surrounded by green space and ample room for automobiles. It strikingly resembles the modern districts in Chicago or Prague or anywhere. Photographer: Thomas Misa.

This chapter addresses these questions with special attention to modern architecture and urban planning. Our volume's framework of circulation and appropriation helps us to comprehend both the character of modernism and its impact on cities. Modernism was a machine that changed the world by imposing a degree of uniformity on modern cities everywhere. At first glance, Antwerp and Chicago have nothing in common, one a North Sea port with a medieval core, the other an inland site of industry and commerce. But in fact the two cities' modern districts are eerily similar in urban form, building type, and lifestyle (see figure 4.1). This is because they both took shape under the sway of modernist concepts of how buildings should look, how cities should be formed, and where residents should live.

To explore these issues, I first examine the *circulation* of modernist ideas about architecture and urban planning. Architects, planners, theorists, engineers, and others who shaped the modern city made study tours, attended expositions, organized town-planning congresses, and published journals, books, and reviews for international audiences. The circulation of modernistic discourse can be mapped with exceptional clarity through CIAM, the Congrès Internationaux d'Architecture Moderne (1928–1959). Although its leading figures claimed their work to be international, even universal, I maintain that CIAM is best understood as a distinctively European institution. The problems it took up, the themes and locations of its major congresses, and its leadership were all European. Modernism was also shaped by the cultures and traditions of individual countries, as can be seen in the *appropriation* of modern architecture and urban planning in two smaller European countries. Examination of the varied fates and distinct inflections of modernism in the Netherlands and Czechoslovakia between the world wars shows that modernism was not nearly the universal juggernaut that its proponents claimed and that its critics bewailed.[1]

This chapter provides a multilayered investigation into the relationship between technology and modernism. Modernism in architecture, as I have argued elsewhere, embraced certain materials and certain forms—steel, glass, and concrete used to make boxy, nonornamented, unbounded spaces for many purposes—because these materials and forms were deemed to express the essence of modernity and mass production. Modernists might use bricks when they were cheap but then smoothed them over to mimic "modern" concrete. Such modernists as Walter Gropius, Bruno Taut, and Ernst May embraced American-style mass production and even transformed large-scale building sites into something that looked like a rational factory. In effect, modernists floated their specific ideas about what materials and forms ought to characterize the modern world on the deeper currents of social and technological modernization. These theorists of modernism offered a linear and largely teleological storyline that leads from modern materials to specific modern forms (Misa 2004, 158–189). Such a technological fundamentalist stance both empirically predicts and normatively requires that modernism be "the same" everywhere—but modernism in fact was always more complicated and open-ended (Banham 1986; Meikle 1995; Lesnikowski 1996; Benson 2002). This chapter documents the universalizing tendencies of modernism, manifested in CIAM, but also describes and accounts for the manifest differences evident within modernism. Modernism never was a deterministic force for change. In Dutch and Czech cities embracing several distinct visions of modernism, we can see how Europeans appropriated a supposedly universal "machine that changed the world."

European Modernism

Founded in 1928, CIAM remained for nearly three decades the modernists' preeminent international organization. Its peripatetic meetings and polemical "charters" were a magnet for modernism in architecture and urban planning, attracting diverse figures into its fold and molding modernist discourse. A cardboard version of CIAM's history has ably served its many latter-day critics. In their view, CIAM was little more than a podium for Le Corbusier. Given his grandiose plan to construct a skyscraper city after razing the center of Paris, as well as his substantial influence on actualized plans for Moscow, Chandigarh, and Brasília, not to mention high-rise public housing in the United States and United Kingdom, Le Corbusier certainly bears a heavy historical burden (P. Hall 1996, 203–240). A second key figure was CIAM's long-serving secretary, Sigfried Giedion, best known today for his popular books *Space, Time and Architecture* (1941) and *Mechanization Takes Command* (1948). Giedion was the glue that held together the diverse members of CIAM, with their conflicting agendas and visions.

CIAM's chief glory was undoubtedly the Athens Charter, a programmatic vision of urban planning that embraced "functionalist" categories. The hegemony of functionalist urban planning led to striking similarities in modern urban districts worldwide. Le Corbusier, functionalism's chief theorist, believed that a city was defined by its major functions—working, dwelling, leisure, and transport—and that these should take place in spatially separate districts. In both western and eastern Europe, urban planners embraced some version of functionalism through the 1960s if not beyond (see Germuska, Heßler, and Jajeśniak-Quast in this volume). This section focuses on CIAM's political and organizational history, while the following sections analyze the relationships of Dutch and Czech architects and planners to CIAM and what is sometimes called the "International Style" of modernism.

Walter Gropius claimed the mantle of "internationalism" in his *Internationale Architektur*, published in 1925, but this phase of modernism was shaped by a distinctly European vision. CIAM took up a resolutely European perspective on defining the problems and possibilities of cities, as Eric Mumford (2000) has recently made clear. While CIAM always had a few members from select countries in Asia, North America, and South America, its membership was preponderantly European and its leadership was exclusively so. A 1936 CIAM map of the world reveals a dense network of European members from Barcelona to Moscow—but only a few other members scattered elsewhere across the globe. All ten of CIAM's major congresses were in Europe. CIAM's vision presumed, for example, that cities were a certain modest geographic size, so that CIAM's standard-scale maps of Amsterdam, Zurich, and Barcelona each fit comfortably on a single standard poster board 1.2 meters square. By contrast, "the scale chosen produced a monstrosity of oversize" when applied to New York or Los Angeles or Detroit (E. Mumford 2000, 84; Sert 1942/1944, 198).

CIAM's European orientation was painfully clear during the years that Giedion spent as a prestigious guest professor at Harvard in the 1930s. Although Giedion hoped to recruit numerous American architects and planners into CIAM, not many paid him any attention. In those years, America's impact on European modernism was much greater than the reverse (Banham 1986; Cohen 1995; Ibelings 1997). Surprisingly few Americans attended CIAM meetings, although several European émigrés did so. One was the Austrian-born Richard Neutra, author of *Wie baut Amerika?* (1927) and designer of the strikingly modern Lovell house (1927–1929) in Los Angeles; he explained to all who would listen that the two continents' accepted standards for elevators, running water, central heating, and housing generally were so different that Americans would never accept CIAM's Europe-focused ideas for low-cost housing. Giedion himself acknowledged that "the problems which concerned us...held no interest for American architects, magazines, or public. Housing, one American member has informed me, was regarded as merely a European problem" (E. Mumford 2000, quote 141; Kentgens-Craig 1999, 17, 48). Even Le Corbusier returned empty-handed from a much-anticipated trip to the United States, despairing about "the land of the timid" (Bacon 2001).

CIAM's organizational structure sharpened its regional focus. The nation-state was a fundamental building block for the organization. Le Corbusier had vigorously sought the commission for designing the League of Nation's headquarters, and even when it went to two nonmodernist architects, he and other CIAM members held up the League as a suitable patron of modern architecture. Because of its notion of powerful nation-states conducting the world's business, the League was an attractive model for CIAM. In its most active period during the 1930s, individuals joined CIAM as part of a group that formally "represented" a specific nation; only in the mid-1950s could individuals per se join CIAM, then distinctly in decline. As we will see, this organizational model deeply affected modernism in the Netherlands and Czechoslovakia.

CIAM's Europe-centered activities extended beyond its well-known focus on functional planning. Its founding theme of "Die Wohnung für das Existenzminimum" (minimum-existence dwelling) responded directly to Europe's severe housing crisis. The 1929 Frankfurt congress showcased over 200 floor plans, fully half of them from that city's active public housing effort. CIAM's quasi-scientific impulse was made clear with the city plans presented on standard-scale boards, labeled only with the city name, wage levels, and apartment areas; strangely enough, no other urban variables, not even land values, seemed to matter. The CIAM architects' efforts to cram a family into an apartment as small as 23 square meters may have been realistic, given the low wages that prevailed in Europe and the frankly appalling housing European workers often endured. Sad to say, the poorest workers could not afford the famous Frankfurt apartments, with their cost-effective design, economical construction, and space-minimizing kitchens; many lower-income families could not pay the high electricity bills (Heßler 2001, 267–307). The fact that many "minimum-existence" buildings lacked bathrooms and central heating, and that some residents carried coal for heating and cooking up as many as five flights of stairs, made these plans irrelevant to those striving toward an "American standard of living" (De Grazia 2005, 75–129). All the same, inexpensive U.S. housing rarely featured the typically European on-site allotment gardens found at Römerstadt and other European public housing (see figure 4.2).

Figure 4.2
The garden in the machine? On-site allotment gardens at Frankfurt's Römerstadt appropriated the hard edge of modernism to Germans' desire for garden soil under the fingertips. Photographer: Thomas Misa.

Often following Le Corbusier's account, the conventional history of the 1933 congress that launched the Athens Charter is fundamentally misleading. Although the congress itself was held, rather grandly, aboard a cruise ship sailing between Marseille and Athens, this venue was a depressing marker of political and economic turmoil. For some years CIAM had planned a major ten-day meeting that year in Moscow. But avant-garde modernism in the Soviet Union was fading fast, with top-down economic planning and the imposition of Stalin's neoclassical style. At the last moment, the Soviet hosts withdrew their invitation and left CIAM scrambling for a site. Germany, too, in the spring of 1933 was no place for modern architects and planners. Already the Bauhaus had been chased out of Dessau, where local Nazis had tacked a sloping roof onto the boxy modern building. Active briefly in Berlin, the Bauhaus was permanently shuttered after Hitler took power in March. Modernist architects were soon purged from the principal national associations as well as from municipal building departments. It was thus something of a miracle that a French art editor, Christian Zervos, arranged the Mediterranean cruise for CIAM. On July 29 the SS *Patris II* departed Marseille for Athens with 93 CIAM members who spoke eleven different languages.

Although Le Corbusier claimed later that the cruise-congress endorsed his Athens Charter, a swirl of controversy in fact prevented a final vote on his proposal. Le Corbusier simply fabricated CIAM's approval, for the voyage actually featured fiery debate on his audacious vision, as well as formal talks by other CIAM members and presentations of scale plans for 34 cities. These included metropolises such as Paris, London, and Budapest; cities of administration such as Rome, Athens, and Utrecht; ports such as Barcelona, Rotterdam, and Genoa; industrial cities such as Detroit and Dessau; new towns including Littoria and Zuidersee; and cities of diverse functions such as Frankfurt, Los Angeles, and Verona. Curiously, the two colonial cities of Dalat (French Vietnam) and Bandung (Dutch East Indies) were ghettoized as "villes de plaisance" (pleasure cities). The cruise-congress ended with numerous rival proposals about urbanism and planning still on the table.

The absence of CIAM's official go-ahead hardly slowed Le Corbusier down. "The Athens Charter unlocks all doors to the urbanism of modern times," his book announced (Le Corbusier 1943/1973, 25): "The destiny of cities will be set right." He forcefully argued for a top-down analysis of cities, set out in such punchy imperatives as "Traffic at high-density intersections will be dispersed in an uninterrupted flow by means of changes of level"; "The destruction of slums around historic monuments will provide an opportunity to create verdant areas"; and "Private interest will be subordinated to the collective interest." Cities, Corbusier argued, comprised the functions of dwelling, work, leisure, and transport, with each function having a distinctive urban logic and separate region. Lewis Mumford never forgave Le Corbusier and CIAM for leaving out of the urban scene such vital functions as politics and culture.

Formally, CIAM represented itself to the world as being above politics. Its technocratic stance posited "one best way" of designing cities—namely, modernist architecture and functionalist urban planning. As an elite, international, avant-garde organization, CIAM would change society. To this end Le Corbusier courted diverse political regimes, from the League of Nations to the Communist Party in Moscow and the fascist Vichy France. He flexibly altered his famous slogan "architecture *or* revolution," with its distinct suggestion that his architecture might prevent a working-class revolution, to "architecture *and* revolution," depending on who was paying his bills (E. Mumford 2000, 48).

Despite its technocratic stance, CIAM had many members who were explicit about their politics; in fact, CIAM was founded by left-wing German, Swiss, and Dutch architects. In his CIAM contributions, the Czech Karel Teige, a fervent Marxist, explicitly advocated a class-based analysis. Germany's modernists received prominent commissions from left-wing city governments to build workers' housing, and in the early

1930s several of them even worked in Soviet Moscow. European modern architecture came to the United States in the persons of Walter Gropius and Mies van der Rohe, who took up distinguished teaching careers in Cambridge and Chicago and trained the influential postwar generation of architects. Both men jettisoned whatever leftist political sympathies they may have entertained in Germany, and instead stressed those formalistic aspects of their architecture that seemed politically neutral. Nevertheless, twenty years of FBI files document their allegedly "un-American activities" (Kentgens-Craig 1999, 180–182, 204–214, 238–244).

Politics decidedly inflected the U.S. appropriation of CIAM's modernism. To begin, there was the *nonreception* of such prominent European figures as Mart Stam and Hannes Meyer, whose strong left-wing sympathies raised a red flag in America (Kentgens-Craig 1999, 124–128). Suspicion of radicals was also evident in 1953 when Harvard's Graduate School of Design searched for a dean. Impressively, the three top finalists were all CIAM members. Oscar Niemeyer, the co-designer of Brasília's functionalist housing blocks, was apparently the first choice, but since he was a card-carrying communist and McCarthyism was at full tilt, his candidacy was a nonstarter. The second choice, Ernesto Rogers, a prominent Italian architect, declined the job. Eventually, the committee turned to José Luis Sert, the Spanish-born president of CIAM since 1947. Sert, one of the authors of Barcelona's "master plan," had been angling for an American job ever since Harvard University Press published his landmark volume, *Can Our Cities Survive?* (1942).

As a semiofficial statement of CIAM's mature ideas, Sert's book drove a curious wedge between the European and the American understandings of urban planning. Oddly enough, though his book was subtitled *An ABC of Urban Problems, Their Analysis, Their Solutions*, Sert all but ignored such substantial U.S. planning efforts as the famous 1929 *Regional Plan for New York and Environs* (even though he pinched several diagrams from it) and the proliferation of New Deal housing and urban planning efforts. Lewis Mumford and Catherine Bauer, from whose *Modern Housing* (1934) he took an illustration of Frankfurt, appear only in the footnotes as sources rather than in the text as influential figures themselves. The vibrant transatlantic housing community anatomized in Daniel Rodgers's *Atlantic Crossings* (1998, 394–408) had evidently vanished. Sert had proposed several illustrations of U.S. steel plants and Tennessee Valley Authority dams, but U.S. military censors removed them (the book appeared in 1942). The book does feature numerous illustrations of slums, traffic snarls, parking lots, and chaotic urban scenes from New York, Detroit, Chicago, and Los Angeles, but these were stark antimodels of urban planning. More positive images of urban designs in Philadelphia, Chicago, and New York were presented without identification or architect's name as instances of universal principles of planning (Sert 1942, 3, 59, 65, 107, 149). For instance, an illustration of Philadelphia's Carl Mackley public housing high-rises—which Rodgers (1998, 403) hails as "an authentic piece of European social modernism brought home"—appears in Sert's book without any identifying caption. Sert's readers might conclude that CIAM's modern architecture and planning was universal, but Sert knew better. As he admitted to Giedion in 1947, "We cannot continue to consider central Europe as the main field of interest for CIAM. Explanation of CIAM ideas both to east and west should be the basis for our program" (quoted in E. Mumford 2000, 185).

Dutch Modernism

The "triumph" of modernism, of course, was continually contested. Indeed, modernism's apologists in CIAM and elsewhere waged an energetic campaign to establish and entrench their particular interpretation of the modern world (Banham 1986; T. Smith 1993). By carefully reviewing the several strands of modernism in the Netherlands, we can recapture some of the contingency and variety that the modernists worked so hard to efface.

In the Netherlands, two rival schools—each claiming to be authentically "modern"—emerged around the First World War. While most international histories emphasize the contributions of de Stijl, the "Amsterdam school" of architects and theorists actually had greater influence on Dutch building projects of the time. The colorful, playful, even whimsical building designs from the Amsterdam school stand in sharp contrast to de Stijl designs, which embraced the machine age, resisted ornamentation, and sought a single impersonal "style." The Amsterdam school's emphasis on individual creativity and intentional lack of method constituted a Dutch version of expressionism. Their journal *Wendingen* appeared first in 1918, a year after the founding of the rival journal *de Stijl* (de Wit 1983; Ibelings 1997; Warncke 1998).

A beautiful example of the Amsterdam school's alternative modernism is the Sheepvaarthuis (Shipping Office Building) built in Amsterdam during 1912–1916. The architects' extensive use of ornament, including stylized human statues at its entrance and along its facade as well as the colorful stained-glass windows inside, stood as a direct affront to the formalism and minimalism advocated by de Stijl. And the architects' determination to design all elements of the Sheepvaarthuis building right down to its interior detailing, including furniture and floor coverings, took up the modernistic determination to create an all-embracing "total work of art," or *Gesamtkunstwerk*.

Michel de Klerk, one of the three Amsterdam school architects of the Sheepvaarthuis, was no traditionalist. He forthrightly embraced the modern age in a surprisingly critical assessment of H. P. Berlage published in 1916. It was Berlage's sixtieth birthday, and de Klerk was supposed to say something pleasant about the famous designer of Amsterdam's Stock Exchange (1903) and author of the city's landmark Plan Zuid (1902–1915)—but no. "In my opinion Berlage has not been to the forefront for decades now," de Klerk wrote in the pages of *Bouwkundig Weekblad*. "He is not sensitive to the tingle of the new, the shock of the sensational, the overwhelming impression

Figure 4.3
Modernism against the machine. In sharp contrast to the boxy machine form modernism of de Stijl, Amsterdam's Spaarndammerplantsoen housing project (1917–1920) featured rounded brick columns and curvaceous waves of brick facade. Colors are shades of purple, brown, and red. Workers came home to these entrance doors. Photographer: Thomas Misa.

which actually characterizes the modern today" (quoted in Ibelings 1995, 23). Other prewar achievements of the Amsterdam school were a set of 16 villas at Park Meerwijk and the public housing complex at Amsterdam's Spaarndammerplantsoen, where rounded brick columns and curvaceous waves of brick facade decorate the entrance doors (see figure 4.3). Amsterdam school architects designed many other social housing blocks, stylish post boxes, and ornamented bridges that can still be seen in the city, as well as the Bijenkorf department store in The Hague (1924–1926). For their part, de Stijl denounced the Amsterdam school's supposedly mistaken approach to the modern.

Not all Dutch architects in these years accepted modernism, of course. Foremost among these was M. J. Granpré Molière, who developed a self-described traditionalism. A leading example was his Vreewijk project in the south of Rotterdam (1916–1919), designed to be a village-like garden city (see figure 4.4). This urban planning concept

Figure 4.4
Appropriation of the British garden city in Rotterdam. Self-proclaimed traditionalist M. J. Granpré Molière's Vreewijk (1916–1919) with pitched tile roofs, exposed bricks, and green space around the semidetached houses. Photographer: Thomas Misa.

came from Ebenezer Howard's Letchworth and Welwyn Garden projects. All such garden cities privileged the individual semidetached house and stood in sharp contrast to the "modern" ideas about city planning anticipated in Berlage's Plan Zuid and realized in several notable districts in Rotterdam under J. J. P. Oud. The modernist schemes designed large city blocks and embraced the monumentality of uniform designs (Taverne et al. 2001).

Dutch traditionalists did not lack for ideas or even prominent commissions, and from the 1920s through around 1940 they were a distinct force in Dutch society. Traditionalists held that architecture should be "timeless" and reflect the traditions of particular regions or countries (rather than capturing the fleeting spirit of the modern age or striving to be uniform and international). They also had the goal of enhancing traditional crafts and using traditional materials. Traditionalists designed such well-regarded public buildings as the town halls of Enschede and Waalwijk and the Museum

Boymans in Rotterdam. In 1924 Granpré Molière become professor of architecture at Delft, where he trained many students including the talented and prolific J. F. Berghoef. And in the journal *Rooms Katholiek Bouwblad* founded in 1930, traditionalists had a prominent publishing outlet.

Traditionalists might have carried their case, but the modernists closed ranks among themselves and presented a strong united front. In this instance, the dynamics of international circulation, namely CIAM's insistence on "national" representative groups, seems crucial. Dutch modernists were not initially united: in the 1920s, the cities of Rotterdam and Amsterdam appeared to form two rival modern architecture movements. Architects in Rotterdam formed the Opbouw in 1920, and, while initially it included a diverse set of members, under the leadership of Mart Stam it soon became dedicated to left-wing politics and a severe functionalism in architecture. Meantime in 1927, a group of Amsterdam architects, distinct from the Amsterdam school, founded "De 8" to investigate and develop their own functionalist approach to architecture.

The rival Rotterdam and Amsterdam factions were brought together through their joint participation in CIAM. The Dutch architects and planners Cornelis van Eesteren, Mart Stam, H. P. Berlage, and Gerrit Rietveld were among the founding members of CIAM in 1928, and van Eesteren even joined CIAM's inner circle of organizers. Members of both the Rotterdam-based Opbouw and the Amsterdam-based De 8 were active participants in CIAM-sponsored congresses and expositions. One highpoint for these groups was CIAM's adoption in 1931 of Dutch standards that stemmed from van Eesteren's systematic studies as chief architect of the Amsterdam public works department. Through these activities Dutch modernists found they had common cause and even began publishing a joint journal, with the portmanteau title *De 8 en Opbouw* (1932–1943).

This architectural consolidation around functionalism came to be appropriated under the name of Nieuwe Bouwen. Nieuwe Bouwen represented the strongest domestic Dutch expression of functionalism in architecture and for a time became virtually a synonym for Dutch modernism. Its members looked to the modern world of machines and mass production for inspiration, sternly rejected historical styles and ornamentation, and conducted many experiments in building with new materials such as steel, glass, and especially concrete (Derwig and Mattie 1995). A signature building of this movement was the Van Nelle factory (1925–1932), located outside Rotterdam. This eight-story, glass curtain factory complex was designed by the prominent firm of Brinkman & Van der Vlugt. In addition, the Amsterdam suburb of Betondorp (literally, "concrete village") was an experiment where architects tried out eleven different types of concrete construction. These experiments—indeed the willingness to embrace innovation—gave these modernists a key rhetorical advantage over the traditional-minded architects who continued to cultivate traditional forms and materials. And yet it is important to note that traditionalists also built with concrete; moreover, the modernists' heavily publicized Betondorp experiments did not generate notable breakthroughs with the new material. Although traditionalists still had many vibrant and valid ideas for detached houses, town halls, churches, and museums, they were not as well prepared to embrace modernism discursively.

Building in the Netherlands largely came to a halt during the war years, and full-scale construction did not resume until the economic recovery of the 1950s. (German occupation authorities disbanded *De 8 en Opbouw*.) Rotterdam had been heavily bombed in May 1940, and the city government cleared additional space after the war by demolishing a working-class neighborhood in the center of the city. Here, under the influence of Dutch modernism, was built the most American city in the Netherlands, with American-style tall office buildings and an American-style shopping mall at its center. As the chamber of commerce affirmed (Taverne 1990, 146), "[we] are now going to renew and modernize the city, whereas we otherwise would have had to make do for a long time with obsolete premises."

The successor firm to the designers of the Van Nelle factory emerged as one of the largest and most influential architecture firms in the country. In Rotterdam alone, Van den Broek & Bakema was responsible for such high-profile projects as a large shopping complex built for three companies (1948–1951) and the Lijnbaan pedestrian shopping district (1949–1953) as well as villas, factories, and churches. The Lijnbaan, inspired by the U.S. shopping centers of Victor Gruen, the Austrian-born designer of the archetypal enclosed shopping mall, showed an "indelible American stamp" (Taverne 1990). In his otherwise gloomy *The City in History*, Lewis Mumford (1961, plate 63) warmly admired Rotterdam's rebuilding and specifically praised the Lijnbaan: "done in modest materials and on a modest scale, meant to house a variety of smaller shops, restaurants, and cinemas, [it] is exemplary in almost every way." Other modernist buildings that went up in Rotterdam during these years were Marcel Breuer's new store for De Bijenkorf, a number of apartment projects in and around the central district, and the modernist Central Station (1950–1957). Modernists had had their eyes on the city's Hofplein for years. A 1927 Opbouw plan, embracing the notion that traffic must be "the foundation of town-planning design," envisioned a great sweeping roadway cut through the square with adjacent apartment blocks featuring rounded facades. In the event, the Hofplein, once a promising area for nightlife, was cleared after the war for a traffic roundabout; other automobile-friendly alterations included the widening of the Coolsingel into a full boulevard (see figure 7.3).

In 1957, in the midst of the postwar building boom, the traditionalist editors of the *Katholiek Bouwblad* were forced to admit that "for the first time in centuries, a universal style representing the culture of an age" had emerged: not traditionalism, but modernism. Yet many Dutch architects were savvy enough to realize that any universal style could become a straitjacket. With this in mind, a group known as Team 10 took determined steps to bury the universalistic pretensions of CIAM and to create modern alternatives to the then-dominant glass-and-steel functionalism. Among their efforts were the founding of the Liga Nieuw Beelden (1955) and continuing lively discussions in the eclectic journal *Forum*.

This ferment brought about a succession of modernist styles. Rotterdam, especially, showcases today a vibrant profusion of modernistic bridges, buildings, and sculptures; clearly modernism, by whatever name, is alive and well there. Modernism as a continuing and confident movement is prominently on display at the Netherlands Architecture Institute, situated in its own modernistic building just across from a street of functionalist villas and opposite the traditionalist Museum Boymans. In its core exhibit, the Netherlands Architecture Institute recognizes no fewer than *eleven* modern architectural styles and stages neatly spanning the twentieth century: realism, the Amsterdam school, de Stijl, functionalism, monumentalism, modernism, structuralism, neorationalism, neomodernism, postmodernism, and supermodernism.

Czech Modernism

In his childhood, architect-historian Eric Dluhosch crossed from traditional to modern architecture. In the early 1930s his family lived in a multistory apartment block in the Malá Strana district of Prague. To the street, the turn-of-the-century building presented a heavily ornamented Renaissance-style facade. But the building's real life hummed inside its courtyard. Clouds of dust filled the building's courtyard on Mondays when housewives and maids attended to the weekly ritual of carpet cleaning. On Tuesdays delivery wagons brought sacks of potatoes to feed the families as well as sacks of soft brown coal to fuel the apartments' heating and cooking stoves. A gypsy knife-sharpener plied his wares on Wednesdays. Even in urban Prague, the weekly rhythms of life came from the village.

After her husband's early death, Dluhosch's mother relocated the family to one of Prague's outlying garden suburbs. There in middle-class Petřín the family lived in a modern two-story apartment house. Gone were the coal stoves, the ornamented facade, the heavy furniture, and the vivid tableaux of the courtyard. A gas range now cooked the potatoes, while steam piped in from a central plant warmed each room. The family's living rooms, as Dluhosch recalled, were done in pastel colors and "without a trace of ornament" (Švácha 1995, xix). A journey of twenty minutes on electric streetcars took the family to the newly opened Bílá Labut' department store, a glass-fronted building that remains visually striking even today. For shoes they went to the equally modernistic Bat'a store on Wenceslas Square. Bat'a shoes came from the company's town-and-factory complex at Zlín, while the flagship Prague store, with its tall plate glass windows, was widely admired for its modern, seemingly "American" design. With its garden suburbs, glass-fronted department stores, vibrant urban life, and profusion of architectural innovations, Prague arguably was every bit as modern as Frankfurt.

We can map the emergence and circulation of modern architecture in Prague and indeed Czechoslovakia through the career of Jan Kotěra, widely hailed there as the "father of modernism."[2] Kotěra's career had numerous parallels with H. P. Berlage's: both men created a distinctive early-modernist architectural style (independent from historical styles and yet distinct from later glass-and-steel functionalism), and they taught a crucial generation of students. After Kotěra finished his engineering apprenticeship in Prague, he began a three-year course of study at the Academy of Fine Arts in Vienna in 1894, the same year that Otto Wagner took up a prominent professorship there. Wagner broke with historicism, experimented with art nouveau, or Jugendstil, then created an early version of modernism with the Post Office Savings Bank (1904–1906) in Vienna. While this building's immense blocks of carved stone on its lower floors were done, as Wagner explained, in the time-honored "Renaissance way of building," its upper floors were clad with thin stone panels in the "modern way of building"; these panels not only weighed one-eighth as much as those of traditional stone (or less) but also suggested "a number of new artistic motifs" (Moravánszky 1998, 158). The Wagner circle created a cosmopolitan center of prewar modernism. Kotěra joined the debating club that evolved into the famed Vienna Secession and befriended Adolf Loos, just back from a three-year visit to the United States. Loos's polemical essay "Ornament und Verbrechen" (Ornament and crime) (1908/1998, 167) famously argued that "the evolution of culture marches with the elimination of ornament from useful objects."

Kotěra's early professional career spanned Europe and beyond. Awarded the prestigious Prix de Rome, he spent six months on a study tour in Italy. Teaching at the Prague School of Applied Arts beginning in 1898, and later at the Prague Academy of Fine Arts, Kotěra actively cultivated an international outlook, gleaning information on the latest architectural developments from professional publications, lectures, and books as well as from study tours to Paris (1900) and to the United States, Holland, and England (1904–1906). Ten years before the Wasmuth volumes made Frank Lloyd Wright famous in Europe, Kotěra had studied and written about Wright's organic architecture. Kotěra's striking design for the St. Louis Exposition (1904) attracted the attention of famed designer-theorist Hermann Muthesius, who soon established contact with the Prague-based Mánes Association, giving a celebrated lecture there in 1910. The most influential of Kotěra's international trips was that to the Netherlands, where he saw firsthand H. P. Berlage's Amsterdam Stock Exchange building. Kotěra promptly praised it in the prominent Czech journal *Volné směry*, and something of Berlage's restrained ornamentation and geometrical structure found its way into Kotěra's most important early building, the city museum in Hradec Králové (1909–1913). One contemporary noted of this period Kotěra's "declaration of a pan-European stand and the acknowledgment of a more universal expression in architecture" (Kotěra 2001, 25).

Kotěra's students and collaborators were leading figures in the Czech effort to digest modernism's "tingle of the new." Fully 15 of 26 leading modern architects in the 1920s and 1930s had studied with Kotěra or one of his students.[3] Other notable members of the Czech modern movement, such as Josef Chochol and Bohumil Hypšman, had been students of Otto Wagner in Vienna, while a third group formed around Antonín Engel, a noted town planner who had also studied under Wagner, at the Czech Technical University in Prague. Yet what is notable about the Czech architectural scene is its profusion of diverse "modern" styles. The singular strength of nationalism, the selective weakness of CIAM, and the distinctive conditions of the capital city of Prague shaped the appropriation of modernism in the Czech context.

Fragmentation plagued Prague's administrative history for many decades. The four districts of Prague's core (Hradčany, Malá Strana, Old Town, and New Town) had been administratively joined in 1784, and between 1850 and 1901 four additional districts (Josefov, Vyšehrad, Holešovice, and Libeň) were annexed to it. But numerous nearby districts remained administratively distinct, often enjoying the independent privileges of royal chartered towns. As late as 1904 royal charters were given to Smíchov and Bubeneč, respectively, immediately south and north from the Hradčany castle district, as well as to Karlín just to the east. These acts of royal meddling convinced many Czechs that Austria-Hungary was not facilitating the rational growth of the burgeoning city but rather was keen on frustrating it. Such administrative fragmentation made it especially difficult to extend transport services across the region, and also hampered the expansion of water, sewage, and electricity services. The sizable area of the city and suburbs, some 40 separate towns and villages in all, was brought under one unified administration only in 1922. It is quite striking that Prague's underground metro system, planned in the 1930s but not constructed until the 1960s, frames a large triangle within the city, while the city's extensive tramways connect to the outlying suburban regions that constitute roughly the greater Prague of 1922.

While such fragmentation probably helped Prague preserve something of its "unspoiled" historical character, the city did not altogether escape modern urban planning. The largest-scale planning schemes such as those that remade great swaths of Paris and Vienna were certainly hampered by the high cost of open land in Prague and by the spatially divided authority over it. The largest clearance targeted the old Jewish quarter. Acting in the name of modern hygiene, Prague's authorities after 1887 opened up the medieval streets in Josefov, knocked down all but a dozen of its original 260 stone buildings, and dramatically transformed the urban fabric. Prague's Jews managed to retain their town hall, six of their nine synagogues, and most of their medieval graveyard (Giustino 2003). Urban planning elsewhere in the city was largely a patchwork of modest-scale efforts, inspired by such varied models as Otto Wagner's rationalistic street networks, the garden city movement, and even urban-block reformers.

Yet even as the suburbs expanded and the problem of connecting them to the city core recurred, the combination of Prague's rivers, its steep hillsides, and an awareness of the city's cultural significance prevented the construction of large-scale motorways to the suburban districts. The singular attempt to deploy full-blown Le Corbusier-inspired leveling of the city core—Josef Havlíček's nutty plan for building three monstrous skyscrapers in the middle of the central New Town district (1937, 1943–1944)—fortunately went nowhere.

The second notable factor shaping modern architecture in the Czech lands was nationalism. Czech nationalism was a complex construct as Derek Sayer (1998) has recently emphasized. A simple storyline of heroic success became dominant, especially after independence in 1918, but in reality nationalism emerged gradually over many decades. For instance, aristocratic patronage and Enlightenment idealism framed several apparently nationalist efforts, such as the plans for the National Museum (formally established in 1818). Even the nationalistic nobility in the Czech lands were in many cases German-speaking Catholics. Would-be Czech nationalists in the nineteenth century suffered under many burdens. It was bad enough that rival Hungary was elevated within the Austrian-Hungarian empire, reducing the Bohemian crown to third rank, but Vienna and Budapest also distinctly overshadowed Prague in cultural prominence. For decades, leading Czech artists, architects, and professionals routinely went to Vienna to study and bring back Viennese theories. All these insults, imagined and intended, needled Czech nationalists.

Even under imperial rule, an early boom in nationalist-inspired Prague civic architecture took form. Evidently Prague hoped to transform some of Bohemia's growing economic wealth into cultural institutions that might chip away at the hegemony of Vienna and Budapest. In 1861, Prague's newly reorganized city council fell to the control of middle-class Czechs. They commissioned a spate of notable buildings much prized by tourists and Czechs today. The very names of the National Theater (1868–1883), Czech Polytechnic (1873–1874), and National Museum (1885–1890) tell the story. In addition, monuments to key figures in Czech and Bohemian history went up in squares and parks across the city. Yet Prague was still under the imperial thumb, as was made painfully clear when Austria-Hungary (after initially promising otherwise) forced Prague to purchase its own city walls at high cost. There would be nothing in Prague like Vienna's monumental Ringstraße.

Czech nationalists after the First World War initiated a number of civic, commercial, and residential building projects. Across the country modernism, in several distinct versions, became something of a banner for independence, especially the modernist department stores noted above and the houses and villas designed by avant-garde architects for middle-class or professional clients. Technology accompanied modernity. As Thomas G. Masaryk, the much-celebrated founding president of the Czechoskovak Republic, wrote of his triumphant entry, "I decided to use a democratic automobile to go through the cheering Prague; I did not want to use an old gilded coach that was so typical of the past age" (quoted in Peichl and Slapeta 1987, 162).

And not all the action was in Prague. The city of Brno, in Moravia, staged a strikingly modernistic Exhibition of Contemporary Culture (1927–1928) to celebrate the tenth anniversary of independence from Austria-Hungary, while other modernist manifestations included its Nov Dům housing exposition (1926–1928) and Mies van der Rohe's famed glass-wrapped, open-plan Tugendhat house (1930). Not far away, in Zlín, the Bat'a shoe company erected a model town complete with a 17-story skyscraper (1937–1938) that was Europe's second tallest. Originally a garden city development, the Zlín complex became better known for its functionalist steel skeleton buildings. Zlín's head of construction, Vladimír Karfík, was yet another poster child for international circulation; after training at the Czech Technical University in Prague, Karfík worked in Le Corbusier's office, made a study trip to England, and then spent four years in the United States (1926–1930) where he worked in three Chicago architecture offices, including that of famed skyscraper designers Holabird and Root, and also with Frank Lloyd Wright.

A unique Czech architectural style melding modernism and nationalism developed, known as rondocubism, and sometimes even called the "national style." Rondocubism was an offshoot of the prewar experiments to design buildings using the fractured forms of cubist painting. Whereas Dutch cubist architecture of the time was square and "cubic," Czech cubists deployed almost any angle but orthogonal ones; Pavel Janák (1911) aptly labeled their cubist style as "the prism and the pyramid." In Prague today one can still see a great number of cubist buildings, such as Josef Gočár's so-called Black Madonna (recently reopened as the indispensable Museum of Czech Cubism) and several streets of striking cubist houses and apartment buildings beneath the Vyšehrad Castle. Rondocubism in the 1920s toned down the distinctive pyramidal shapes but retained the repeated geometrical designs, often triangular, adding allegorical statues or historical friezes to cement national glories. Buildings such as the Czechoslovak Legiobank (1921–1923) or the department store Adria (1922–1925) are easily identified as rondocubist. The facade of the Legiobank is heavily ornamented with various geometrical elements, diverse sculptures, and a frieze depicting the heroic suffering of peasants, workers, and soldiers during the First World War. And in case the sculptures and friezes weren't nationalistic enough, the rondocubist designs were connected to motifs in popular Slavic folk traditions and even featured the Czech national colors, red and white. Many villas, apartment buildings, banks, department stores, and crematoria in Prague, Hradec Králové, and elsewhere were built during the 1920s in this popular "national style."

Paradoxically, CIAM had only modest impact on Czechoslovak modernism, despite the high profile of Czech figures within CIAM. Indeed, the Czechoslovak national delegation to CIAM incompletely reflected the country's architectural diversity and obscured its vigorous debates about modernism; instead, the CIAM delegates largely represented the left-wing advocates of "scientific functionalism" that coalesced around the singular figure of Karel Teige. Teige launched his career as an artist, writer, avant-garde organizer, and hyperactive editor after graduating from a *reálné gymnasium* in Prague in 1919 and attending lectures in art history at Charles University. It is entirely in character that Teige, a founding member of the left-wing arts group Devětsil, canceled the group's first meeting in December 1920 because "it is inadmissible to recite poetry at a time when the police are shooting workers in the streets of Prague" (Dluhosch and Švácha 1999; Teige 2000, 351). Two months later, he presided over the group's first sizable event, held in the Revolutionary Stage theater. Subsequently, he founded, joined, or promoted innumerable avant-garde movements and publications.

As this sketch of his life suggests, Teige lived in a vortex of international activity and left-wing activism. His early writings included evaluations of Italian futurism, French cubism, and Russian constructivism, as well as several manifestos on international proletarian art. In 1923 upon taking up editorships at *Stavba*, the prominent monthly publication of the Club of Architects, and at the newly founded *Disk*, a Devětsil-sponsored review of art, architecture, and design, Teige launched his campaign for a certain constructivist interpretation of modern architecture. Two years later, he was the founding editor of the *Revue Devětsil*, which typographically expressed its politics on the cover as "RED" (Švácha 1990). He quickly established personal friendships with such leading figures as Walter Gropius and Le Corbusier, internationalizing the architectural scene in Prague by inviting these figures and others to lecture. In the mid-1920s he traveled extensively across Western Europe and in fall 1925 spent a month in Soviet Russia on a cultural exchange. Teige's international profile made him a natural choice among the CIAM's central organizers (Giedion and Le Corbusier in Zurich) to create a Czechoslovak national branch.

The founding members of Czechoslovak CIAM were all either members of Devětsil or otherwise advocates of Teige's program of left-wing politics, "scientific" functionalism, and collective housing. Czechoslovak CIAM comprised such figures as Jaromír Krejcar, the leading Czech functionalist architect in the 1920s, a devotee of the machine aesthetic, consultant in the mid-1930s to Moscow's town planning institute, and designer of an impressive glass-and-steel pavilion in Paris (1936–1937); Jan Gillar, who worked on minimum-existence and collective housing projects, then succeeded Krejcar as the country's leading left-wing architectural theorist; and Josef Havlíček, the co-designer of several high-rise office and apartment buildings, including the 13-story General Pensions Institute (1932–1934), and author of an unrealized plan for imposing Le Corbusier's functionalist planning on Prague. All three were also members of Devětsil. Other early Czech CIAM members who were not active in Devětsil, but cut from the same left-functionalist cloth, included Adolf Bens and Ladislav Zák.

There is a simple reason for the close correspondence of Devětsil and Czechoslovak CIAM. In autumn 1929, just before the first full-fledged CIAM congress in Frankfurt, Teige was deeply involved with organizing the Levá Fronta (Left Front) as a unified organization of left-wing Czech intellectuals. Within the Levá Fronta Teige also created an architectural section (ASLEF) that gained prominence for its imaginative use of architecture in social criticism. Teige simply "chose the members of the Czechoslovak CIAM Group from the membership roster of ASLEF." He reported to Giedion that this group would focus on the problems of housing and the sociological aspects of architecture. These Prague-based activities culminated in 1932 with a Congress of Left Architects, which Teige boldly proclaimed as an international force in its own right, fully comparable to CIAM (Spechtenhauser and Weiss 1999, 223–224).

For a time Teige had substantial influence on CIAM. He placed a wide-ranging social and economic analysis of housing shortages on CIAM's agenda. He even debated with the formidable Le Corbusier, and his own social-reform stance helped frame CIAM's work on the minimum-existence dwelling. The Czechoslovak group under Teige gave well-regarded papers to CIAM's congresses and provided detailed data for its national surveys. However, one consequence of CIAM's 1933–1934 crisis, which framed the Athens cruise-congress discussed above, was a complete falling-out with Teige's left-wing crowd in Prague. Teige himself hastened the break by angrily blasting CIAM's "trivial and second-rank questions on analysis, solution, and realizations" and its failure to take the battle against poor housing and chaotic city growth to the root of "the elimination of the capitalist system itself" (Spechtenhauser and Weiss 1999, 235). Czechoslovak CIAM was soon reconstituted with new members from the Moravian city of Brno. Nonetheless, as mentioned earlier, CIAM itself had only modest impact on Czech modernism. The international circulation of Czechs as well as the visits to Prague by many leading architects, outside of CIAM official meetings, had already internationalized Czechoslovak architectural modernism.

The Baba model housing project (1932–1940), built in a northern district of Prague, is an apt time capsule of Czech modernism. From a certain distance, the 30-odd houses with their boxy modernistic profiles appear to be straight out of the famous 1927 Stuttgart housing exposition, a mythic origin point for architectural modernism. Indeed, as the thirteenth in the Werkbund's series of housing expositions held across Europe, Baba was a direct successor of Stuttgart. Members of the Czechoslovak Werkbund, organized in 1920, secured a three-hectare hillside site in Dejvice overlooking central Prague and announced a competition for "minimal houses" in both row house and free-standing configurations. Numerous architects, Czechoslovak as well as foreign, including Adolf Loos and Mart Stam, submitted designs in hopes of finding a patron, but a house was actually built only if a Czech owner was willing to pay for it. The exact course of these architect-patron negotiations will never be known, for the original records of the housing estate were destroyed after the communist takeover in 1948 because its resolute modernism went against the grain of the mandated neoclassical social realism. Recently, Stephan Templ (1999) has rescued the Baba project from historical obscurity by using unconventional historical sources, including an amateur silent film shot during construction.

On close inspection, it is clear that Baba was no carbon copy of Stuttgart. Baba took form exactly when the Czech modernists were vigorously debating the merits of Teige's strict scientific functionalism as compared with a rival "emotional" functionalism that took a wider view. Adherents of scientific functionalism believed that architectural designs should reflect logical, objective, and functional requirements. Emotional functionalists agreed with the need for mass housing and standardized designs, but they aspired to design buildings to sustain a human who "not only works but also sings" (Švácha 1995, 275). Furthermore, the members of Prague's artistic, political, and professional community who paid for the houses pointedly declined to have them conform to mass standards. Diversity of functionalist designs was the immediate result (see figure 4.5): the detailing, internal plans, and even external materials of the Baba houses vary widely. At least three distinct construction methods were used: reinforced concrete skeleton, a massive system using large molded bricks, and a hybrid combination technique. The high-rise experts Havlíček and Honzík even proposed a prefabricated house using industrial materials, but no client took them up and that house was not built.

Several distinct modern architectural styles can be discerned in a walk through the Baba streets, the totality of which can be said to represent the Czech appropriation of modernism. Pavel Janák and Josef Gočár capably represented the older generation. With an echo of cubism, Gočár covered part of the Villa Glücklich's exterior with an irregular stonework pattern, while an incongruous stair-step design frames the Linda house by Janák. A severe functionalism, sometimes known as purism, inspired other buildings that most closely conform to the strict "white functionalism" pioneered in Stuttgart. A distinct variant on functionalism, using rounded forms, inspired two houses done by Ladislav Zák. Find a house with sweeping round corners looking like a steamship's bridge, and you have located Zák's visually striking Čeněk house or his economical Herain house, the smallest of the Baba houses at just 63 square meters. A rounded stairway parapet identifies the Zadák house by theater architect František Zelenka. Zák's third Baba house features a daring cantilever design for its upper-floor terrace. All the Baba houses were certainly "modern" and reflected Czechoslovak architects' active participation in the European scene between the wars (Mart Stam was the only foreign designer whose Baba house was actually built). Yet it is striking that Czech modernism, at Baba and elsewhere, was not merely a foreign import from Le Corbusier or the Bauhaus or anywhere else but rather was a product of Czechoslovaks' domestic appropriation of international modernist ideas.

Figure 4.5
Varieties of functionalism at the Baba housing estate (1932–1940). While "scientific functionalists" aimed at mass housing and standard designs, "emotional functionalists" designed expressive forms for a human who "not only works but also sings" (Svácha 1995). Early in his career Antonín Černy built severe functionalist apartments and office buildings, but for Baba he designed this emotional functionalist house with rounded forms (1939–1940). Photographer: Thomas Misa.

Modernism and Europeanism

This chapter has documented at least three significant dimensions of diversity within modernism. This fact alone is noteworthy because modernists publicly proclaimed there was just "one best way" of making cities. Clearly they did not agree on what this singular way was. First, within CIAM itself there were heated debates and substantial disagreements about the height of buildings and the shape of cities. Not even the commanding figure of Le Corbusier carried the day during debate on the so-called Athens Charter. I believe that CIAM, despite its pretensions to universality, is best understood as a specifically *European* institution. Americans, for example, saw the housing question, as CIAM framed it, "as merely a European problem." And, curiously enough, by masking the distinctive U.S. urban planning activities, Sert's landmark volume *Can Our Cities Survive?* set the stage for urban planners to *believe* they were applying truly universal principles. Finally, while modernism flourished in Europe between the wars in a context of left-wing politics and proactive municipal housing projects, the movement was stripped of its early social-reform stance and political bent when it came to the postwar United States.

Second, surprising diversity within modernism can be located in the Netherlands—an observation supporting the importance of appropriation processes for the development of differences between cities. The Amsterdam school of expressionist architecture as well as the machine-aesthetic followers of Berlage each claimed they were authentically modern while denying that mantle to the rival. Even within the restricted realm of functionalist architecture there were two distinct schools, centered on Rotterdam and Amsterdam, that for a time contested this concept too. CIAM's largest impact on Dutch modernism came when it forced the functionalist rivals to collaborate and paved the way for the joint journal *De 8 en Opbouw*. The functionalist architects with this strong consolidated base thus eclipsed the self-proclaimed traditionalists, whose leading journal was eventually forced to admit that modernism had become a "universal style representing the culture of an age."

Third, diversity within modernism flourished in Czechoslovakia also. Modernism came early to the Czech lands, before the First World War, in the form of Picasso-inspired cubism that resulted in a distinctive cubist architecture. Nationalism obviously influenced monumental buildings even before independence from Austria-Hungary in 1918, but it was in the 1920s that a distinctive "rondocubism" emerged and was known, for a time, as the national style. It distinctively combined modernism with nationalism: these buildings featured abstract but folk art–inspired geometrical designs, historical statues and friezes that commemorated national glories, and color schemes that signified the new nation. What is more, Czechoslovak architects vigorously debated "scientific" versus "emotional" functionalism in the 1930s. In part because of this home-grown ferment, CIAM had only modest impact on the Czech scene. The Czechoslovak national representatives to CIAM, both before and after the falling-out with Teige in 1935, were really only a selective subset of the active Czech modernists. Little of the debates *within* Czechoslovak modernism filtered its way into the Europe-wide realm of CIAM.

The image of modernism as a powerful juggernaut flattening all in its wake was carefully cultivated by certain imperialistic modernists, but is hardly an accurate understanding of the movement, as these three examples show. How, then, can we account for these similarities and differences in the texture of cities and the shape of space? Orthodox partisans of modernism might wish they had a single yardstick to measure whether any given building or city conforms to an authentic modernism. But who can say that the expressionist Amsterdam school practiced an inauthentic modernism?

After all, Michel de Klerk fully embraced the "tingle of the new," and his Spaarndammerplantsoen housing project resulted from the same urban, activist left-wing politics as did Frankfurt's much-celebrated Römerstadt. A similar observation is valid for Czech cubism and rondocubism in architecture, even though these were later eclipsed by functionalism. Any explanatory scheme presuming there was "one best way" to the modern seems destined to fall flat.

With the concept of *circulation*, we direct attention to those historical processes and institutions where concepts, practices, artifacts, and systems that appear to "change the world" are located. CIAM is one such institution; other studies can be made of international standard-setting bodies, consulting firms, science and engineering as intellectual disciplines, transnational corporations, and the profusion of intergovernmental agencies that constitute European integration. In like measure, with the concept of *appropriation* we can account for the culturally specific ways in which supposedly universal concepts and practices are taken up, modified, or even rejected. In this view, Czech rondocubism was hardly a diverting side path in the triumphant march toward modernism, but on the contrary a culturally significant effort to meld modernism with Czech nationalism. A different case is the continuing and ongoing dynamism of Dutch modernism evident even today. One can situate this too, not as a vague "cultural preference" that stands above historical analysis, but as the consequence of efforts by specific Dutch architects in the mid-1950s to replace the ossifying CIAM, with its universalistic pretensions, with new institutions that would reinvent modernism several times over.

Finally, we need to recognize that full awareness of the historical value of European cities emerged slowly and fitfully among full-blown modernists—and may even today be contingent. While Rotterdam, as we noted, enthusiastically embraced modernism during its postwar rebuilding, the central districts of Prague were little scathed by modernistic reconstructions until very recently. Today the east side of Prague, for a part along Wilsonova near the old city wall, is the new site of a massive elevated-highway bridge. Even Le Corbusier acknowledged the valuable historical heritage of Europe's cities. If rather narrowly, he did allow that "architectural assets must be protected... if they are the expression of a former culture and if they respond to a universal interest ...and if their preservation does not entail the sacrifice of keeping people in unhealthy conditions...and if it is possible to remedy their detrimental presence by means of radical measures, such as detouring vital elements of the traffic system or even displacing centers hitherto regarded as immutable" (Le Corbusier 1943/1973, 86–87). One hopes that Prague's planners take a wide view of their city's "architectural assets" to include not only the well-known tourist treasures but also the city's less-well-known modernistic heritage. It would be a pity to preserve baroque Prague but sacrifice the vibrant modernist diversity of cubist, rondocubist, and functionalist Prague.

Notes

1. For their expert guidance on Dutch and Czech modern architecture, I am greatly indebted to Elisabeth van Meer, Kimberly Elman Zarecor, Arjan van Rooij, Dick van Lente, and Liesbeth Bervoets. I benefited also from the subtle nationalist interpretation of Teige in Zarecor (2002). Remaining errors are, of course, my own.

2. The terms "Czech," "Bohemia," and "Czechoslovakia" need to be used with some care. Generally when modernism was most active in Prague, I have used "Czech modernism" (although "Bohemian modernism" might equally be appropriate); later, especially after 1934–1935, when the Moravian city of Brno became more active as an international center, one might speak of "Moravian modernism." For institutions representing the new nation-state, I have used (e.g.) "Czechoslovak CIAM."

3. Kotěra's students included Adolf Bens, Josef Gočár, Pavel Janák, Jaromír Krejcar, Otakar Novotny, and Kamil Roškot.

II Representation and Reform

5 Spectacles of Sociability: European Cities as Sites of Consumption

Paolo Capuzzo

The history of urban technologies has often been understood in functional language, in the sense that they did things to cities. It certainly is true that gas and electricity networks provided residents with light and heat; omnibuses, streetcars, subways, and automobiles transported people within and between cities; water pipes and sewage lines made urban life more hygienic and more comfortable. This functional description is, however, just one side of the story. Urban technologies also had important consequences on the symbolic and ideological levels, as several contributions to this volume show. Prototypically modern technologies helped create a sense of modernity, with cities as the principal site. For instance, Noyan Dinçkal shows that, besides carrying passengers, Istanbul's streetcars also carried value-laden notions of "modernity" and "the West." And Per Lundin shows how European traffic engineers in the 1950s brought American values, experiences, and expectations back to Europe with the "car-friendly" city.

Similarly, the history of urban consumption cannot be understood in purely functional terms. Everywhere, markets combined economic transactions with social intercourse, and shops displayed products in an inviting and attractive manner. The technical networks of transport and communication sped the circulation of consumer goods and consumers, but new patterns of consumption were not merely adjustments to change in these technical structures. In fact, this chapter argues that an economic history of the modern city must take up an explicit cultural dimension: cities are not only sites of markets and commercial transactions, but also places of sociability and spectacle.

The city played a central role in overthrowing the ancien régime of consumption (Roche 1989). The globalization of commerce, particularly in the northwestern part of Europe in the seventeenth and eighteenth centuries, led to the introduction of new products and new habits that transformed European society. The diffusion of coffee, tea, cocoa, tobacco, Indian calico, and chinaware changed daily habits, and created new places and modes of sociability (Burnett 1999). These products were less burdened with social and legal restrictions, and their use resulted in new cultures of consumption that eroded older forms of symbolic power. Cities were the places where these novel products arrived from overseas, and where people met, exchanged habits, and learned to use them. The commercial cities provided a space that favored this connection between commercialization and cultural exchange (Curtin 1984). At the eve of the modern era, cities blossomed as a space of cultural cross-fertilization because the globalization of commercial networks widened the available material resources, and the deregulation, expansion, and commercialization of urban space facilitated contacts among different cultures. Nearly everywhere, cultural and technological innovation percolated through the urban scene.

Scholars studying consumption have identified the commercial city as a workshop of the modern culture of consumption, and they propose a theoretical framework that allows the interpretation of certain characteristic social processes that, although not yet the European norm, were well established in some cities already in the seventeenth and eighteenth centuries. In the nineteenth and early twentieth centuries, cities wielded influence throughout Europe, spreading new forms of consumption and new habits that shaped and reshaped social and power relationships. Cities furthermore played a leading role for individual consumption through at least the 1950s. More recently, consumption has taken a different spatial form due to increasing geographical mobility, changing ideas about urban planning, and the consequences of new media, which together recast the boundary between city and countryside (see Heßler, this volume).

The precocious consumer culture of seventeenth-century Amsterdam or eighteenth-century London should not be understood as the necessary "first step" toward modern consumption as it later developed throughout Europe. The diffusion of new consumption patterns was influenced by wide-scale changes in commercial geography, in transport and communication technologies, in production systems, and the prevailing political and social cultures. There was certainly the circulation of a normative urban middle-class culture of consumption in the nineteenth century, but the historical literature directs us to pay close attention to changes, peculiarities, and adaptations. Of course, there were leaders and followers in European patterns of consumption, but, as elsewhere, European modes of appropriation were not passive imitations and they occurred in distinct urban contexts that need investigation.

By modifying the organization of urban space, commercialization opened new symbolic and aesthetic dimensions. Consumption scholars influenced by the Marxist analysis of commodity fetishism aim at a "critical analysis" of aesthetic forms of commercial culture and urban imagery. The task of such a critical analysis is to overcome the common separation between urban economy and urban aesthetics, and to interpret these phenomena together. This theme animated a well-known volume by Wolfgang Fritz Haug (1971), but its most influential presentation can be found in Walter Benjamin's Parisian *Passages* (1982). Influenced by the surrealists, particularly Aragon's

Paysan de Paris from 1926, Benjamin developed his analysis of the "commercial urban *rêverie*" engendered by the new world of consumption as a critical tool directed against capitalist society. His aim was to break down the dreamlike experiences created by commercialization. Later formulations of this approach, such as Guy Debord's (1992) concept of "spectacle," also intended to subvert daily life and its symbolic codes, but these attempts tended to develop in a more pessimistic direction. They emphasized the ability of consumption to develop independently of the material processes that constitute it as well as its ability to swallow up subjective experiences.

This theoretical approach helps frame the relationship between economy and aesthetics in the urban environment. Nevertheless, from this perspective, the consumer as a subject remains in the background or is totally absent. So, while helpfully analyzing the social semiotics of commercial cities, these scholars frequently discuss "meaning" and "radiance" as though they were autonomous forces able to determine the social fabric completely, while social actors vanish behind the symbolism of objects. This approach pays insufficient attention to the active role of consumers in the key processes of reception, selection, and interpretation of symbols and messages.

Recent contributions from anthropology and ethnography add to our understanding of urban consumption by showing the subjective tactics that guide the practical and discursive appropriation of objects (Friedman 1991; D. Miller 1987; Sassatelli 2004). For these authors, the subject actively builds meaning at the level of practices and discourses. De Certeau (1980) considered the consumer as a *bricoleur* who combines meanings to create a new relation between the self and objects. In this chapter, I regard consumers, shoppers, and tourists as active subjects rather than passive spectators, even though they do not act in a completely free environment but within power relationships that connect them to strategies of production and commercialization. Thus, the process of subjectification through consumption can "fail" and create a sense of discomfort, frustration, and alienation in the face of objects that do not appear to be expanding the self, individual consciousness, or actors' knowledge (cf. Marcuse 1964). In sum, then, consumption has to be seen as a practice that is shaped by many forces, material as well as symbolic.

The Networked City and the Geography of Consumption

Urbanization and the increasing division of labor strengthened the role of the market in supplying society with goods. Commercial and technical infrastructures enhanced the influence of markets. Water and road networks in the eighteenth century, and railways in the mid-nineteenth century, widened the scope of markets and increased commercial competition by making the movement of goods faster and cheaper, more punctual and reliable. The flood of goods into cities and the new consumption culture they brought about profoundly influenced urban geography, urban morphology, and the urban landscape. Forms of commerce changed by degrees. Medieval institutions such as annual fairs—typical of a rural, poorly networked economy—waned markedly or dramatically changed form (De Grazia 2005). The image and function of urban markets were also transformed: city markets became junctions of commercial networks based on the regional specialization of production integrated by new transport technologies (see Disco, this volume). Thus, urban marketplaces were continually reorganized—as testified by several new forms of regulation introduced in the nineteenth century—to keep pace with new needs. Urged by reform movements that struggled to improve health conditions, the state intervened to guarantee food quality and to regulate the handling and processing of food in such places as shops, stores, slaughterhouses, butcher shops, bakeries, and dairies.

Peddlers, too, changed in revealing ways. By the middle of the nineteenth century, British peddlers had abandoned isolated rural areas with weak commercial infrastructures. Overall, the need for peddlers had decreased, since the number of shops and the connections between retailers, wholesalers, and producers had increased greatly (Alexander 1970). Instead, peddlers began to exploit their mobility and flexibility to focus on the commercial spaces opened up by rapid suburbanization. The relationship between suburbanization and peddling seems to be a structural one. Even in the 1960s in the Milanese *banlieue*, for example, the suburban function of peddlers was strikingly similar to that in Britain a century earlier (Capuzzo 2003).

In addition to the functional changes in the older forms of commerce, entirely new places of commerce and consumption emerged in the nineteenth century. Arcades represent the first visible evidence of consumption practices and places absent in the typical city of the ancien régime (Geist 1969). They were private spaces that were open to the public, where new building materials and technologies—glass, iron, gas lighting, seductive displays, and comfortable heating—created an environment that combined consumption with sociability. In short, they appeared to fulfill Benjamin's description of an urban daydream. Benjamin saw in the arcades a resource for re-enchanting the world. The aesthetic dimension of the world of commodities, the seductive experience of consumption, and the commercialization of social life—all these contested what Max Weber viewed as the "disenchantment of the world."

The capital of this form of urban consumption, as for much else, was Paris, where the first *passage* was built before the revolution: the Jardins du Palais Royale (1780). Yet the golden age of the Parisian arcades was the first half of the nineteenth century. Such famous *passages* as the Panoramas (1800), Grand Cerf (1825), and Verdau (1846), as well as the Galérie Vivienne (1826), created urban places that blended social life and commercial culture. As intentional investments of private capital in the urban morphology in order to promote commercialization, arcades represented visual and material evidence of the way in which the culture and economy of consumption transformed the modern city.

Figure 5.1
London's Burlington Arcade, built in 1819 for shopping, sociability, and spectacle, and still open today with up-market shops and consumers. Photograph © Andrew Dunn, 8 June 2005 ⟨www.andrewdunnphoto.com⟩; used under Creative Commons Attribution ShareAlike 2.0 License.

Arcades spread throughout Europe, although the international circulation of this distinctive urban form remains to be studied in detail. In London, the Royal Opera opened a walkway of shops (1816–1818) opposite the theater entrance, while the Burlington Arcade opened a full promenade with 47 leased shops in 1819 (figure 5.1). The Corridor opened in Bath in 1825, promoting a relationship between shopping and vacationing that became a typical feature of tourist cities. The Galeries Royales Saint-Hubert soon became the cornerstone of urban life in Brussels. In Vienna the Freyung Passages were built between 1856 and 1860, simultaneously with the grandiose Ringstraße. One exception to the otherwise pervasive pressure of commercial forces was the Galleria Vittorio Emanuele II in Milan, built in 1867, the result of public initiative celebrating Italian unification.

Special exhibitions and world's fairs also promoted new forms of consumption and helped transform the urban landscape. Besides the most famous exhibitions in London and Paris, medium-sized cities also hosted exhibitions. And even if their buildings were temporary, exhibitions had an incalculable impact on the public's imagination and expectations. The circuslike environment helped sell the public on new technologies. It is no accident that promoters of path-breaking technological innovations showcased them at the great expositions: the telegraph in London in 1851; the elevator and reinforced concrete in Paris in 1867; the sewing machine, typewriter, and telephone in Philadelphia in 1876; the refrigerator and electric light in Paris in 1878; the phonograph in Paris in 1889 (Greenhalgh 1988).

In addition to these spectacular innovations, the great expositions displayed the tableaux of modern life. They made the public aware of the commercial applications of technology with a direct impact on daily life, from personal body care to house appliances, from food products to hobby items. Engineers and artists were commissioned to design spectacles that melded technology with exoticism. Several exhibitions featured suggestive architectural structures, most famously the 1851 iron-and-glass Crystal Palace in London, or the Eiffel Tower and the Galerie des Machines built for the Paris exposition in 1889. These workshops of the modern imagination outlined an entire iconographic apparatus of modernity.

To host an exposition, cities had to develop their transport infrastructures to handle the deluge of visitors traveling to the city and circulating within it. Railways brought thousands of people each day to central terminals, while tramways, omnibuses, metros, and urban railways took them out to the special districts typically established for an exhibition. In both Europe and the United States, these exhibition districts often developed into major leisure, recreation, or commercial areas.

We can see this long-term transformation at work with exceptional clarity in Vienna. The Prater district, traditionally a local Viennese leisure area, had long been the site of popular festivals, such as Saint Brigida Day, at which social classes melded together in a carnival-like environment (Capuzzo 1998, 391–392). The Prater was the site of Vienna's 1873 exposition and consequently developed a more diverse mix of commercial infrastructures. The district's excellent transportation facilities led in 1895 to it becoming the site for the theme park Venedig in Wien (Venice in Vienna), which attempted to reproduce Venice by combining the exoticism of nineteenth-century fairs with new technological attractions such as photography, electric lighting, phonographs, and cinema (Rubey and Schoenwald 1996). This simulation of Venice—complete with buildings, canals, bridges, shops, coffeehouses, even Venetian *gondolieri*—proved surprisingly popular for five years before it was dismantled. It started a new kind of amusement that was followed by later simulations of Japan, Spain, and Egypt. Venedig in Wien and its successors thus created a new space for reconfiguring the public of consumers on the basis of mass entertainment. One experiences a latter-day version of Venedig in Wien, complete with imitation *gondolieri*, in Las Vegas today.

These turn-of-the-century multimedia spectacles did not long survive the arrival of moving pictures, however. The champion of Venedig in Wien, Gabor Steiner, went bankrupt in 1912, a sign of the decline of mass entertainment that had concentrated in a single space elements taken from different urban subcultures, in this case revues, ballets, cabaret, and the Viennese *operetta*. Initially, movies were a mixed entertainment that took place within music hall and fairs as one attraction among others, but soon movies became a specific, stand-alone form of entertainment. Promoters built special movie theaters within the city that modified urban sociability and the urban imagination (Charney and Schwartz 1995). Cinemas reinforced the commercial and amusement specialization of particular city areas, while other leisure activities like football and race courses developed in the outlying urban areas that were linked to the city core by electric streetcars or trolleys. These developments involved a much wider social stratum. The poorer class of workers had previously had a localized social life, but the new forms of cheap entertainment created new spaces of sociability for them.

Palaces of Consumption

Department stores such as Paris's Bon Marché were once understood to be the sine qua non of modernity and consumption, but recent authors have questioned whether in fact they were so distinctive as commercial institutions. Commercial techniques taken to be characteristic of the department store, for instance, were already part of the commercial culture of European cities' luxury shops and bazaars in the first half of the nineteenth century (Walsh 1999). What seems indisputable is that department stores were a striking discontinuity on the symbolic and the everyday, practical levels. The birth and diffusion of department stores remains an important turning point in a cultural history of consumption (R. H. Williams 1982; Lancaster 1995). In the mid-nineteenth century, as Jeanne Gaillard (1975/1997) put it in her classic book on Paris, the larger European cities changed from being "introverted" to being "extroverted." The department store was a catalyst, a result, and an accelerator of this transformation process, so it is necessary to consider its birth and development within the context of general urban change.

Private capitalists in Paris invested in the urban structure to promote commercialization, urban transportation, tourist infrastructure, commercial institutions, and various services (Harvey 1994). The transformation of the transportation system in Paris began in the early nineteenth century, led by the needs of traffic and commerce. As the city allowed for better circulation, a powerful omnibus company began to operate in Paris. Thanks to the company's managerial and organizational concentration, and despite the regime of monopoly, urban transportation became affordable for the expanding middle classes. The outcome was a social geography that tended to separate affluent people living in the city center from the lower social strata that found themselves increasingly in the outskirts. Department stores flourished in this context. Paris and its many department stores gained a leading role in urban imagery through popular novels such as Zola's *Au bonheur des dames* (Crossick and Jaumain 1999; M. Miller 1981). The French capital maintained its leadership in urban transport facilities until the end of the century, when Berlin and other cities took better advantage of the opportunities opened by the electrification of transportation.

Department stores in city centers contributed to huge rent increases while being at the same time the result of such increases. Consequently, they had to maximize selling volume at low unit profits. Given the escalating land prices, they became a force toward the verticalization of the urban form, a process that was most pronounced in the United States. Only in the 1920s did the commercial pressure toward vertical development become so strong in Europe that building regulations that had limited vertical structures were successfully challenged.

Department stores were seldom, however, at the forefront of innovation in architectural styles. Jugendstil department stores in Paris, Antwerp, and Brussels contributed to the renewal of the urban landscape, as did modernistic department stores in Prague somewhat later; but most department stores adopted the tried-and-true monumental historical styles. Sometimes monumentality was developed according to a sober rational design, as in the Wertheim department store on Berlin's Leipziger Platz, but even in Germany the turning point in department store architecture occurred only in the 1920s with the building of the Schocken department stores designed by Erich Mendelsohn (James 1999). Mendelsohn's architecture emphasized the accelerated rhythm of the modern city, reinforcing the communicative potentials of the urban stage. His dynamic functionalism combined the simple geometry of American factories with agile structures wrapped by wide, rounded windows (see also Misa, this volume). The work of other architects of the Neues Bauen movement also contributed modernistic commercial designs.

The diffusion of the department store by and large followed the geography of European urbanization. At the beginning of the twentieth century, almost all big European cities had at least one. Where the middle class or the skilled working class was sufficiently affluent, department stores developed also in smaller cities. In Central and Eastern Europe and in Italy, where the level of consumption was relatively low, department stores were primarily to be found in big cities. On the eve of the First World War, for example, the Bocconi department store reached nine cities and in Naples it even competed with the Mele firm (Amatori 1989). In big cities with several department stores, customers were segmented by economic class. In Berlin, for instance, Wertheim catered to affluent people, Tietz targeted the middle classes, and A. Jandorf & Cie served the working class.

Figure 5.2
Erich Mendelsohn expressed the rhythms of the modern city in the Schocken department stores of the 1920s. This store in Stuttgart wrapped rounded windows around the simple geometry of American factories. While his famous Einstein Tower (1919–1921) was in Potsdam, near Berlin, Mendelsohn's other modernist stores, in Nuremberg, Breslau, and Chemnitz, pointed out that not all the action was in capital cities.

Looking carefully at the chronology and geography of department stores in Europe, one can easily spot distinct patterns of diffusion. Paris, which absorbed financial, demographic, and symbolic energies like no other capital city, was strategically important as a center of fashion and consumption. So in France the diffusion of the department store radiated outward from the primacy of Paris. Britain largely followed the Paris model, with London at the center. By contrast, in Germany peripheral companies first developed a branch system in middle-rank cities and then moved to gain control of stores in the metropolis of Berlin (Coles 1999). Italy, as noted above, also had a remarkably decentralized pattern of department stores.

Department stores took advantage of the "extroverted" city. Invariably, department stores were located in streets with large or newly increased traffic flows, such as those in

Haussmann's Paris or the Kensington area in London. Another key site was the intersection of streetcar lines, such as the Mariahilfer Straße in Vienna, the Leipziger Straße in Berlin, Piazza Duomo in Milan, or along the ring that replaced the old city walls in many European cities. In general, the stores contributed to the increase of prices in nearby areas because they attracted masses of visitors and encouraged other commercial developments. Despite the fierce opposition of small shopkeepers, department stores often had a positive influence on the commerce of the nearby areas; but they also accelerated the marginalization of other, more distant areas (Jaumain 1996; Morris 1993). Since department stores were powerful actors in the urban space, it is not surprising that they aroused fear and concern. In the first phase of their development, department stores aimed at conquering central areas, and their image was linked to the premises on which they made their greatest investments in building up their symbolic meaning and appearance. Later, in the first decades of the twentieth century, some department stores began to follow the middle classes into the suburbs.

Department stores became catalysts of modernity in part by introducing technological innovations meant to fascinate customers as well as to stage a seducing theater for the display of goods. They also became retailing machines organized according to rational managerial principles that regulated labor, marketing, advertising, and selling. By the 1880s the pneumatic tube sped up the labor-intensive collection of money during the day, while the cash taken in could be transferred rapidly and safely to the banks. The elevator, introduced in the 1850s, rapidly diffused at the end of the century, allowing the vertical increase of department stores; escalators spread in the interwar period, initially causing some discomfort among customers not used to this mechanical movement. Air conditioning was also introduced during this period to make stores more attractive in the hot summer months.

Early on, lighting represented a decisive element in the display of commodities. Visitors to London even in the eighteenth century gazed at shop windows dramatically lit by candles and mirrors. Gas lighting spread throughout the city in the nineteenth century, while department stores rapidly adopted electric lighting in the 1890s (see Schott, this volume), allowing designers to develop sophisticated theatrical effects in the display of goods. The American experience of New York and Chicago, documented in the journal *Shop Windows*, arrived in Europe with the opening in London of the U.S.-designed and -funded department store Selfridges in 1909 (Leach 1989). In the interwar period, the profusion of brightly lit signs outside the department store further transformed the urban landscape.

The appropriation of arcades and department stores was simplified by the establishment of restaurants, hotels, coffee bars, hairdresser's salons, theaters, cinemas, tea rooms, and many other services on a rapidly changing urban stage. Sometimes these services were located in the department store, but more often they were situated in the surrounding area. These commercial areas became a new sort of urban public sphere, although they were created to make profit rather than to promote civic discussion. Businessmen attracted people by improving the aesthetic standards of the urban environment and creating new attractions.

In *Shopping for Pleasure,* Erika Rappaport (2000) links the building of London's fashionable West End to the formation of a public sphere for middle-class women. Mostly from the suburban districts of London, these women flocked to the new, clean, and reasonably safe shopping areas in the West End, a sharp contrast to the dirty, ambiguous, and even somewhat dangerous traditional marketplaces such as Covent Garden. Prior to the establishment of this commercial space, unaccompanied women had been regarded as a problem. A woman alone in the street was assumed to be lower-class, or even a prostitute. In London, then, urban transportation helped establish a new connection between the domestic private sphere and the urban public sphere. The interaction of the commercial strategies of businessmen with the social practices of women made the West End an area of urban sociability, an outcome favored also by a remarkable presence of aristocrats and the unusually affluent living there. Thus, even in Victorian London, the construction of gender among the middle class occurred in the public sphere and not only in private space (Walker 1995; Yaffa 2001).

Commercial strategies, social practices, and urban spaces obviously interact in complex ways. Commercial strategies are developed by business people, advertisers, and architects who make use of specific intellectual and material resources. When applied to urban space, these strategies are confronted with other forces, such as production and housing, technological change, building and planning culture, land use regulations, and, last but not least, social and individual practices that contribute to build meanings and images of this space (Falk and Campbell 1997). Even in places that architects intended to be areas of commerce and consumption, people met for many other reasons, sometimes even openly in contrast with the area's original intentions.

Leaving the City

The close connection between commercial culture and urbanity began to fray after the Second World War. In the United States the development of out-of-town shopping increased dramatically beginning in the 1930s. Large shopping malls outside the cities flourished with the mass use of cars, the spread of suburban residential patterns, and the development of motorways. These new shopping infrastructures could outcompete traditional urban shopping centers in price because of their lower land costs and their economies of scale. Europe had a different experience because the mass use of cars began only in the 1960s, delaying such processes by a whole generation. Moreover, European city centers remained attractive because of their morphological features and the lifestyle they allowed. Urban planners seeking to identify "livable" cities have often noticed these patterns. Still, out-of-town shopping malls spread in Europe, too, in the 1970s and 1980s, although at a slow pace and with different geographical features. These developments do not seem to have led to the same catastrophic consequences as in some American city centers. Even in Britain, which has urban patterns quite similar to American ones, shopping malls modified the spatial organization of consumption but did not destroy traditional city centers (Gayler 1984).

These urban and suburban changes are closely related to mobility and residential patterns. We could say that in the first part of the twentieth century commercial structure depended on mass public transport, while later—as cars became the main mode of transportation—difficulties in reaching central urban areas pushed customers toward more accessible suburban locations. In the 1950s and 1960s, even in Europe, the watchword of planning was to adapt cities to the requirements set by the mass use of cars—creating the "car-friendly city," as an influential German traffic planner phrased it—yet, due to the structure of many European cities, this was not an easy task (see Lundin, this volume).

Shopping malls continued the commercial spectacle that began with department stores, but on a wider scale and with more sophisticated technological equipment. Moreover, they played on safety concerns and perceptions that were becoming problematic for many old city centers. Ethnographies of the shopping mall often point out attitudes and symbolic practices that strikingly resemble literary descriptions of the nineteenth-century consumer euphoria. Shopping malls were even the site of active appropriation of commercial space for distinctly noncommercial purposes, for instance, youth socialization. Nonetheless, suburban shopping malls were more standardized and more predictable than their urban counterparts. Nineteenth-century commercial culture had organized space and created technologies of circulation that contributed to the homogenization of the European urban areas of consumption, but cities remained differentiated and unpredictable entities in which the culture of consumption interacted with active social cultures and a dynamic urban fabric.

Building a commercial infrastructure on entirely new premises, with no symbolic or historical background, was a different matter. Here the consumer cultures separated from the urban dimension tended to become standardized, but the *non-lieux* (literally, nonplaces) that were supposed to make the space more comfortable turned out not to be very attractive in the long run (Augé 1992). Since the 1980s, citizens and planners have become more sensible of the damage represented by the decline of old urban centers. New urban strategies have aimed at the relaunch of city centers, while new urban mobility policies finally reconsidered the role of cars in the city, invested in public transport, and created walking areas (Capuzzo 2004). Thus, shopping has been reintegrated into the set of social and morphological resources offered by the city. Out-of-town shopping malls and urban commercial districts will probably coexist by exploiting different economics and social patterns of consumption.

Cities to Consume: Tourism and Urbanization

The revival of traditional European city centers testifies to the continuous strength of factors like sociability and spectacle in the area of consumption. After the fall of the Berlin wall, cities like Prague and Kraków reemerged as spectacular sites of commerce and culture where visitors mingle in famous coffeehouses and behold various tourist attractions. In fact, tourist cities can be seen as the ultimate stage in the development of urban consumption. You do not go to Rome to visit Benetton stores but to experience the city as a whole.

Prague, Kraków, and Rome are, of course, old cities with a long history of visitors and, as they later came to be called, tourists. In other areas, brand-new towns have catered to visitors. Often enough, such cities emerged in coastal regions, or at least in areas with natural beauty or calmness. Increasingly, extreme forms of environmental exploitation have accompanied such developments, especially in areas that have become seaside resorts. Organizing the environment in order to make it an object of consumption represented a decisive step for some backward European regions that found in the tourist industry their main source of private and public investment. The success of a seaside resort depended on the quality of its natural resources, on the investments made, and on the ability to build up a symbolic attraction, a sort of representative glamour. The spectacle staged at seaside resorts depends on delicate interrelationships among the perishable resources of land, air, and water.

Many cities and regions in Europe have come to depend on consumption and tourism as their main sources of economic, demographic, and architectural development. Tourism financed urban growth and modernization of these "cities to consume." Technology in these leisure cities had the same double function as in the department store: such cities contributed to the "re-enchantment of the world," in that they stimulated new fashions and attractions to catch a public looking for fun and amusement; and the development of tourism as a main economic engine of urban development required technical improvement of the city's services and infrastructures, its planning regulations, and skillful management in both the private sector and local administration. Tourist cities often represented the leading agencies for the economic development of entire regions.

Railways modified the geography of tourism in Europe, as they contributed to the social broadening of holiday-makers. The aristocracy still tended to set the overall pattern, but interesting opportunities for commercial developments came from new, middle-class customers. Cheaper and quicker transportation was a fundamental condition for broadening the customer base, and the proximity between port cities and seaside resorts helped the latter to put up a railway service, although they were not able to finance it with their own resources. In parallel with strategies for appealing to the masses that reinforced homogenization, the rise of middle-class tourism also gave cities stiff incentives to adopt countervailing strategies of differentiation and product segmentation.

A recognizably modern tourist industry began with the shift from private country houses to entire cities that offered special touristic services. Spas, which had a long tradition in Europe, developed into modern tourist centers in eighteenth-century England, as they began to commercialize their natural resources and make their urban environment attractive to customers (Hembry 1990). The success of Bath rested on its building up an efficient thermal establishment, creating leisure facilities such as theaters, coffeehouses, restaurants, and hotels, and organizing leisure activities such as balls, concerts, and other performances. Shopping became a central feature of leisure cities, which might even offer attractive luxury shops comparable to those of London or Paris. Bath and Chester were famous for their shops, as were Nice and the many *Kurorte* (health resorts) in Central Europe in the nineteenth century.

The development of tourist and commercial infrastructures around hot springs spread on the Continent, too. The Belgian town of Spa became a fashionable holiday resort for the European upper classes after 1830, and provided a meeting place for the Continent's aristocracy. Private companies and local authorities cooperated to strengthen commercial facilities and carried out city planning and architectural projects to imitate the Bath pattern. In Italy and Germany, as well as in the Habsburg monarchy, the *Kurorte* were at the forefront of urban modernization and became a symbol for the spirit of old Europe. Its crisis was memorialized by a novel set in a fictionalized *Kurort*, Thomas Mann's *Magic Mountain* (Prodi and Wandruszka 1996).

Italy had an ancient thermal tradition, and many of its resort cities commercially exploited their natural resources. However, resorts such as Montecatini and Salsomaggiore, for example, emerged only at the beginning of the twentieth century because of the general delay in Italian industrialization and urbanization. The success of these cities depended—as in other European cases—on the interaction of several actors: a municipal government able to invest in the urban layout to create spaces for leisure (squares, promenades, boulevards, parks) and to provide modern facilities; entrepreneurs, local or foreign, who promoted tourist structures and entertainment; and nearby universities that guaranteed high-quality medical care. These developments came together in Salsomaggiore, an economically depressed salt-processing center that became an internationally well-known spa resort. Salsomaggiore exploited its healthy waters, proven to be so by the University of Parma, and a direct railway connection to Milan, which provided the city with affluent customers as well as financial and entrepreneurial resources. These new investors built a tourist infrastructure able to attract people from across Italy and indeed from abroad, deeply modifying the urban landscape (Bossaglia 1986).

In the nineteenth century, tourist cities began to compete with each other by targeting an affluent and sometimes even famous public to build up their image as fashionable holiday resorts. Cities that managed to attract this group of people also drew in the celebrity-minded and status-conscious middle classes. Sometimes this strategy resulted in a well-regulated broadening of their customer base, but other times in a devaluation of the resort's glamour. These marketing techniques are still used today to promote and publicize holiday resorts, but it is not easy to manage them in the long run because fashion cycles inevitably move from one site to another. Bath is a good example. Its fame began to decline in the middle of the nineteenth century, and the city became a spa for the elderly, leaving to other cities the leadership in European fashion resorts.

In the early years, Bath served as a model for British bathing resorts. The fashion of a cold bath—encouraged by medical knowledge—favored the development of bathing establishments in northern Europe (Corbin 1988). Cold-bath resorts developed on the Baltic and the North Sea, where the upper classes of central northern Europe gathered. These cities were often promoted by public servants, as in the case of the Belgian city of Ostend. There the *Kursaal* was located on the beach and extended into the sea by means of a pier (Lombaerde 1983). In time the British city of Brighton became the leading city for this new fashion, after extensive investments in land, forests, and pavement in effect created a new coastal architecture, complemented by the city center's attractions and entertainment for holiday-makers. Brighton symbolized a transition in tourist cities: health care for the aristocracy gave way to amusement for the middle class.

The main change in the geography of tourism occurred with the rise of winter tourism on the Mediterranean coasts in the late nineteenth century. These areas, especially the Côte d'Azur and the Ligurian Riviera, had been very poor and economically dependent on fishing and agriculture, but in a few decades they became fashion resorts famous throughout the world, with cities like Cannes, Nice, Monte Carlo, and Sanremo. *Les hivernants* began to patronize Nice in the late eighteenth century, as people from northern Europe spent some months there between November and April (Pemble 1987). Initially holiday-makers were attracted by the mild winter and healthy air, but soon Nice developed an exciting social life in hotels, cafés, music and dancing halls, and gambling houses. After the establishment of the railway increased Nice's accessibility, the city grew at a tremendous pace at the end of the nineteenth century. At first, British holiday-makers were predominant (the main avenue was named Boulevard des Anglais), but soon enough tourists came from across Europe. To meet the requirements of affluent customers, Nice's municipal council carried out a scheme for rapid urban modernization that included entirely new water supply, sewer, and energy networks. New hotels anchored the region's tourist industry and became the main driving force

Figure 5.3
Tourism brought modern urban life to parts of the Mediterranean coast, such as the Promenade du Midi (ca. 1900) in Menton, in southeast France, about 15 miles by rail from Nice. Along the seashore was the tourist town of hotels and foreigners, while the separate native town was perched on the mountainside, with narrow, steep streets inaccessible to automobiles. Source: Library of Congress, Prints and Photographs Division, Washington, D.C. 20540, USA. LC-DIG-ppmsc-05934 ⟨hdl.loc.gov/loc.pnp/ppmsc.05934⟩.

of a general economic boom. Such tourist development attracted substantial foreign capital, which certainly aided the modernization of Nice and indeed the whole coast.

Yet Nice's moment in the sun did not last forever. Other Boulevards des Anglais developed along the Côte d'Azur, and car racing became the new modern attraction in the area. In the 1920s the image of the Côte d'Azur was renewed by Americans, and the exciting and somewhat reckless time spent at the Riviera was epitomized in F. Scott Fitzgerald's autobiographical *Tender Is the Night* (1934). In the 1930s the Côte d'Azur became a year-round resort. In time, however, the international competition grew so that Nice, like Bath before it, lost its position as a fashion leader and became a quiet resort for pensioners from northern Europe.

A further turning point in the European history of tourism occurred in the interwar period, as medical knowledge began to support the healthfulness of the sun and the warm seas, while a new aesthetic celebrated tanning (Triani 1988). These new trends dramatically expanded the demand for summer tourism in the Mediterranean. Cities in the region adopted different marketing strategies to promote their image through such media as literature, movie, radio, and posters (Meller 1998). These decades were important also for the social history of tourism. Mass tourism first appeared in Britain at the beginning of the twentieth century, initially driven by commercial forces, but also supported by the state in the interwar period. In some countries, state activities took place within a democratic framework, as in the case of workers' holidays organized by trade unions in Britain and France (Cross 1993). In other countries, such as Italy and Germany, mass tourism was part of a general policy of integrating the masses into the totalitarian project of fascist dictatorships (Baranowski 2004; De Grazia 1981). Here, mass tourism was not in the first instance based on private initiative, but a phenomenon orchestrated by the state to control the masses and increase their loyalty toward the government.

Notwithstanding the expansion of tourism, the 1920s and 1930s were of course decades of economic stagnation and international conflict. The explosive development of mass tourism set in only after the Second World War. Due to its enormous environmental impacts, mass tourism decisively altered the relationship between natural resources and tourist investments. The mass use of cars, organized bus touring, and airplane charter flights made Mediterranean seaside resorts easily accessible to reasonably affluent customers from northern Europe. Italy, Spain, and Greece exploited these new opportunities, and their backward regions received financial and technological inputs through enormous investments in the tourist business.

The Costa del Sol readily illustrates the forces that mobilized a new kind of urban dynamic. This southern coast of the Spanish Mediterranean, stretching roughly from Nerja to Gibraltar, represents an extreme case of urban modernization, environmental transformation, and technological diffusion driven by touristic consumption. As early as the late nineteenth century, the city of Málaga offered a cheaper alternative to the Côte d'Azur for winter holidays. Still, tourism in Spain was slow to develop because of inadequate investments and a mediocre railway system that left the Costa del Sol on the margins of the main flows of European tourist traffic. An exception to this trend was San Sebastián, not coincidentally on the opposite end of Spain (Walton 1996). Already a destination for domestic tourism by the mid-nineteenth century, San Sebastián flourished as an international destination with the establishment of a railway connection to Paris. San Sebastián imitated Biarritz, but was cheaper than its French counterpart. At the beginning of the twentieth century, San Sebastián increased its international fame owing to an efficient local administration that fostered its architectural development, established modern urban services, and favored investments in such sport facilities as golf, yachting, and tennis. By the 1930s and 1940s it was one of Europe's most important seaside resorts, a center of shopping and entertainment.

The joint action of public authorities and foreign tour operators created conditions for a successful tourist sector in Spain. In the 1920s, during the dictatorship of Primo de Rivera, the government improved the infrastructure and created publicly owned tourist hotels in cities like Madrid, Málaga, and Seville. At the beginning of the 1950s, Torremolinos and other coastal villages attracted American celebrities, including Anthony Quinn, Rita Hayworth, Ava Gardner, and Ernest Hemingway. Tourism transformed the economy of these coastal villages, which had been based on fishing and agriculture. The opening of the Spanish economy led to a binge of hotel building throughout the Costa del Sol. The real boom in Spanish tourism occurred in the 1960s, when Spain, still under the repressive Franco regime, abandoned autarchy and turned explicitly toward mass tourism (Marroyo 2003, 329ff.). The political economy promoted by the Francoist Catholic technocracy developed the tourist industry to help the balance of payments. German investors and customers played an important role in the Balearic Islands, promoting the rapid urban growth of Palma de Mallorca, while varied international investments developed the Catalonian coast and above all the Costa del Sol. Above all, the Spanish state was absolutely crucial: it built tourist hotels, controlled the quality and prices of their services, promoted professional training, and favored the construction industry and the implementation of new infrastructure.

The decisive technological step toward mass tourism in this region was the rebuilding of the Málaga airport. In 1953 charter flights from northern Europe began bringing masses of tourists to the Costa del Sol. Cheap charter flights and attractive holiday packages made the region competitive on the international tourism market. In the 1970s trips to Spain represented the most important business of the big European tour operators: in 1980, the traffic to Spain from Britain and Germany accounted for 35 percent of all package flights in Europe. Urban coastal development occurred with few obstacles and little regulation. The timing and geography of urban developments precisely followed the increasing distance from the Málaga airport. In order to extend tourist urbanization eastward from Málaga, a motorway had to be built to bypass the bottleneck of the inner city, opening up for exploitation the area leading to Nerja.

The accessibility of the Málaga airport, the closeness to the sea, and the presence of an existing village that could provide shops, amusements, and basic services favored established towns like Torremolinos, Benalmádena, and Fuengirola. This area experienced a building boom thanks to its low land costs: in the early 1960s, a flat in Torremolinos cost half as much as a similar flat on the Italian coast and a quarter of the price of one on the Côte d'Azur (Barke and France 1996). But these massive construction investments occurred without any effective city planning regulations. Even sizable urban centers that exceeded 10,000 inhabitants, such as Benalmádena and Mijas, had no urban plan, and there was no institution responsible for overseeing, let alone controlling, the rapid pace of tourism-driven sprawl (Roura and Bernier 1978).

Urbanization, tourist facilities, and the associated infrastructures spread rapidly up and down the coast, permanently altering the Andalusian coastal landscape. Profit-maximizing investments in the tourist business favored intensive building activities and oversize constructions. The building of the huge complex of Playamar in Torremolinos, with its 21 17-story tower blocks that can lodge 6,000 people, became the architectural symbol of these transformations (see figure 5.4). A lively building dynamic remade Benalmádena, too. The building of a marina, finished in the late 1980s, turned this old port into a modern residential, commercial, and tourist district with suggestive postmodern architecture, while the building of an amusement park on the coast provided a family entertainment anchor. By contrast, Marbella tried to target affluent customers, following the traditional presence there of Spanish and international

Figure 5.4
Modernism and mass tourism run amuck in Torremolinos, Spain. On the Mediterranean coast at the Playamar complex, residents of the 21 virtually identical 17-story high-rise buildings generate heavy environmental impacts. Photographer: Juan Antonio Ruiz.

aristocracy. Although the aristocracy itself gradually left the city, they were replaced by a new elite composed of show business personalities, sports stars, and wealthy visitors from the Arabian Gulf. For a time Marbella stagnated in the 1980s, but it recovered to become the coast's leading golf destination and managed to keep a flavor of distinction and exclusiveness, in no small measure because the nearby city of San Pedro de Alcántara neatly segregates the region's construction and tourist workers from the region's tourists.

Technological investments and the exploitation of natural resources have thus created a new urban region on the southwestern fringe of Europe. Urbanization on the Costa del Sol depended on natural resources, such as the coastal landscape and the sea; on the initiative of business people and politicians; and on the waves of visitors that appeared after the Second World War. Since the late 1950s, well-paid British blue-collar workers began to spend their holidays in Spain, taking advantage of the cheap peseta. A holiday on the Costa del Sol became a longed-for annual custom that allowed people to escape from daily routines, to break taboos and habits in a hot, exotic climate refreshed by the sea, to be intoxicated by alcohol and enticed by opportunities of unusual sexual encounters. This kind of escapist holiday had already been pursued by working-class youth in the interwar period (Fowler 1995). Often, money saved from the whole year was spent on two weeks' holiday, a significant attribution of value to an experience that allowed these groups to transcend their given social roles and their everyday commitments, offering them the possibility of playing with a different identity.

The vacation image of the Costa del Sol took on a cheap and working-class character in the British media. But from the 1970s onward, a new flow of people began to settle down on the coast. The emerging class of "rich retired" who benefited from the social insurance introduced after Second World War increasingly invested in bungalows or flats in southern Spain. These retired people from northern Europe, particularly Britons who had enjoyed the coast as tourists in the 1960s and 1970s, decided to spend their retirement years there, attracted by the low cost of real estate, the mild weather, and the ability to escape their destiny as working-class pensioners (O'Reilly 2000).

The actual results of these attempts by consumers to change their identities are difficult to evaluate. Their encounter with new realities and opportunities may have resulted in a free, uninhibited, and peaceful new life. Yet perhaps it created a sort of artificial social emptiness, where the absence of a real life bordered on the delirious, as in James Ballard's novel *Cocaine Nights* (1996), set on the Costa del Sol. In either case, these ambitions and desires strongly contributed to the urbanization of southern Spain. It is not easy to determine the British presence on the Costa del Sol, because official censuses are not reliable. It has been estimated, however, that by 1990 fully 75 percent of the inhabitants in Mijas were British and that half of them were older than 60 (Jurdao Arrones 1990). These expatriates hardly integrated into Spanish society; in fact, most of them do not even learn to speak Spanish and, as a further sign of their lack of assimilation, stores selling imported British food flourish there.

Due to the arrival of foreign retirees, the building business has moved since the 1970s from hotels to chalets and flats for older or retired foreigners. Most of the building firms have also been foreign. They bought the land from Spaniards, developed the lots, and sold flats. Real estate investments, including the purchase of flats, were particularly profitable for foreigners thanks to a complicated system of fiscal evasion. The flats were bought not by individuals, but by international companies whose legal seat was usually in Gibraltar, Panama, or the Isle of Man. The residents were represented as investors in a company that conveniently enough provided them with a flat. In this way the "purchaser" could evade property taxes, sales taxes, and death duties, because a purchase, bizarrely enough, was recorded not in Spain as a real estate acquisition, but abroad as an exchange of shares not subjected to fiscal deductions.

In the last fifty years, the Costa del Sol has been exposed to a new set of demographic pressures. Besides tourists and "foreign residents," the region has attracted many Spaniards in search of job opportunities. The combined impact on the natural resources and landscape has been tremendous, forcing the Costa del Sol to rethink its commitment to unregulated tourism. The first victim of the region's wild urbanization was Málaga itself, since its beach and coastal landscape have been destroyed and the city has been increasingly polluted. The coast's image has evolved over the decades: an uncontaminated heaven in the 1930s through 1950s; a low-budget mass holiday place in the 1960s and 1970s, with an active beach scene and an exciting nightlife; the ambiguous fame acquired in the 1980s and 1990s due to standardized holiday packages, a destroyed environment, crime and drug dealing, and the stagnation and decline of the tourist presence.

Local governments are trying to invent a pattern of environmentally friendly tourism, encouraging visits to the Andalusian hinterland and promoting sport activities, particularly golf. The attempt is to target the most affluent tourists, reducing the number of visitors without reducing the profits from tourism. Yet it is doubtful that the reduction in the number of tourists alone will guarantee a better natural environment. At the center of this controversy are golf players, who are usually active consumers of tourist services and spend much more than the typical beach visitor. However, golf courses put pressure on Andalusia's natural resources, since a single golf course requires about 40 hectares and untold volumes of scarce water (Doblas 2001).

In addition to these negative effects on the natural environment, social tensions have also emerged. The presence of the British enclaves has motivated some Spanish scholars to raise charges of neocolonialism because the land was sold to foreign investors who have exploited and destroyed the natural environment, making local people dependent on foreign investments and subject to volatile speculation. Spanish peasants became construction workers for a time, but with the end of the tourist boom, they will be compelled to move elsewhere. Britons are accused of not being willing to integrate, they refuse to learn Spanish, and they do not pay taxes, while at the same time expecting health care, which is particularly expensive for a population of retirees.

Thus, the experience of the Costa del Sol shows the ambiguous nature of an unregulated urban and technological modernization of environmentally attractive areas. The pressure of consumption and desire can mobilize investments capable of changing a landscape completely, but the result can be the destruction of the main natural resources and the collapse of the social and ecological environment.

Consumption and Exploitation of Space

The ecological footprints of cities are enormous, as is their role as centers of commerce and consumption. When citizens socialize downtown, or when tourists view freshly renovated churches and plazas, they do not always reflect on the ecological impact of their lifestyle. Urban imagery is based on other premises and tends to follow the logic of business and engineering. By means of new technology, a spectacular urban selling machine was designed that consumers appropriated according to their own social and cultural needs. "Economic rationality" and "cultural enchantment" interacted in building up the urban scene of consumption. At work has been an interplay of forces and practices in conflict and negotiation: the management of consumption requires that social forces are put to work and natural resources are used. Consumption is not an autonomous force but has to be understood within the complex social fabric in which it occurs.

Mass tourism contributes to the creation of uniformity throughout Europe (cf. Hård and Misa, this volume). As middle-class urban consumption patterns become the norm all over Europe, it becomes increasingly difficult for individual towns to find ways to distinguish themselves. Typical in this regard is the development in the German *Ruhrgebiet*. After coal mines and steel plants were gradually closed down, almost all of the former industrial towns in the area have redefined themselves as sites of consumption and culture. Germany's largest shopping mall is to be found in the former coal town of Oberhausen, and some of its most exciting museums are in the former steel town of Essen. But that is another story.

6 Progressive Dreams: The German City in Britain and the United States

Mikael Hård and Marcus Stippak

In the mid-1890s the town was one of the first in Germany that designed a publicly owned and managed electric tram network. In all other areas of modern town management it has proven its progressive character—in industry and in social life. Among its most notable achievements count the construction of a first-class water supply and sewage system, a slaughterhouse, an electric power plant and a gasworks, as well as an imposing swimming hall.
Stein (1913, vii)

With these words, Darmstadt's lord mayor proudly presented his city's achievements in industry, finance, hygiene, social policy, and technology. Because a truly "modern" town was not merely an industrial one, the latest developments in urban technology played an important part in his campaign to create a modern image of the town—as did the fact that Darmstadt hosted "one of the best equipped and most frequented institutes of technology in Germany." Darmstadt, a mid-sized town in the southwestern part of Germany, today markets itself as a "science city," highlighting its many research and development institutes. A century ago the town's industrial character was more pronounced: for example, the local pharmaceutical giant Merck was then considered a "chemical industry of world fame" (Stein 1913, viii). Darmstadt already had manufacturing sites, a centralized water supply system, and excellent railroad connections, but to be a truly modern town it also needed hospitals, museums and galleries, music halls and theaters, schools and recreational areas.

Darmstadt also promoted itself as a "town of art and culture" with an "extraordinary cultural and technical infrastructure" (D. Schott 1999a, 295, 297). The municipal authorities tried to bring together a number of seemingly incompatible functions: industry and art, culture and nature, labor and leisure, garrison and commerce. They also, quite naturally, attempted to accommodate diverse social groups. The city designated new land along the railroad tracks for industry, while simultaneously attracting wealthy senior citizens, members of the middle classes, and a resident colony of famous Jugendstil artists by designing new residential areas and turning a part of town into an avant-garde village.

By presenting the many advantages of their city, Darmstadt's leaders hoped to increase the number of wealthy taxpayers, and by representing their town in terms of modernity and progress, they celebrated their own deeds. According to a narrative of rational planning and cultural advance, the town's attractiveness resulted from the local government's innovative policy and competent administration. Darmstadt's lord mayor and civil servants shared a conviction with most German colleagues that only proactive politicians and strong public authorities could deliver the many services required by a modern city. Their aim was to guarantee all citizens a certain basic standard of living, a goal still known today as *Daseinsvorsorge* (caring about someone's well-being), by means of *Leistungsverwaltung* (service-oriented public authorities and utilities) (cf. Blotevogel 1990).

Compared to the "urban jungle" of British and U.S. cities—condemned by writers such as Charles Dickens and Upton Sinclair—German cities were to all appearances idyllic and harmonious. Through various initiatives and interventions, German civil servants had seemingly controlled urban growth and created humane living conditions for the masses. In the English-speaking world, the German program became known as "municipal socialism," and many commentators on both sides of the Atlantic, even those who were usually repelled by socialism, judged the achievements of German city authorities favorably (cf. L. Mumford 1938, 425).

Municipal socialists embraced modern technology. Appropriating up-to-date technologies would help solve the many problems that plagued rapidly growing cities in the industrial era. To finance and manage large urban technological systems, they believed in regulating business and in expanding the influence and competence of local administrations. Claiming that private initiatives in areas such as street cleaning and the emptying of cesspools left much to be desired, municipal socialists argued that a strong public administration would be much more far-sighted and efficient. Local politicians argued that public ownership of utilities would guarantee the maintenance and timely expansion of water, gas, electricity, and public transportation. If left to private interests, so the story line went, urban technologies would be unevenly constructed and remain inaccessible to many who needed them.

Despite the apparent lead of German local governments in creating a "modern" city, British reformers had advocated municipal socialism even earlier. During the 1880s, the British Fabian Society had argued for the socialization of important branches of industry, as well as for public ownership of energy provision, water supply, and transportation. German reformers then appropriated their ideas about a peaceful kind of socialism. The lobbying activities of the Verein für Socialpolitik (Society for Social Policy), a leading association of university professors, civil servants, politicians, industrialists, and journalists, brought about a "change in economic policy" on both the national and local levels (Krabbe 1979, 269).

The apparent success of these changes motivated urban reformers to look to Germany rather than Britain as the leader in municipal administration. Beginning around 1900, municipal engineers, architects, and planners from many countries envied the self-consciousness and optimism of their German colleagues. As Andrew Lees (1985) and Daniel Rodgers (1998) have pointed out, German cities functioned both as promising models and realized utopias for many Anglo-Saxon urban reformers. The professional character of German urban planning and the way in which German citizens identified with their towns impressed progressive reformers in the United States and Britain. In *Atlantic Crossings*, Rodgers (1998, 141) observes that, according to one U.S. reformer, "in Germany one found 'organic' cities, possessed of 'communal self-consciousness.'" In 1914, British journalist William Harbutt Dawson explicitly praised German cities for their progressive property policies, visionary planning concepts, and humane welfare facilities, and discussed the advantages of publicly owned utilities in gas, water, sewage, and electric power. Even after the outbreak of the First World War, the legendary British planner Patrick Geddes commented on the innovative and efficient character of German city authorities.

In this chapter we anatomize the campaigns that produced and propagated this new image of German cities. The success of the German model among urban reformers did not have a simple material background but was the outcome of concerted discursive strategies. In the following, we use the concept of "discursive framework" to analyze the circulation of ideas and concepts between and among Germany, Britain, and the United States (Hård and Jamison 1998). We claim that urban reformers in North America and Europe argued within a common framework of interpretations and representations. Rodgers (1998, 159) has pointed out that progressive reformers on both sides of the Atlantic framed solutions within "a shared language of germs and sewage, gas prices and streetcar fares." At home, the reformers appropriated the international experience to fit the particular needs of their respective cities.

This chapter does not seek to challenge Lees's and Rodgers's observations on the importance of the German city in Anglo-American discourse, but it does question Lees's assumption (1992, 147) that reformers in Germany, Britain, and the United States made up a united *Gesinnungsgemeinschaft*, a community with a common temper or turn of mind. While German and British reformers certainly shared some general ideas and visions, differences between Germany and the United States were in many respects so pronounced that it makes more sense to talk about a *Diskussionsgemeinschaft*, a community with a common language, a shared way of talking and discussing.

We describe how the circulation of urban technologies was mediated by conferences, visitations, journals, and translations. As Noyan Dinçkal does in his chapter on Istanbul, we investigate the symbolic character of technology in urban discourses. Whereas Turkish reformers attempted to make their capital more "European," Anglo-Saxon progressives aimed at making their cities somewhat more "German." Our focus is not so much on the workings of urban technologies, but rather on their meanings, how these meanings circulated on an international scene, and how they changed when appropriated by various actors. We study the role "modern" technology played in the visions developed by municipal engineers, architects, and planners on both sides of the Atlantic. Our concern is the ways in which discursive frameworks spanned national boundaries. Our focus, then, is on the narratives that actors created about technological developments in their own countries or cities, as well as their narratives about foreign achievements or mistakes.

In a more abstract sense, our interest lies in representations of the self and of others in the area of urban technology. With this focus on meaning and representation, we align ourselves with cultural historians and anthropologists who understand that material objects are loaded with meanings, and that artifacts are more than mere tools lying around for us to use as we see fit (Hård and Jamison 2005). Such objects convey symbolic meanings, carry values and norms, and are objects of desire as well as contestation (Hård 2003). This insight also pertains to urban technologies, city structures, and buildings.

A complementary cultural perspective also trickles through recent urban history. We find cultural analyses of city planning concepts and discussions of the aesthetic meaning of buildings, as well as discussions of "high" culture and various urban subcultures. Current histories of urban planning treat scenarios and visions, discourses and images. These analytical concepts also inform recent writings on progressive reformers, garden city advocates, and urban ecologists (Stradling 1999). Our contribution to this literature is an explicit focus on the circulation of imagery and the discursive appropriation of urban technologies. We are interested in how German cities were represented around 1900, and we wish to understand the role that urban technology played in such representations. To this end we work out the narrative forms in which the achievements of the modern German city were told, paying particular attention to the meanings attributed to various technological networks and systems.

This chapter covers the first two decades of the twentieth century, spotlighting the ideas and views of ordinary municipal engineers, mayors, and public officials. Ferdinand Rudolph, chief engineer of the Darmstadt waterworks, is our typical hero. Our approach is intended to form a counterweight to the many urban histories that discuss the cultural appropriation of urbanity on the basis of a very limited number of world-famous cities, such as Paris, Berlin, London, and New York, which appear over and again in this literature. Additionally, most studies of urban debates have overlooked the contributions of engineers and other technicians. Intentionally or not, this literature perpetuates the image of the objective, disinterested engineer who sees no need to become involved in political issues. On the contrary, we show that rank-and-file engineers also contributed actively to the symbolic representation of cities.

Urbanization: An International Challenge

Since the early days of industrialization, British industrial cities received bad press. The young Friedrich Engels added to a flood of critical comments when, in *The Condition of the Working Class in England* (1845/1987), he observed how the urban environment sickened the working class. Prostitutes, alcoholics, and orphans filled the streets, and even those who were lucky enough to have a job and a place to live could not count on a long life. Industrial work was unsafe, and residential areas were overcrowded, with damp, cold dwellings.

Reformers and charitable organizations faced an overwhelming task. Despite their pressure, the British Parliament was slow to pass reform laws. Investigations by the Poor Law Commission sometimes even diverted attention away from basic problems. The Commission's chairman, lawyer Edwin Chadwick, claimed to have empirically shown that the less people ate, the healthier they were. In his *Report on the Sanitary Conditions of the Labouring Population in Great Britain* (1842), Chadwick downplayed the need for higher wages and instead proposed that installing water closets and cleaning the streets were the best means to help the working classes (Hamlin 1998). Education and health propaganda were key ingredients in Chadwick's controversial recipe for reform.

Germany experienced similar problems when it began to industrialize on a large scale in the late nineteenth century. Industrial cities especially grew very quickly. The Ruhr area with its coal and steel towns stood out, but even in greater Berlin the number of inhabitants increased almost sevenfold between 1850 and 1910 (Reulecke 1985, 203f.). Untold thousands were housed in notorious *Mietskasernen*, or "tenement barracks." These disreputable buildings typically covered a whole block and often contained up to six privy-packed courtyards inside. Many of the apartments were suffocating and dark, and residents suffered in humid basements during winter or in hot attics during summer.

German authorities confronted the "social question" more actively than their British counterparts did. German mayors had sizable administrations at their disposal, and in most German states they controlled resources that were unheard of in the Anglo-Saxon world. Full-time working professionals such as engineers, medical doctors, and lawyers managed German cities. The legal guidelines for housing were successively tightened, so that toward the end of the nineteenth century, German cities forbade the building of apartments in which living rooms, bedrooms, and kitchens lacked windows of a certain size. Strict rules also specified the height of buildings and the distances between them.

Conditions in U.S. cities differed substantially from those in Britain and Germany. Many American cities had literally grown out of nothing in a relatively short time, expanding at such a breathtaking rate that it was virtually impossible to guarantee acceptable humane living conditions or create widely accessible services or utilities. The turnover of residents was usually much higher than in European cities, and immigrants at the lower end of the social ladder suffered health problems due to filthy housing and dangerous working conditions. In his best-selling novel *The Jungle* (1906), Upton Sinclair famously documented Chicago's chaos and filth.

In the United States, private enterprise—often operating by means of municipal franchises—had a larger impact on urban technologies than on the European continent. Although municipal authorities planned and directed urban expansion to some extent by laying out streets and railroad tracks, designing public parks, combating street debris, and erecting waterworks and public baths (Hoy 1995), private interests played a bigger role in the energy and public transportation sector (Roberts and Steadman 1999, 174). Urban problems triggered the U.S. reform movement of the 1890s. These reformers, called progressives, were middle-class professionals—typically journalists, medical doctors, and civil servants—who aimed to alleviate the worst forms of social inequality and reverse what they considered a steady process of urban decay. Progressive reformers in the United States eagerly turned to Europe for inspiration. Jane Addams, for example, founded her famous Hull House in Chicago after having toured large parts of western and southern Europe. European cities such as Frankfurt, Dresden, Birmingham, and Glasgow became influential models for Addams and other U.S. reformers. Impressed by the apparent cleanliness and orderliness of these cities, they praised the municipal authorities who took responsibility for urban developments. Dazzled by the achievements of Europeans in such areas as public transportation and street cleaning, U.S. reformers tended to accept almost everything European as positive. Although some realized that the political system in a country like Germany was not particularly democratic, U.S. reformers praised the German authorities' control over construction and planning, as well as what the reformers saw as the German authorities' disinterested and noncorrupt character.

In the decades around 1900, Germany became a model for British urban reformers as well. British and American reformers looked favorably upon the fact that German mayors were salaried and elected for relatively long terms, and that civil servants were permanently employed experts. Also, German cities could plan effectively because they owned relatively large areas of land, and the authorities had considerable freedom in confiscating private real estate for the public good. According to the reformers' view, these conditions enabled planners to act more independently of vocal interest groups and to make plans with long time horizons. This same administrative-legal framework also facilitated the orderly expansion and management of water and sewage systems, electricity networks, and tramlines. In his book *Networks of Power* (1983), historian of technology Thomas P. Hughes documents the role model function of German cities—Berlin, in particular—in the area of electrification.

Frequent study trips and numerous conferences bonded the reform discourse community together. Translation of books and pamphlets facilitated the circulation of information and knowledge, while informal talks and formal lectures reinforced intellectual ties with social ones, thus facilitating the emergence of a transatlantic discourse community. Members of the Boston Board of Commerce, for example, toured several European countries in 1911. In the same year Johann Heinrich von Bernstorff, the German ambassador in the United States, gave talks about German cities to the City Club of New York and the Board of Estimate and Appointment, arguing that experience and expertise were indispensable for the management of modern cities. Bernstorff also focused on the necessity of centralized power and planning capacity. In his view, the German system of municipal government and administration "requires the whole working time and all the powers and thought of able men who have acquired special knowledge of the problems of town administration by a long experience of the work" (Bernstorff 1911, 7). In addition, the ambassador argued for the municipalization of technical works and elaborated upon the extensive possibilities of tax revenue that German cities had at their disposal. The City Club applauded Bernstorff and formally approved both instruments.

The German Cities' Exhibition: The Role of History

Germany's emergence as an international model was no coincidence or accident, but the outcome of German authorities' conscious policy of self-representation. Not least because the rapid expansion of public administration and urban technological systems required enormous economic resources, mayors and civil servants felt a need to legitimate their deeds to their own citizens. Especially after 1900 they also felt they had an international mission to fulfill in urban development, and were eager to show foreign visitors their cities' achievements. Municipalities chose a form of representation that was common at the time: the German Cities' Exhibition of 1903 was modeled on the large world's fairs that had taken place in London (1851), Paris (1889), Chicago (1893), and other leading industrial centers.

The German Cities' Exhibition provides the point of departure for our analysis of the symbolic character and meaning of urban technology. First, we examine the picture that local politicians, engineers, civil servants, and physicians drew of German cities in the two-volume work *Die deutschen Städte* (German cities). This publication offered "immediate intellectual encouragement" to everyone with an interest in the modern city and its future development (Wuttke 1904, I: xliv). Based on this source, we identify the ways in which the city and urban technology were represented. We also connect the various essays to form an overall picture of the German city and its urban technology.

Die deutschen Städte comprised at least four main lines of argumentation. The first presented an entirely uncritical, *progressive* stocktaking that praised the latest urban developments; the second, *historical* line of argumentation placed the modern city and its authorities at the pinnacle of a long tradition of municipal development, its origins dating back to the late Middle Ages; the third drew a clear line of *demarcation* between German cities and foreign (especially English and French) ones; and the fourth revealed visions of the role that the German city and urban technology ought to play in the *future*. In summary, *Die deutschen Städte* echoed the incessantly optimistic tone of the contemporary sanitary literature, where "things moved forward, things moved upward" (Rath 1969, 81). And, as can be seen in the example shown in figure 6.1, its numerous illustrations placed the "improvements" of German cities in a favorable light.

In introducing the first volume, Robert Wuttke targeted three social groups. Civil servants in smaller communities were expected to learn from the experiences of the larger, pioneering cities. The exhibitors, most of whom represented larger cities, aimed to convince the second audience—civil servants on the national level—that cities were the true champions of progress and modernity. The third audience was international: representatives from 13 countries participated in the opening ceremony of the German Cities' Exhibition, and exhibitors wished to convey to them the grandeur of modern German cities. Yet Wuttke drew a challenging picture: "Shadows darken the shining picture of the cities and their development. In contrast to earlier times, in which the middle classes carried the responsibility for local politics, they are today isolated. Since the 1870s our political party scene has changed dramatically. Powerful parties pursue objectives that are contradictory to those of the cities" (Wuttke 1904, I: xxii). Therefore, the exhibitors also hoped to revive, or create, a common middle-class identity among urban citizens.

Similarly concerned with the shifts in German politics at the time, Wuttke also commented on the problems caused by the rapid turnover of urban inhabitants and the loss of shared identity: "Today the urban population is a restless, agile crowd; following the economic situation, it flows into the city, it flows out of the city.... In the last century the interrelations between the city and its individual residents changed fundamentally.... Among the masses we must convey an understanding for the cities' great cultural activities; the bourgeoisie has to gather again under the banner of the cities" (Wuttke 1904, I: xix, xxii).

Wuttke's narrative was one of historical progress: against the backdrop of a dark past, the latest urban developments ushered in a new golden age. His line of argument was easy to follow. Initially, he observed that German cities after 1875 experienced swift population growth. As a result, urban municipalities were forced to define a new self-image and to reform the structure of local administration. Wuttke, a historical anthropologist, suggested that these changes in urban life were fundamental and historically unique. Cities were under such great pressure that they did not have the time to find out whether their new fields of activities would develop in the expected way. Despite these difficult circumstances, Wuttke pointed out that cities obtained a wealth of experience that enabled them to look optimistically toward the future. He concluded that

Figure 6.1
The progressive character of technical change was underlined by art nouveau–style illustrations, in this case of "traffic improvement" in Nuremberg. Source: Wuttke 1904, II: 43.

German cities had developed their own abilities and strengths, and that these guaranteed continual progress.

Although Wuttke dismissed most of the urban past as useless because it did not offer any relevant experiences and procedures, he thought it essential to historically legitimate the modern city. He identified urban technologies as crucial elements of municipal activities and approved the ways in which mayors and their staff had taken on responsibility for urban technologies and matters of public health. The pervasiveness of urban technological systems meant that small and mid-sized cities could skip the stage of experimentation and move straight to state-of-the-art solutions. This evidence seems to support Ralf Stremmel's suggestion (1994, 247) that cities' histories, or rather their pasts, were not a central element in their self-imagery of the time. Nonetheless, Wuttke and his colleagues thought it critical to point out the historical roots of the modern city, to give its municipal authorities and its technological facilities a kind of historical legitimation. In a rather blunt attempt to "invent a tradition" (Hobsbawm and Ranger 1983), Wuttke drew a direct connection between what he called the medieval middle classes and the contemporary middle classes. Furthermore, he declared, cities and their middle-class leaders were the main carriers of a national consciousness and proponents of the German empire. From the early nineteenth century onward, cities had established themselves as the new centers of power within the German nation.

At least rhetorically, other actors supported Wuttke's efforts to legitimate the modern city through the selective use of history. During his visit to the Exhibition in 1903, Bernhard Fürst von Bülow, prime minister of Prussia and chancellor of the German Reich, praised the cities and their achievements—in fact, drawing the lines of continuity back to the tenth century. History thus played an important role in the representation of cities and urban technology around 1900. History's role was one of identity-building and justification, and the intention of the historical narrative is obvious: from the Middle Ages on, Wuttke's middle-class *Bürgertum* eagerly accepted responsibility for political, economical, intellectual, and moral developments, within the city and beyond. Urban technologies had been managed by a progressively minded, but tradition-conscious local elite that applied technology for the general public's benefit. In other words, there was no reason to worry about the future development of the city and its technological facilities, because both political and technical experts were at work.

The German Cities' Exhibition: Nationalism and Nature

With his third argument, Wuttke drew a line of demarcation between German and foreign cities. The objective was to allay doubts about the potentials of the city and urban technologies. Conceding that, in the past, English and French cities had offered the most important models for improving the urban environment, Wuttke admitted that

German authorities had initially proven unable to cope with industrialization and urbanization. The superior public finances of England and France in the earlier phase had left no choice for German cities but to become their apprentice. By the turn of the twentieth century, however, the Germans viewed themselves no longer as students but as teachers.

In Wuttke's nationalist narrative, German cities now had the power to become exemplars for the world. He claimed the German Cities' Exhibition demonstrated this role reversal and showed the new independence of German municipalities, which had liberated themselves from foreign influence and created their own, characteristically German way of managing urban life. Wuttke even compared the cities' development from 1880 to 1900 with the effects of the coalition wars against Napoleon at the beginning of the nineteenth century. Using quite belligerent words, Wuttke called this emancipation process a "fight for liberation" in which German ideas prevailed (1904, I: xxi).

Wuttke's was not an isolated voice for nationalism. In slightly less militaristic language, Dr. Otto Wiedfeldt, councilor of Essen, described how German cities had cut the cord from their English and French counterparts. Until the last third of the nineteenth century, Wiedfeldt (1904) suggested, German municipal authorities blindly followed a "Manchester-like attitude." Due to England's far-reaching political and economic influence, almost all German cities were fascinated by what Wiedfeldt believed was the irrational idea that the promotion of private interests would automatically generate public benefits. German civil servants disposed of this liberal ideology and overcame their reluctance about running larger facilities only after they had realized that private owners would rarely support waterworks and gasworks or electric power stations for the public well-being (Wiedfeldt 1904, I: 182f.).

Dresden's lord mayor, Gustav Otto Beutler, echoed these sentiments. In his inaugural address to representatives from Paris, Vienna, St. Petersburg, Prague, and other foreign cities at the opening of the Cities' Exhibition, Beutler described the development of German cities as a historically unique process. Like Wuttke and Bülow, Beutler found the recipe for German cities' success in the middle classes and their involvement in urban affairs. With a typical *Sonderweg* argument, Beutler claimed that one would not find anywhere in the world a stronger and more fruitful cooperation between municipal authorities and the middle classes than in Germany. He was convinced that the municipal administration's increasingly professional character and the middle classes' strong commitment to voluntary activities created an unparalleled combination. Wuttke, for his part, boasted that German cities were unique in having managed to establish "close contacts" between scientific knowledge and political and technical practices (Wuttke 1904, I: xxxvi, xxi).

Ernst Grahn, a well-known German civil engineer, and a certain Professor Nowack, a physician, expanded this line of interpretation. Using more matter-of-fact language,

Grahn (1904, 305) reviewed the considerable achievements of the Deutscher Verein für öffentliche Gesundheitspflege (German Association for Public Health) since its founding in 1873. This association, he explained, had successfully established itself as a mediating body between scientists, engineers, and civil servants—a viewpoint supported by recent scholarship (e.g., Hardy 2005). In an even more enthusiastic manner, Nowack (1904) signaled his unanimous approval of this narrative by underscoring city authorities' policy of adopting sanitary arguments based on firm scientific evidence.

The pivotal role assigned to experts is entirely characteristic of the international discourse community. When German representatives stressed to foreign audiences the scientific character of German urban management, they were drawing upon similar arguments made about German manufactured goods. In the 1870s, German commentators had apologized for the low quality of industrial products; indeed, the leading engineer Franz Reuleaux had famously admitted that German products were "cheap and bad." But as German industry gained export shares, its self-consciousness increased. Something similar happened in German urban planning. Albert Südekum, member of the German Parliament and editor of *Kommunales Jahrbuch*, an important urban reform annual, explained to a group of U.S. visitors how scientific investigation and expert knowledge had helped German cities achieve greatness:

City administration in Germany is becoming the science of community living.... The city is far more than a business affair. It is much more than a political agency. It is an agency of social welfare with unexhausted possibilities. Our cities are trying to utilize art and science, the improvements of steam and electricity, in the service of the people. We are aiming to socialize industry and knowledge for the common good. (quoted in Howe 1913, 114)

Wuttke's fourth argument revealed visions of the role that the German city and urban technology might play in the future. Interestingly, this narrative contributed to defining German cities in terms of "machinery." Whereas Wuttke himself stressed the pervasiveness of engineering knowledge within the urban administration, Wiedfeldt put forth the "municipal utilities" as central actors in the formation of the modern German city. Modern technology, he said, would make its way even into those communities still lacking any kind of facilities. The circulation of information between cities—not least through their annual progress reports—would generate such tremendous competition that no city, however small, would dare lag behind and risk being accused of withholding the best technical solutions from its inhabitants (Wiedfeldt 1904, I: 181, 186). Similarly, Nowack (1904, I: 455) highlighted the ways in which the modern city depended on technical works: "Facilities such as waterworks, sewage systems, waste disposal and housing control, food and epidemic control... are characteristic features of the modern city. Their erection guarantees to give man unspoiled, natural conditions of air, water, and soil." He claimed that such facilities enhanced the well-being of a city's inhabitants and helped preserve their full abilities and develop society's full potential.

Municipal building surveyor Hermann Klette also addressed the process of optimizing nature and making it subservient to social needs through technology and engineering. He cited numerous achievements: road tunnels cut through mountains, streets paved at every altitude, and bridges erected to connect cities, economic areas, and countries. Within cities one could find impressive railway stations, docks, tramlines, and suspended railways. Below the surface, technical networks of gas and water supply, sewers, telephone lines, and pneumatic delivery tubes competed for limited space. Each network formed one part of the urban organism and placed "terrific forces" at the disposal of cities (Klette 1904, I: 385). According to civil engineer Grahn, technical utilities, especially waterworks, had become the most important field of municipal activity at the beginning of the twentieth century. Yet Grahn simultaneously diagnosed a growing indifference toward waterworks. Somewhat resignedly, he lamented the cursory attention people paid to the function and history of the local water supply system—except when faced by a serious water shortage. Nonetheless, Grahn (1904, I: 302, 344) firmly expressed his belief that the public could be convinced of the necessity of continuing to mechanize the city in general and to improve the water and sewage systems in particular.

Klette shared Grahn's optimism, though he admitted there were many shortcomings in developing a satisfactory sewage treatment technology—one of the most discussed environmental problems at the time. Even if engineers and physicians were still far from developing an acceptable method, Klette nonetheless conceded that the wealth of experience and specialist knowledge that urban-based civil engineers had collected would ensure a speedy end to the "era of storm and stress" (1904, I: 419). In addition, Nowack announced that the modern city—both mechanized and hygienic—was nearly completed and, despite any shortcomings, the moment had even come to export the blessings of urban technology and urban norms of cleanliness to filthy villages: "While the cities invested a great deal of money to drain off their excrements and waste water ... the villages remain a serious threat for the city dwellers' health, because of their dirty houses and yards, their cesspits and the polluted soil, the contaminated drinking water and the often unbelievable carelessness about germs" (Nowack 1904, I: 458).

"The Example of Germany": The British Picture

Geddes and Dawson were not the only British urban experts impressed by German cities. In 1907 planner Thomas C. Horsfall announced to an international audience that the German state of Saxony had the "best housing law in the world" (quoted in Albers 1997, 144). Horsfall was chairman of the Manchester and Salford Citizens' Association for the Improvement of the Unwholesome Dwellings and Surroundings of the People, which worked to improve living conditions for the working classes and the urban environment in general. His book *The Improvement of the Dwellings and Surroundings of the People* (1904) reflected the concerns of the Manchester reformers.

The subtitle of Horsfall's book, *The Example of Germany*, reveals its focus and laudatory tone. To the author, German towns ranked higher on every possible score than their English counterparts. The air in German cities, with the possible exception of Essen, was much cleaner than in any English industrial town. Most German cities had large parks where citizens could relax and regain health and strength; their streets were cleaner; and the houses in working-class districts were better maintained than in England. Even German apartments were tidy and neat: "English observers who visit the homes of German workpeople are generally surprised at the high average of orderliness and apparent comfort which they find in them; and, on the other hand, Germans who visit the homes of English workpeople in the poorer districts of our large towns are surprised to find so many dirty and neglected dwellings" (Horsfall 1904, 162). Horsfall traced these striking differences to city-sponsored education initiatives that made the German population more conscious of personal and public hygiene. For example, German pupils were taught the advantages of cleanliness and an intimate relationship with nature, and they had access to warm showers and school gardens. Like his middle-class counterparts in Germany, Horsfall was convinced of the benevolence of this kind of policy.

The strategic way in which Horsfall used German self-representations provides evidence of the close connections between members of the reformers' discourse community across national borders. In his plea for municipal socialism, for example, Horsfall drew heavily on a 1902 report by city authorities of Cologne. Their "Report on the Condition and the Management of the Communal Affairs in the Financial Years 1891 to 1900" was an explicit effort at public relations and self-justification, showcasing the improvements that resulted from socializing private companies. The report's authors adopted the well-known discursive strategy of presenting their particular interests as general interests: "There is now an imperious demand that those undertakings, which supply the needs of the whole community, shall be managed by the municipal authorities, as they know better how to protect the general interest than traders, whose first care is to make profit" (quoted in Horsfall 1904, 139). The Cologne authors listed the numerous areas now successfully under public control: the gasworks, the local electricity company, tramlines, parts of the railway, the waterworks, slaughterhouses, the harbor, and the wharves. In all of these cases, they argued, public ownership and management had proven advantageous to the inhabitants of Cologne, whereas a liberal system driven by private interest might not guarantee the greatest happiness to the largest numbers.

Horsfall appropriated the achievements of Cologne in an effort to reshape the administration of English cities and towns. Claiming that the English administrative system was plagued by amateurism, Horsfall argued that it needed to be replaced by a modern system in which professionals had a decisive say. In order for mayors to be fully devoted to their tasks, make informed decisions, and resist corruption, they had to be elected for longer terms and given decent salaries. Equally important, modern mayors needed large staffs of experts to provide decision makers with plans, calculations, and other documents. As Horsfall (1904, 31) saw it: "The active management of municipal affairs in German cities is very largely in the hands of a special class of technically trained officials, who apply scientific administrative methods to a degree unknown in other countries."

According to this argument, modernity required expertise, a factor that was desperately lacking in English cities. As urban technologies spread, ever more professional experts were needed to maintain and administer the various networks, systems, and projects. The outcome would be a kind of technocratic system in which scientific knowledge and methods were paramount. The "necessities of modern life" called for professional administration (Horsfall 1904, 35).

Horsfall's book, obviously, must not be read as an objective investigation. His extremely positive representation of German cities can be understood only against the background of what he considered to be the intolerable situation in many English towns. Horsfall's discursive strategy—to make changes come about in his home country—may be illustrated by two quotations that juxtapose England and Germany. At the outset of his book Horsfall (1904, 4) declared why his fellow countrymen ought to take German cities as their model: "German towns [are] much pleasanter to live in than English towns." And, when he later discussed the concrete advantages of the expert system, Horsfall suggested that the building plans that German authorities produce "fill an Englishman ... with surprise and with admiration" (ibid., 27). The reformer's basic narrative was one of contrast and juxtaposition, and the role attributed to German cities was one of role model and pattern.

"How the German City Cares for Its People": U.S. Appropriations

We are not the first to observe this rhetorical strategy. Daniel Rodgers (1998) devotes two chapters to urban policy and planning in his all-encompassing treatment of the transatlantic circulation of reform ideas. Although he uses different terminology, Rodgers clearly shows that this process was one of active appropriation rather than simple transfer of ideas. Significantly, he notes that U.S. observers concentrated on just those factors they considered better in Europe than in the United States: "No wretchedly paved streets and packed streetcars, no corrupt city bosses, no waste, no lawlessness, no anarchy" (Rodgers 1998, 142). Furthermore, U.S. progressives adopted European ideas they could use at home and did not necessarily convey balanced representations of Europe. Albert Shaw, a U.S. journalist, explicitly said that he did not aim at objectivity in his city surveys: "my task in all these matters is not to find out the faults of foreign people and foreign systems, but to find out their merits in order that we may profit by their wisdom and be stimulated by their good performances" (quoted in Rodgers 1998, 133). In the U.S. progressive discourse, "the European city" was both a point of comparison and a source of inspiration.

Rodgers treats at some length Frederic C. Howe, a lawyer and politician-turned-journalist who actively preached the European gospel to his fellow U.S. citizens. An active member of the Public Ownership League of America, Howe published several books on European urban policy and wrote numerous articles in popular magazines. Together with the Boston Chamber of Commerce and other organizations, Howe helped to arrange study trips around Europe. In several speeches, subsequently published as *How the German City Cares for Its People* (1911) and *European Cities at Work* (1913), Howe illustrated the points made above. When he described European cities, Howe usually compared them directly with U.S. cities. For example, he pointed out that instead of giving concessions to private companies to run public services, European authorities organized these services themselves and often made a handsome profit: "In England and Germany, the city owns these profitable enterprises, and protects its people. We think of the sacredness of private privilege. European people think of the sacredness of human life" (Howe 1911, 17).

When Howe described the achievements of German city planning, one may rightly ask whether he had visited Germany or paradise: "The City [authorities] laid out the streets for miles around the city. They laid out broad boulevards with parkways and prominades [sic]. They made them artistic and beautiful. They made the side streets spacious and comfortable. Within a short walk to every man's house they placed a play ground or a garden, and they made the garden as artistic as possible" (1911, 18).

Like other U.S. progressives, Howe turned to European cities to gain inspiration and develop arguments that could help his cause. After his journeys to European cities, he returned home with images of community and consensus, harmony and happiness. Somewhat naively, especially for a former politician, Howe supposed that serious conflicts never accompanied European urban planning. In Europe, he claimed, there "was no conflict between the city and the railway over streets, stations, or terminals"; "tracks rarely disfigure the city" (Howe 1913, 106).

Juxtaposition, inspiration, and instrumentalism run through Howe's writings. His chief message is that European, and especially German, cities had been planned for the common good rather than for profit, and that harmony, beauty, and health were the outcome rather than the chaos, ugliness, and filth found in the "accidental" cities of the United States. Howe's discursive strategy illustrates well what philosopher John L. Austin, in his speech-act theory, once called *How to Do Things with Words* (1962/1976). Howe very consciously picked the observations he thought would prove effective in the U.S. debate, and in doing so changed the terms of this debate. He was aware, for instance, that U.S. observers often criticized the German election system for not

Figure 6.2
A typical reformer's illustration, depicting the "spacious street construction" of a grand boulevard in the Charlottenburg district of Berlin. Source: Howe 1913, 94.

being particularly democratic. To meet such criticism, Howe attempted to transform the meaning of "democracy" by emphasizing the achievements and performance of strong local government rather than questions of suffrage and ownership. As Rodgers (1998, 143) puts it, Howe's writings "were part of a struggle to socialize the language of democracy—to balance its rhetoric of rights and privileges with a new rhetoric of services, outcomes, and results."

In large part, Howe and other progressives preferred a strong public sector because of the need to maintain the technical systems modern cities depended on. Private ownership would, in their view, lead only to substandard results; indeed, it was unprofitable for society and the economy as a whole. Howe focused on public transportation, maintaining that publicly owned streetcar systems were usually of higher quality, provided travelers with better comfort, and covered greater areas.

Admittedly, Howe's juxtapositions were not particularly fair or objective. When comparing German industrial cities with those in the United States, he stated bluntly: "German industrial cities are not mean and tawdry" (Howe 1913, 12). Drawing a direct comparison between the steel town of Gary, Indiana, and Düsseldorf on the Rhine, Howe not very appropriately stated that "Gary was planned for the making of iron

Figure 6.3
A U.S. reformer's enthusiasm for the Dresden railway station's "splendid proportions," with its tracks "built for permanence and beauty." Source: Howe 1913, 12.

and steel; Düsseldorf was planned for people" (1913, 43). Düsseldorf's many poor inhabitants stuck in dreary housing would hardly have agreed. According to Howe's narrative, even men of industry and commerce were convinced of the superiority of public ownership and administration of urban services. Frankfurt businessmen reportedly considered "the city as a big collective enterprise, with assets which comprise all the wealth and property of all the people" (Howe 1912, 73). He saw their city as a harmonious whole, an organism whose survival depended on the subordination of private interests to the interests of the community.

Organic images prevailed in Howe's writings. Both urban politics and the technical networks that penetrate modern cities are described in organic terms. In *European Cities at Work*, Howe writes that "Street railways, gas, electric light, and water are treated as the cities' vital organs to be owned and operated for service, comfort, and convenience. They are owned by the city because they control its life, its growth, its development" (1913, 106). In other words, the urban machine that characterizes modern cities has developed organically and, like every higher organism, must be governed by a brain that has a broad overview.

Contexts of Conflict and Corruption

The appropriation processes analyzed in this chapter unfolded in a charged political, economic, and social context. The German Cities' Exhibition can be understood against a backdrop of conflict and struggle on the one hand, and uncertainty and ambition on the other. Like the U.S. progressives, German reformers had a strong sense of mission. In the nineteenth century, public works and municipal engineering emerged as important areas of technical development throughout the industrialized world, and urban planning and hygiene became academic subjects in many countries. The foundation of water, gas, and electricity works was critical in this regard and, to meet cities' needs, new fields of engineering emerged on both sides of the Atlantic. Private and public utilities were set up to provide urban residents with energy and water, and to remove their garbage and sewage. These achievements not only changed the urban fabric fundamentally, but they also required enormous investments. Legitimizing these investments was a major function of the German publications analyzed in this chapter. German pamphlets and reports that U.S. observers read as heroic stories of pioneering technological achievements (and a writer such as Stremmel [1994] misreads as purely descriptive accounts) originally had a political purpose.

Since the early days of municipal engineering, cities made enormous investments to address sanitary and urban environmental problems, ensure continuous supplies of water and energy, and provide efficient means of transportation. European city authorities built public baths and installed underground water pipes and sewage systems, they effectively treated waste and refuse, and they issued laws that were meant to improve housing conditions. Furthermore, cities laid out gas and electricity networks, and made streets broader, straighter, and better lit. At the beginning, these developments were frequently contested, even if they might seem natural in hindsight. Homeowners and others who had to finance the construction and maintenance of the water supply system initially opposed even running water. This is one reason Rudolph, the head of Darmstadt's waterworks, described the advantages of public waterworks with such fervent enthusiasm. Instead of recapitulating ten years of heated political debate, intense discussion, and countless expert hearings, Rudolph chose to smooth things over. Public opposition and inefficient municipal administration were conveniently forgotten, and instead Rudolph obliquely mentioned only the "diverse efforts" of the participants (Stein 1913, 154). Rudolph's narrative was one of continuous technical development and progressive municipal decision making (Stippak 2007, 157). Sure enough, although he and his colleagues had helped improve the living conditions in Darmstadt, heavy investments accompanied this accomplishment. To defend these against budget-minded critics, Rudolph and others thought it advantageous to represent urban technology in an affirmative way by applying the standard progressive narrative, "things moved forward, things moved upward."

Self-images functioned both as inward legitimization and external representation. Outside observers, impressed by these concerted acts of demonstration, transformed these self-conscious narratives to fit their own domestic conditions. In effect, British and U.S. commentators transformed German achievements in the area of urban development. In the Anglo-Saxon world, the German ideal of service-oriented public authorities and utilities (*Leistungsverwaltung*) became a force in a political struggle against amateurism, vested interests, and corruption. American progressives in particular regarded urban planning as a means to install a kind of expert-informed democratic order and make cities more livable and beautiful (Rodgers 1998, 160–162). As we have seen, juxtaposition was their most common strategy of appropriation: whereas German commentators tended to compare present with previous conditions, foreign observers contrasted German conditions with those in their own countries. Images circulated widely in the transatlantic discourse community that evolved after the turn of the twentieth century, but these images and ideas took on quite different meanings in their respective national discursive frameworks.

Considering the slower means of transportation and communication in those days, the coherence of the community is astonishing. Urban experts and commentators traveled extensively, and they were able to exchange information in several languages. Although progressivism and other urban reform movements did not persist as a tight discourse community beyond the 1920s, the transatlantic connections that we have called attention to in this chapter did not dissolve (as Rodgers [1998] distinctly hints). As parts III and IV of this volume show, after the Second World War ideas and images about cities tended to move *to* the European Continent from the United States or from Britain. In traffic and urban planning, Anglo-Saxon solutions would have a decisive impact on experts and politicians on the Continent. Like Howe and Horsfall, continental planners traveled abroad for inspiration, and on their return home, they made extensive use of images and calculations to convince their fellow citizens that the future rested on the western side of the Atlantic Ocean.

7 Constructing Dutch Streets: A Melting Pot of European Technologies

Hans Buiter

"The city of Pest with its boulevards and technical infrastructures leaves...a spacious and grand impression on the visitor and should be a point of attention for engineers,"[1] reported a Dutch engineer to his colleagues in 1895. M. Sijmons had just returned from the Eighth International Congress on Sanitation and Demography in Budapest, where engineers from all over Europe had discussed water supply, sewer systems, and other means to improve public health. Since 1876, this congress had been a regular event that set in motion the pan-European circulation of knowledge among sanitary and urban engineers.

International congresses were not the only way that people, information, and technical concepts circulated across Europe. Books, journals, and personal visits served the same purpose. Sijmons had performed a feat of technical tourism by meeting fellow public works engineers—not only in Budapest but also in Dresden and Vienna. In 1896 Budapest hosted a full Dutch delegation after their grand tour of European cities to study streetcar systems. As can be seen in this volume, there are innumerable examples of urban engineers traveling internationally to seek inspiring models and usable advice on urban practices.

During these same decades around 1900, urban streets throughout Europe were transformed from multifunctional, relatively empty, and rather disordered places into well-ordered and explicitly zoned spaces in which artifacts, systems, and humans were each given their own place. Streets today are multilayered spaces that may have sidewalks for pedestrians, safety islands for street crossings, defined lanes for buses and streetcars and sometimes also for bicycles, entrances for the subway, and parking lots for automobiles, as well as special areas for markets and outdoor cafés (see figure 1.3). Signposts, traffic lights, pavement markings and patterns, and even cameras are meant to control the users of the street. By means of inclusion and exclusion, the character of the street as "public space" has changed. This chapter explores these changes with particular attention to the ways in which the street has been redefined through technical means. My focus is on the relations between the varied functions of the street: transportation, energy supply, water and sewage systems, and sociability. To show how streets have been technically and socially co-constructed, this chapter will examine the planning and building of separate and combined sewers around 1870, the electrification of streetcars around 1900, the early adaptation to the automobile around 1930, and finally the automobile's emerging dominance after the Second World War. At the center of my analysis are the coevolution of streets, sanitary systems, and traffic networks in the leading Dutch cities of Amsterdam, Rotterdam, The Hague, and Utrecht.

Since there was a constant flow of technological knowledge, concepts, and artifacts from other European cities, the story of Dutch street innovations has a strong European flavor. Examples from other European cities were not merely sources of inspiration for Dutch engineers and city officials, but also strong rhetorical elements in discussions about the introduction and adaptation of this technology. European solutions often influenced the design or improvement of Dutch cities, but sometimes the flow of information and knowledge went in the other direction. This chapter pays attention to the actors involved and the platforms they used, and it also reflects on the "Europeanness" of this flow of ideas and practices. To a large extent, the chapter is based on my book *Riool, rails en asfalt* (Buiter 2005a).

Streets as Public Spaces

For Jürgen Habermas (1962/1989) the accessibility of public space is one of the outstanding qualities of liberal bourgeois society. However, authors such as Lyn Lofland (1973) and Richard Sennett (1977) have made clear that Habermas's public space is not always accessible for everybody and that the public character of urban streets is full of paradoxes. There is little disagreement that urban public space expanded during the seventeenth, eighteenth, and nineteenth centuries. In many Protestant areas, religious complexes were opened up to the public and many new streets were constructed for general use (Olsen 1986). Public parks were laid out for recreational purposes, in sharp contrast with enclosed private gardens. Local municipal authorities in most locations took over the organization and maintenance of urban public space.

Because city officials took charge of this expansion in public space, urban residents lost influence over what was happening outside their front doors. Rich and poor residents had used the street for a variety of purposes, contributing to the vibrant tapestry of street life celebrated by painters such as Jan Steen (1626–1679) and William Hogarth (1697–1764). The new municipal experts curtailed the customary private use of the public street to store a business's goods or a family's possessions. Practicing one's profession in the street or trading goods there were likewise restricted. Furthermore, loud singing, heavy drinking, rough sports, brawling, and animal fighting were also restricted or outlawed. These manifold attempts to increase the respectability of public space and to discipline its users coincided with the rising dominance of the middle class. As Madeleine Hurd (2000) points out, the public space emerging in the nineteenth century had a masculine middle-class character, and working-class activists had to struggle to gain access to it.

Urban historians have conducted a lively debate on several aspects of the street. Donald Olsen's *The City as a Work of Art* (1986) placed the development of streets in a broader context and addressed topics such as the replacement of gutters, underground sewers, and the introduction of macadam to facilitate cross-town traffic. Olsen interpreted these changes in relation to the dominant values that governed life, distinguishing different styles in housing and streets in London, Paris, and Vienna. His observation that private space was much more valued in Britain than in France accounts for the more spacious houses and poorly valued streets in London, contrasted with the small apartments but broad, luxurious, and lively boulevards in Paris. Clay McShane (1994) also claims that shared cultural ideals of housing shaped the designs of specific streets. He points to the Anglo-Saxon preference for unpretentiousness and suburban living to explain the transformation of city streets from lively, multifunctional spaces into transit lanes for trolley and automobile traffic that carry workers between the suburb and the city center.

Peter Baldwin's prize-winning *Domesticating the Street* (1999) also views the street as a contested space. Baldwin focuses on the development of public space in Hartford, Connecticut, between 1850 and 1930. He acknowledges that some streets were transformed into lifeless traffic arteries, but stresses that others were still vibrant in 1930 and remain so through today. He rejects the bleak picture painted by Davis (1988) and others: "The street has not died—but streets have changed, some more than others" (Baldwin 1999, 4). While traffic was dominant in some streets, others had playgrounds and remained social gathering places, that is, classic multifunctional urban spaces. Spatial specialization and segregation are the key words in his story. Baldwin recounts attempts to ban newspaper-selling children from the streets, to relocate peddlers at special markets, and to suppress prostitutes in central areas. Local reformers campaigned to purify the Hartford public space by founding parks and playgrounds where children could play and families could enjoy recreation, and by removing street vendors, drunks, and prostitutes from the streets. Modern city planners and traffic engineers followed in the footsteps of these reformers, and after 1900 the streets of downtown Hartford were transformed from multifunctional spaces into orderly traffic lanes for streetcars and automobiles. Baldwin points out that the segregated urban space that resulted was an unintended outcome rather than deliberately planned.

Whether or not they have retained their character as public space, streets have been transformed into technical ensembles that contain artifacts, systems, and networks. Sewers, water supply systems, streetcars, energy and communication networks exist at, over, or under the street surface. Curiously, despite the large historical literature on urban systems and networks, not much attention has been paid to the historical development of the street as a *place* where these systems and networks coevolve (Tarr and

Dupuy 1988). Streets can be seen as conjunctions of different technologies, each with its distinct traditions and regulations. For example, the introduction of integrated sewer systems has to be seen in light of the discussions on public health and the ways to avoid cholera and typhoid epidemics in the second half of the nineteenth century. Values of cleanliness and tastefulness played important roles in the competition between different technologies. And the application of these transport and sanitary technologies must be set into the broader reformist drive for more beautiful and respectable public spaces.

The Rise of the Sanitary Street

Rotterdam was the first city in the Netherlands to discuss an integrated sewer system. Around 1840, when it was the country's fastest growing city, its city officials decided to upgrade the existing sanitary networks (Noort 1999; Nieuwenhuis 1955). At the time, Rotterdam, like other Dutch cities, applied different systems for the removal of household waste, feces, and surplus water. Feces were collected in brick cesspools that were supposed to be emptied every year or two by special contractors with the city government. These contractors, who were also responsible for collecting and removing stove ash and other household waste, carted or shipped the wastes out of town, where they were processed and sold to farmers. Wastewater from households, rain, and industry was transported out of town by an interconnected network of small private sewers, gutters, ditches, and canals.

According to the miasma theory, which dominated all public health discussions prior to the acceptance of Louis Pasteur's germ theory in the 1880s, smelly waters and stinking alleys were very suspect. The many slow-flowing canals that penetrate Dutch cities made the possibility of miasma a particularly pressing concern. Since the canals, ditches, and cesspools could harbor disease, and because they expelled terrible odors (the miasmas believed to cause illness), urban engineers tried to develop new systems for improving public water supplies and removing feces. The devastating cholera epidemics that periodically afflicted Dutch cities from 1832 on forced urban elites and officials to take these problems seriously.

In 1841 the newly appointed city architect W. N. Rose designed a waterway system at the edge of Rotterdam to facilitate water transport and the flushing of the inner-city canals and the ditches in the newer parts of town. For economic reasons, however, the Rotterdam city council did not approve his scheme. Only after several deadly cholera epidemics did they accept Rose's plan for a "water project." In 1856 work began on a new half-moon-shaped canal system to connect the ditches and the canals in the polder city (built on reclaimed land) and to transport this surplus water, by means of steam-driven pumps, to the river Maas.

Although the water project's goal was to transport wastewater out of the city, it also created a new neighborhood. The experienced father-and-son architects J. D. and L. P. Zocher designed a canal zone, complete with prestigious villas and pedestrian boulevards in British landscape style. The inspiration for transforming the former city walls into green zones for recreation stemmed ultimately from Paris, where in the 1780s the former city walls were transformed into "boulevards" (A. Jacobs et al. 2002, 74). ("Boulevard" in fact came from the Dutch word *bolwerk*, or fortification.) After his victories in Germany, Napoleon had imposed the boulevard model there to weaken the German cities' defenses (T. Hall 1997, 300). Later the approach circulated to Dutch cities, nicely illustrating the sometimes convoluted international circulation of urban models. After Napoleon's brother became king of Holland, the border town of Arnhem was in 1808 the first in the Netherlands to lay out *wandelingen* (walks) along its former town walls. From the 1820s on, the Zochers created several *wandelingen* in Arnhem, Haarlem, Rotterdam, and other Dutch cities, thus appropriating the eighteenth-century French boulevard model (Cremers et al. 1981).

While the water project was under way, the existing methods of transporting feces were also reconsidered. In the aftermath of the cholera epidemic of 1853, and with strident input from the urban committee of public health, the Rotterdam city council reconfirmed the ban on dumping feces and other waste into gutters, sewers, and public waters. Hygienists dominated the public health committee, and they advised the city council to introduce *fosses mobiles* to replace the existing cesspools. In its 1856 report, the committee referred specifically to Paris, where such barrel-like containers were used to collect and transport feces. However, after a newly built quay collapsed into the river Maas, city architect Rose was removed from the job, and the new head of public works, A. W. Scholten, looked for other models. Despite the opposition of Rose—who owned a firm that collected feces from the urban cesspools and sold the content to farmers—Scholten went to Hamburg to study its water and sewer system. In his 1858 report to the urban Committee on Local Works, he proposed a Hamburg-like solution, observing that Rotterdam was situated at a river of about the same size as the Elbe and that the number of inhabitants in the two cities was almost equal. Scholten stressed that Rotterdam should not follow the experience of London, which had too large a population to safely dump its sewers into the relatively small river Thames.

Scholten's choice of Hamburg as a role model instead of London, the birthplace of the integrated sewer, was a shrewd tactical move. At that moment—the second half of 1858—the London sewer was not a very attractive model. Due to a shortage of water that summer, as well as the increasing load of sewage flowing into the Thames owing to the newly popular water closets, the stench of the Thames became so terrible that the meetings of Parliament had to be suspended. The "Great Stink" of 1858 remains a staple in London's colorful social history.

Hamburg's 48-kilometer-long interconnected sewer system was unique in continental Europe. At the time of Scholten's visit there was simply no other water carriage system of that size. After a great fire had burned down a large part of Hamburg, the British civil engineer William Lindley proposed building an integrated sewer system to transport wastewater and feces to the Elbe. Lindley had just completed a railway line from Hamburg to one of its neighbor cities, spoke German fluently, and was held in high esteem. Like his mentor, the famed but erratic Edwin Chadwick, Lindley believed that improving urban sanitation would stimulate the local economy and stabilize society. To this end, he introduced in Hamburg bathing and washing houses modeled on those in the English city of Liverpool, and in cooperation with William Chadwell Mylre he designed a water supply system using an inlet two kilometers upstream (Evans 1987, 133–135; Simson 1983, 61–88). Thus, Hamburg drew upon state-of-the art English practices, but without the taint of London.

Back in Rotterdam, Scholten proposed a sewer system modeled on Hamburg's. Just as in the Elbe city, a fresh-water system was part of the Rotterdam plan. While in Hamburg the water of the elevated river Alster was used to flush the city sewers, Scholten planned to make use of the water in Rotterdam's defense canals, located a few meters above the new parts of the city. After long discussions, the city fathers in 1863 accepted his plans, banned cesspools in the inner city, and abolished regulations that had previously forbidden inhabitants to flush their feces into the urban canals. The city also then constructed sewers, filled in inner-city canals, repaved streets, and constructed sidewalks. Sadly, Scholten would never see these results; in 1861 he succumbed to typhoid, one of the very plagues he had tried to eradicate (Noort 1999).

The new wastewater measures had several unexpected effects in Rotterdam. While the sewers in the inner city successfully transported wastes out to the river Maas, the waters at the *wandelingen*, which were supposed to improve the city's image, instead started to smell nasty. The well-heeled inhabitants of the Zocher villa zone complained bitterly. In light of these problems, the city experimented with the barrel-like *fosses mobiles*, used to transport feces out of the city. But the problems were overcome only in the 1880s with the introduction of yet another sewer scheme designed by the Rotterdam public works. Engineer De Jongh's system consisted of major extensions of the urban sewer system and the building of new steam pumping stations, modeled after those in Berlin (Buiter 2005a, 118–119).

Meanwhile, in Amsterdam, public works engineer J. G. van Niftrik studied ways to improve the flushing of the existing canal network and to construct an integrated sewer system (Niftrik 1869, 1906). Amsterdam was the only big Dutch city at that time with a fresh-water system. Although it was less severely hit by the 1866 cholera epidemic than other urban centers in the Netherlands, the problem was still serious enough to put urban sanitation at the center of public debate. Under the influence of the miasma theory and the ideas of Munich professor Max von Pettenkofer, Amsterdam's focus was on preventing the pollution of groundwater, soil, and air. A top priority was speeding up the transportation of human feces out of town. In 1867 the city's commission on public health published a plan to fill in a large number of central canals and replace them with underground sewers, as well as to improve the flushing of the urban waterways and sewers into the Zuider Zee.

At his own expense, engineer Niftrik traveled to London in 1869 to get firsthand information about the London main drainage. After the 1858 "Great Stink," London had constructed an expensive integrated sewer system to transport rain, wastewater, and feces downstream into the river Thames. Niftrik was highly impressed by the detailed tour he was given by Joseph William Bazalgette (1819–1891), the engineer who supervised construction of the city's main drainage sewers, which were enclosed by the five-mile-long Thames Embankment (Porter 1993). Great Britain now became a model for Dutch engineers. (London would become a popular model for American urban engineers roughly a decade later.) Also in 1869 the architectural periodicals *Bouwkundige mededelingen* and *De opmerker* published articles on systems operating in London, Cheltenham, Coventry, and several other British cities. The London system inspired Niftrik to design an integrated system with steam-driven pumps and egg-shaped sewers made with Portland cement; and in 1870 he published a proposal to replace the existing cesspools, gutters, and canal network with his system. He further proposed the filling-in of ten centrally located canals.

Despite the mayor's support of Niftrik's proposal, city council members favored an alternative model brought forward by Charles Liernur (Jager 2002). Liernur was a Dutch engineer whose biography reads like an adventure novel. He was born Hermann Carl Anton Liernur in the Dutch city of Haarlem in 1828, trained as an architect and hydraulic engineer, and went to the United States at the age of twenty to take part in the construction of railways and lighthouses. As an officer in the Confederate army during the American Civil War, he designed the fortifications of Mobile, Alabama, the last Southern harbor city that held out against the Union army. When deported from the United States after the war, he went to London. There Liernur worked as an editor for the leading journal *The Engineer*, in which he published his ideas on pneumatic sewage systems in 1866. Liernur proposed a pneumatic system to forcibly pump feces to central stations from which the waste would be transported to processing factories and sold as fertilizer to farmers. In addition to this pneumatic system, Liernur favored separate underground sewers to transport wastewater (by contrast, London's integrated system combined rain runoff with sewage and transported the whole together in one pipe). On this score, he followed in the footsteps of Edwin Chadwick who had earlier advocated the construction of so-called separate sewers for London (Tarr 1996, 135–136).

In 1867 Liernur moved to Amsterdam to facilitate his marketing activities in the Netherlands. In the small town of Breda he tried out a prototype of his pneumatic system at the local army barracks. Meanwhile, his system was installed in barracks in Prague on behalf of the Austrian ministry of war. A hospital in the German town of Hanau followed suit, and the Galleria Vittorio Emanuele in Milan planned to use the Liernur system (Zon 1986; Liernur 1873). These projects served as showpieces to interest city officials and businessmen. Liernur also wrote numerous promotional pamphlets and brochures. His publicity strategy was successful, and he soon closed deals with the cities of Rotterdam, Dordrecht, Leiden, and Amsterdam. His proposals to introduce the pneumatic system generally provoked fierce debates between councilors supporting him and the rival advocates of water removal systems following the British combined or integrated models. A host of pamphlets and newspaper articles debated costs, values of cleanliness, and the embedding of these systems in the existing urban matrix. Liernur's opponents pointed to the London main drain and other integrated sewers such as the one in Hamburg that were successfully flushing both wastewater and feces into the river Elbe.

Figure 7.1
The pipes and drums of the pneumatic "Liernur system" in Amsterdam (1885). Source: Collection of Hans Buiter / SHT–Eindhoven, Netherlands.

The city council of Leiden decided to install a pneumatic system to transport feces and wastewater and to fill in the local canal, while the council of Dordrecht accepted a proposal to introduce Liernur's system for a part of the inner city. His system had its greatest impact in the Dutch capital. In 1870 the Amsterdam council decided to experiment with Liernur pipes in the inner city around the Looiergracht where a small canal had been filled in. A sewer for rainwater and wastewater was constructed next to the pneumatic pipes that formed the core of the Liernur system. The system's unveiling in February 1872 drew visitors from all over Europe, especially from Germany, including the mayor and aldermen of Berlin. Two officials from the city of Bremen reported on the system in the leading German journal for public health.

Although its mechanical functioning had some problems, the weakest point of the Liernur system was the commercial exploitation of the collected feces. Customers in the hinterland of Amsterdam were not satisfied with the black dung offered to them, and the municipality thus made less profit on the system than Liernur had predicted. Notwithstanding this commercial failure, the director of public works decided to expand the pneumatic system into several new parts of the city. Despite protests to the contrary, the majority of the Amsterdam city council accepted that it would be too expensive to construct an integrated sewer system. An important argument against such a British-style system was that it required the installation of water closets which, besides having contributed to the Great Stink in London, were regarded as too costly for the majority of Amsterdam's inhabitants.

In the end, the Amsterdam city council adopted three parallel systems: an improved version of the traditional canal system to transport wastewater out of town, the Liernur system, and the barrel-like *fosses mobiles*. In the inner city the majority of the canals were maintained to transport the waste and rainwater out of the city, while new locks and steam-driven pumps were installed to improve flushing. To prevent inhabitants from throwing their feces in ditches or canals, the municipality introduced *fosses mobiles* in some poor neighborhoods.

In the next 25 years (until 1902), more than 120,000 residents in Amsterdam's inner city were connected to the pneumatic network (Zon 1986). Inspired by the interest shown by foreign observers, Liernur decided to move to Germany. After a brief stay in Frankfurt, he moved in 1881 to Berlin, where he started the journal *Archiv für rationelle Städteentwässerung* (Archives for Rational City Drainage). Still, German cities never put his pneumatic system into practice on a large scale, although German sanitary engineering textbooks included positive evaluations of his system well into the twentieth century. The Dutch engineer Sijmons, mentioned above, cited the Amsterdam experience with the Liernur system at the 1894 Budapest conference on hygiene and demography. During the subsequent discussion, one city council member testified encouragingly that in St. Petersburg the Liernur system was operating according to plan.

Despite this varied support, Liernur's days were numbered. The proceedings of international sanitary conferences are excellent sources for measuring the acceptance of various water and sewage systems. From the second sanitation congress in 1876 through the Paris sanitation congress of 1900, the combined and separate systems were compared to the pneumatic system and *fosses mobiles*. Finally, the Paris congress of 1900 formally recommended the combined water carriage and sewer systems (Simson 1983, 172). From then on, the British-style combined system became the international standard. Accordingly, the city council of Amsterdam decided in 1902 to replace the municipal Liernur system (Reesema 1902). An integrated sewer system was designed in the outskirts of the city to get rid of rainwater, wastewater, and feces all at once (*Rapport van de commissie*... 1906; Bos 1915). In the inner city, the existing canal system kept its function as a means to transport wastewater and feces.

By the turn of the century, the three largest Dutch cities employed a similar sewage system. The last remaining above-ground gutters had been replaced by underground sewers, and ditches and canals had been filled in. The streets were regularly cleaned by fresh water to improve hygienic standards, and municipal sanitary technologies were installed, including wastebaskets and public urinals (see figure 1.5).

The Mechanical Street

In *Streetcars and Trolleys* John McKay (1976) describes how streetcars based on the American and German models spread throughout Europe. Since McKay did not deal with the Netherlands, this section examines whether the Dutch case differs from this general pattern.

It was Belgian and British entrepreneurs who in cooperation with local businessmen set up the first streetcar companies in Dutch towns in the 1860s. In The Hague a Belgian-owned company operated horse-drawn lines out to the seaside resort of Scheveningen, and a few lines through the central parts of town. In Amsterdam the first proposal for a municipal streetcar network also came from Belgian investors, whereas in Rotterdam it was a Paris-based firm.

When electrification of streetcar services became an option roughly two decades later, Dutch entrepreneurs and engineers went abroad to learn from developments in other European cities. Stimulated by the competition from two railway companies running steam streetcars, The Hague's horse-drawn streetcar company introduced electrical streetcars with batteries in 1890, taking Paris as its role model (Blok 1993, 37). Owing to its modestly higher speed, the battery-driven streetcar proved a success. In 1895 the firm asked for permission to introduce overhead-wire electricity on their streetcar lines throughout the whole city. These plans were more or less blueprints of initiatives taken by streetcar companies in the United Kingdom, Germany, and Belgium. The leading German firms AEG and Siemens & Halske had improved the American method of connecting the streetcar engines with the overhead wires. Although the growth of streetcar lines in the 1880s and 1890s was much faster in the United States than in Europe, German streetcar technology, rather than American, became the role model in Dutch cities (McKay 1976, 51–83).

The Hague initiative was not immediately embraced, however. Members of the city council strongly opposed overhead wires. In particular, they feared that this system would damage the aesthetic values of the monumental and representative parts of the capital. Engineers also argued that the construction of overhead lines for the electricity system could interfere with the overhead lines of the urban telephone system. To settle the matter, the council appointed a committee that sent the director of public works, I. A. Lindo, together with two railway engineers, on a study trip to various European cities.

The task that the Lindo delegation set itself was to gain firsthand experience of various streetcar systems. In all, they visited eight German cities in addition to Budapest, Brussels, London, and Paris (Jenniskens 1995; Put 2001). The underground-circuit streetcar in Budapest, constructed by Siemens & Halske, and the gas streetcar system in Dessau were of particular interest. Still, Lindo concluded that none of the alternatives to overhead electric traction—pneumatic streetcars, gas streetcars, and streetcars with underground circuit—was reliable enough. Not surprisingly, suppliers of gas streetcars strongly opposed his findings, and argued that Lindo had not been visiting the right cities.[2]

The issue remained unsettled until The Hague's city council decided in 1902 to set up its own electric power station. The council directed N. J. Singels, an engineer working for the centrally located Siemens & Halske power station, next to the main shopping arcades, to sketch a new municipal lighting network. Together with Lindo, Singels proposed to electrify the streetcars to get a better load factor for the new municipal electricity enterprise (see Schott, this volume). The city council was won over quickly, and an agreement was worked out with the private streetcar company. The company would continue to run the urban streetcars, and the municipal power station would provide the necessary electricity. The remaining opposition in the city council was overcome by pointing to the fact that overhead traction was already employed in many other European cities. The editor of the Dutch engineering periodical *De Ingenieur*, R. A. van Sandick, even claimed that overhead electricity wires would stress the modern character of The Hague and mark the central role of traffic in the modern city.[3] Objections of the national government against the construction of overhead traction in the narrow lanes of the parliament district were overcome by the promise to construct a new, broad traffic artery between the Spui and the Buitenhof, north of the Dutch parliament buildings. To facilitate the operation of the streetcar network and emphasize the representative character of the parliament district, the Hofweg, a 25-yard (22-meter)-wide boulevard, was cut through a densely built inner city

Figure 7.2
Allegory of the inhabitants of The Hague trapped in the struggle between a stork (the city of The Hague) and a "tram monster" in 1901. Many contemporaries feared the impact an electrical tramway would have on the streetscape of the Dutch seat of government. Source: Collection of Hans Buiter / SHT–Eindhoven, Netherlands.

neighborhood consisting of small alleyways, warehouses, workshops, and pubs. Councilor J. H. Warneke hoped to build the boulevard as wide as possible to give the Dutch seat of government an international flavor (Buiter 2005a, 80).

The Hague case was not exceptional. From Amsterdam another delegation went to Germany and Hungary to inspect streetcar technologies, and engineers from Rotterdam and Utrecht undertook similar grand tours. One outcome of this extensive circulation of people and information was a striking degree of uniformity. In all four of the large Dutch cities, German firms were responsible for the layout of an electric-traction system with overhead wires, including electric power stations. Accordingly, in the years between 1900 and 1907, German streetcar technology became the Dutch standard and persisted for many years thereafter. Streets were modified to suit the new, faster mode of traffic, and sidewalks were built to protect pedestrians. New traffic regulations forbade pedestrians and others to occupy the streetcar zone when trams were approaching. At some street corners, buildings were even pulled down to improve traffic safety. Because of the higher speed of the electrical streetcar, safety precautions were considered indispensable. In fact, the coming of electrical streetcars meant that the open and informal accessibility of the street actually diminished for everyone.

On the other hand, electrification was meant to democratize the use of public transportation. The Hague's city council took determined steps to achieve this aim, requiring the streetcar company to lower the standard fare from ten to five cents and to offer special low-fare rides for workers in the early morning hours. Between 1905 and 1910, in fact, passenger trips on the city streetcar more than doubled from 12 million to 31 million (Schmal 1990). For the first time, a means of transportation was truly public in character. This democratization went hand in hand with an effort to discipline passengers. Instead of offering stops wherever passengers preferred, the electric tram would stop only at well-defined places. Furthermore, municipal regulations prohibited passengers from smoking, spitting, or cursing while traveling. Such rules were partly a response to complaints by the middle class about the behavior of their fellow passengers (Dalen 1979, 456).

Although the Amsterdam elite generally had fewer doubts than their counterparts in The Hague about overhead wires for traction, one problematic site in Amsterdam was the Dam Square, then and now the most monumental public space in the city. To win over the city council, the leading Amsterdam streetcar engineer J. H. Neiszen pointed to the numerous German and other European cities where overhead traction had become the standard. In the harbor city of Rotterdam, by contrast, there were no vocal objections against overhead traction. Rotterdam's streets seem to have been more easily adapted to streetcar traffic than those in The Hague and Amsterdam.

The streetcar transformed urban space in several ways. In Amsterdam, plans were put forward by the director of the municipal streetcar company to fill in a canal (Reguliersgracht) to create a faster connection between the center and the southern parts of the city. Due to strong opposition, however, only the widening of another street actually occurred. Schemes to transform the Damrak (an important central canal) into a broad boulevard were also halted because the national government considered these plans too costly (Wagenaar 1998).

As before, changes in urban space were easier to bring about in Rotterdam. In 1909 Mayor A. R. Zimmerman presented an ambitious scheme to demolish the lower-income central Zandstraat neighborhood and to fill in the adjoining main canal to create a Paris-style boulevard. Zimmerman drew on plans by public works director G. J. de Jongh to construct a ring road at the spot of the Coolvest and the earlier filled-in Goudsevest, with financial support from the national government. The argument that the boulevard would give the fast-growing harbor city a respectable cosmopolitan center

Figure 7.3
The Coolsingel as Rotterdam's "world city boulevard" in the 1930s, with new town hall, electric tramway, and broad sidewalks. Each mode of traffic had its own zone on the street. Source: Collection of Hans Buiter / SHT–Eindhoven, Netherlands.

carried the day. Although some critics asked whether the broad avenue, 90 yards (80 meters) wide, would really fit into the structure and character of Rotterdam's inner city, the majority in the city council accepted the mayor's modernizing proposal.

Between 1913 and 1921, the canal Coolvest was filled in, the milk market was relocated, and a monumental town hall, post office, and exchange building were erected to create a "world city boulevard." In their drive to modernize, the city council even decided to pull down the impressive 200-year-old windmill on the Coolsingel. The results generally impressed contemporaries. By 1930, the Coolsingel, with its sidewalk cafés and neon lights, department stores, cinemas, restaurants, and dance halls, was regarded by many as the most modern street in the Netherlands.

The Electrified Street

Compared with New York, Chicago, and Munich, the streets in large Dutch cities were electrified rather late. Only after the First World War did electric streetlights replace gas lighting. With this shift, the number of street lamps and their intensity both increased. Brighter lights brought increased safety and altered the urban aesthetics. The growing amount of high-velocity motorized traffic was another motive to improve street lighting. When the number of bicycles and automobiles increased on Dutch streets after World War I, there were strident calls to adapt the streets to the rising wave of individual traffic. Traffic safety emerged as a specific concern in newspaper articles and meetings of local city councils.

Along with the increase in traffic, new technical concerns arose, such as the damage trucks did to the street surface and the need for a place to park automobiles. Beginning in 1923 the public works departments of the largest Dutch cities started to lay bitumen on major traffic routes to protect the street surface from the impact of heavy motorcar traffic. Although Amsterdam and other cities had been experimenting with the use of this tarlike substance since the 1870s, the inspiration for new types of concrete-bitumen (an early version of what is known today as asphalt) came from Great Britain. An excursion in 1923 by engineers from ten Dutch cities hosted by British Petroleum proved decisive. Once more, international contacts framed the Dutch developments.

Parking lots were designed in several Dutch cities to accommodate the automobile. In Amsterdam a scheme was advanced to fill in the greater part of the Rokin to make space for parking. This effort was the result of the combined efforts of automobile organizations and engineers of the public works department. In typical Dutch fashion, the city council decided to fill in the biggest part of the canal, but left a third of it unaltered to placate the plan's opponents.

Traffic regulation included a number of diverse measures, with Amsterdam often at the forefront. In 1856 Amsterdam had six streets with one-way traffic, and in 1902 no fewer than 84. The aim of these regulations was to facilitate maintenance, increase

Figure 7.4
The Rokin in Amsterdam in the 1930s, once a watery canal. The newly created space was used to park motorcars and bicycles. Source: Collection of Hans Buiter / SHT–Eindhoven, Netherlands.

traffic flow, and to improve traffic safety (Buiter and Staal 2006). Safety was a recurrent topic, as the mixture of bicycles, trucks, and automobiles caused an alarming number of serious accidents. As a result, most big Dutch cities organized inspections of motorized vehicles and constructed more pedestrian safety islands and "sidewalks." While Amsterdam imposed strict speed restrictions as early as the nineteenth century, Rotterdam did not impose speed limits until 1926.

Amsterdam took the lead in other areas as well. In 1912 a special police traffic corps started to regulate traffic at junctions, based on the model established in German and Danish cities (*125 jaar verkeerspolitie*, 1951). To design this traffic police corps, an Amsterdam delegation visited Hamburg, Cologne, Dresden, and Leipzig, as well as Copenhagen. Inspired by the Danes, stop lines were drawn on side streets in Amsterdam, and special hand-held rotating stop signs were set up to simplify the communication between policeman and motorists.

Rotterdam was the first city in the Netherlands to introduce traffic lights. In 1926—just two years after Europe's first electric traffic light was installed in Berlin's busy Potsdamer Platz (see figure 1.4)—a lamppost was installed to regulate traffic at the junction of the Coolsingel and Van Oldenbarneveldstraat. The Rotterdam police had concluded that using gestures to direct the burgeoning traffic on the broad avenue was no longer sufficient. The officials initially tried the American design with red, yellow, and green lamps, but this solution was not satisfactory. Instead, they decided in 1928 to install a four-lamp system that projected the word "STOP" to halt drivers.

Traffic lights were introduced also to improve the flow of traffic. The first phased traffic lights in the Netherlands were set up in Amsterdam on Leidsestraat for exactly this reason. This ten-yard-wide shopping street was the main artery between the central part of the city and the new quarters in the southwest. Traffic counts in 1930 revealed that each day nearly 5,000 automobiles, 25,000 streetcar passengers, 30,000 bicycle riders, and 33,000 pedestrians passed through the street. To improve the flow, traffic policemen were stationed at the nine junctions where the Leidsestraat crossed side streets. Furthermore, the municipality banned horse-drawn carts and handcarts there during the busiest hours of the day. Most radical was probably the implementation of one-way sidewalks. Proposals by the association of shop owners to improve the maneuvering space for car owners by blocking the traffic of cyclists and trucks during the afternoon shopping hours were not approved by the municipal officials, who feared that cyclists would take over the sidewalks.[4] Suggestions by several motorist organizations to replace the streetcar lines running along the Leidsestraat by motorbuses also went nowhere. Instead, the police developed an intricate bell system to coordinate the individual traffic policemen.[5] When the system was put into practice in December 1929, the maximum speed at the Leidsestraat was set at 20 kilometers per hour. The bell system was meant as a low-cost experiment to see if it enabled traffic to go at a higher speed without significantly increasing accidents.

Knowledge in the field of traffic regulation also circulated between Dutch cities. For example, the Amsterdam police asked their Rotterdam colleagues to pass on information on how their traffic systems functioned. The Rotterdam police evidently thought their traffic light system was too confusing for many street users, and they were still searching for a better system. Consequently, the Amsterdam police decided to collect reports from police departments in Germany, Great Britain, Denmark, and the United States. Finally they decided to introduce the Siemens lighting system using green, yellow, and red lamps, which they believed would become the international standard. Soon, however, a drawback appeared: the red lights burned also when there was no crossing traffic. In London a vehicle-actuated system had been introduced to tackle this problem, and Amsterdam adopted the system in 1936.

The introduction of traffic signals actually increased the driving speed, although some cab drivers thought otherwise. On the Leidsestraat, the original speed limit was

12 kilometers per hour, but after the phased light signals had been introduced in 1934, the speed limit rose to 40 kilometers per hour, allegedly without any loss of safety (Leerink 1938, 406). Of course, higher speeds for vehicles were possible only when bicycle riders kept to the side of the road and pedestrians stuck to the sidewalks. Authorities and traffic organizations set up public campaigns to get street users to adapt their behavior to the new situation.

The Motorized Street

When Lewis Mumford visited Rotterdam in the mid-1950s, he was very impressed by the work done on urban reconstruction after the devastating Luftwaffe bombing of May 1940. One of the most striking features of the new Rotterdam was the Lijnbaan, which opened in 1953 (L. Mumford 1963, 44). This pedestrian shopping mall was the first in the Netherlands and replaced the bombed-out shopping arcades along the Coolsingel. Segregation and specialization were the keywords of the Rotterdam reconstruction activities. The Lijnbaan pedestrian district was surrounded by a brand-new urban highway system linked to the planned national highway network. One of the first cities in the Netherlands to have a highway constructed through it (Provoost 1996), Rotterdam served as an outstanding example of the adaptation of Dutch cityscapes to the automobile in the post–World War II period (see Misa and Lundin, this volume).

In the 1950s street regulations tended to privilege motorized traffic. During the German occupation, motorized traffic had been given priority over nonmotorized traffic, and this regime was preserved even after liberation. Municipalities started adapting their street networks to the automobile, increasing the number of traffic lights and other traffic-guiding artifacts, changing the profile of streets, and implementing extensive road-building programs. The ANWB (the Dutch tourist organization) played an important role in this process. From 1952 on, it published a journal, *Verkeerstechniek*, and organized courses and lectures for policemen and engineers. The adaptation of urban streets to the rising tide of mass motorization was one of the main topics of the journal.

Rotterdam with its bombed-out tabula rasa was one of the first cities to design urban highways, and other cities followed suit. Ring roads around the city center were important, yet often controversial, features of these efforts. In Amersfoort in the late 1950s, a ring road was constructed on a filled-in urban moat, directly adjacent to the town's medieval walls and towers. But in nearby Alkmaar, the city council rejected a similar proposal to replace the city moat with a ring road. Numerous Dutch architects and urbanists went to the United States as part of the Marshall Plan, but generally they were wary of the American urban freeway system (cf. Lundin, this volume). Jane

Jacobs's anti-automobile *The Death and Life of Great American Cities* (1961) was even hailed by Dutch public works officials (Provoost 1996, 83–84).

The strongest inspiration for the postwar Dutch traffic schemes was found in Germany. Traffic engineering was a largely unknown field in the Netherlands at the time, and German traffic knowledge and traffic engineers were much in demand. In 1955 officials in Utrecht contacted the ANWB on how to adapt their city to the expected increase in motorized traffic. The ANWB sent the Utrecht mayor to German traffic engineer M. E. Feuchtinger, a leading international expert, who in due course organized a comprehensive traffic survey and delivered a set of proposals in October 1958 (Buiter 1992). Like other German traffic engineers who rose to prominence in the 1950s by reconstructing war-damaged German cities, Feuchtinger had been heavily influenced by American traffic engineering even before the war (cf. Lundin, this volume). With unintentional insight, the journal *Nieuw Utrecht Dagblad* stressed the "American flavor and quality" of his work (Buiter 1992).

The American character and German origin of the project prompted quite a stir in Utrecht. When the municipality published Feuchtinger's traffic scheme in November 1958, residents were horrified. A citizens' committee collected thousands of signatures, organized a protest meeting, and published a booklet. Student associations marched along the urban moat targeted to be filled in and made into a highway, and in front of the city hall they even hanged Feuchtinger and the mayor in effigy. One protester asked cynically why a German traffic engineer was asked to restructure Utrecht's street network: "In Rotterdam in May 1940 the Germans did a very good job as well in demolishing the old street structure" (quoted in Buiter 1992, 17). Facing this pressure, the city council rejected the plan to fill in the moat and construct a ring road and a grid of wide streets in the inner city. After long discussions, it decided to hire the Rotterdam urbanist J. A. Kuiper to suggest alternatives. In 1962 the council decided to fill in *half* of the moat as a compromise, a decision that was challenged by the national minister of culture. The municipality eventually got permission from the national government to fill in a *quarter* of the moat, a plan that included the construction of a huge shopping mall and a public transport hub called Hoog Catharijne between the ring road and the central railway station.

Hoog Catharijne is a typical case of the local appropriation of internationally circulating ideas. It was based on the model of the pedestrian shopping mall surrounded by parking garages as developed in Detroit and a few other American cities in the 1950s (Hardwick 2004). But whereas in the United States most malls were constructed at the outskirts of the cities, Hoog Catharijne was located squarely downtown. The development firm, Bredero, got information about shopping malls from its Australian branch, whose architects had attended courses at the University of Hawaii. In Australia Bredero was rather successful in developing shopping malls, offices, and parking garages at

Figure 7.5
The distinctive city-center Hoog Catharijne shopping mall in Utrecht spans the traffic artery, which follows the course of the old town moat. Pedestrians can walk from the old town (to the left) through the shopping mall to the central train station (just past the bus station, upper far right). Source: Collection of Hans Buiter / SHT–Eindhoven, Netherlands.

central railway stations. Amsterdam turned down Bredero's proposal to redevelop the area around its central railway station, but its ideas were received more enthusiastically in Utrecht. The concept suited the wishes of the Dutch Railways to redevelop Utrecht's central station and to introduce parking garages. The main features of Hoog Catharijne are a pedestrian area above ground and a shopping street connecting the Vredenburg Square with the railway station and a new central business district at the rear of the station. Bredero did not manage to sell the Hoog Catharijne concept in other Dutch cities, but was more successful abroad; the model was exported to Hannover, Aberdeen, and Epson (Buiter 2005b).

Conclusion

We can now distinguish several different patterns of exchange of information and practices among European cities. The international flow of sewer engineering was much more complicated than the exchange of information on streetcars. Compared to the streetcar, sewer technology was more of a basic technology that defined the urban fabric, was more connected with assumptions about urban life, and was more strongly contested. The exchange patterns visible in the case of street electrification resemble the streetcar pattern. On the other hand, the discussions about pedestrian streets and urban highways were as lively as the sewers debate had been eighty or a hundred years earlier.

Pamphlets, textbooks, journals, congress minutes, and local reports on available European technological choices were remarkably international in scope from the mid-nineteenth century on. Often, these sources present clear overviews of available alternatives and standards on a European scale. They indicate that urban engineers had frequent and intense contacts with officials and engineers in other European cities. The technological tours that engineers took of other European cities are an especially strong indication that some kind of "Europeanness" existed in the minds of the actors. Barriers of distance and language were overcome rather easily, and the lively debate that was stimulated by this exchange in Dutch cities reflected a typically European discourse.

Technology applied in other European cities served as models for Dutch engineers, but in many cases it also had a strong rhetorical power in local discussions. This was clearly the case with sewer and streetcar technologies. However, foreign solutions could also be used as cautionary examples. In the Dutch discussion on urban highway construction, for example, the American pattern was regarded with great skepticism outside Rotterdam. While Dutch engineers and politicians rejected an American approach to urban traffic, they welcomed similar models developed in Germany.

Although the orientation of urban engineers and other professionals was very much European in nature, the concepts and artifacts were appropriated to fit local circumstances. This was the case both with the adaptation of the sanitary systems to the existing canal and street networks and with the streetcar. Dutch boulevards like the Hofweg in The Hague and the Coolsingel in Rotterdam reflect the European orientation of their initiators, and also the modest means and pretensions of Dutch urban societies. Hoog Catharijne is another example of an international concept adapted to a local situation, and in fact this concept was implemented only in Utrecht, and in no other Dutch city. I argue that the local appropriation of sewer systems, electrical tramways, boulevards, and malls worked in two opposite directions. On the one hand, with the introduction of these artifacts, the Dutch townscape became more similar to that of other European cities. On the other hand, since these artifacts and concepts were not introduced in all city centers, one could also say that the differences between various Dutch cities increased. The construction of the Dutch street under the influence of European technologies was never a one-way street.

Notes

1. *De Ingenieur* 10 (1895), 271.

2. Records of meetings of the city council of The Hague, Enclosure 27 (1897), The Hague Municipal Archive.

3. Ibid., 4 March 1904, p. 97.

4. "Report on the Traffic Situation in the Leidsestraat," 7 May 1932, police archives, no. 5627, Municipal Record Office, Amsterdam.

5. "Instruction How to Handle the Automobile," 7 December 1929, police archives, no. 5623, Municipal Record Office, Amsterdam.

III Industry and Innovation

8 Empowering European Cities: Gas and Electricity in the Urban Environment

Dieter Schott

The hopes tied to the development of electrical engineering are gigantic. If we succeed in redeeming part of the sin which the Age of Steam has inflicted on mankind in the Age of Electricity... this will be an unprecedented achievement in world history.
Internationale Elektrotechnische Ausstellung 1893, 35

With this extravagant rhetoric gracing the opening of the 1891 International Electricity Exhibition, the mayor of Frankfurt expressed his hope that electricity might bring about fundamental social and economic changes. With more than one million visitors, the Frankfurt exposition was a major event and an apt starting-point for considering the "age of electricity" (Steen 1991). At Frankfurt some 450 companies demonstrated the major applications of electric energy in domestic and public lighting, industrial and commercial power, and urban transit. Three major technological systems competed in the famous "battle of the systems." Direct current, single-phase alternating current, and polyphase alternating current were each represented by a range of companies that exhibited appliances and carried out practical demonstrations (Hughes 1983). Exhibition-goers were plainly captivated by the transmission of high-voltage current more than 100 miles (175 kilometers) from the river Neckar. The distant river was transfigured at the exhibit grounds into an artificial waterfall. This practical demonstration showed that distant hydropower could be sent to urban and industrial centers. It also showed that energy generation and energy consumption could be spatially separated, fanning an atmosphere of enormous enthusiasm and technological utopianism.

On a more practical level, the Frankfurt exhibition attracted engineers from across Europe and beyond to its comprehensive overview of electrical technology. As such, the exhibition was a key node in the international circulation of municipal engineering knowledge and practices. In hosting the first major assembly of German town councilors, it also brought together representatives from more than 150 German municipalities to consider the problems of urban electrification and to identify their common interests. This event marks the beginning of a national organization of

German cities, which was to culminate in the Dresden City Exhibition of 1903 and the foundation two years later of the Deutscher Städtetag (German Town Conference) (Beckstein 1991).

Frankfurt staged the exhibition in part to address its own problems: it hoped to reconcile the need for a power generation plant with the desire to minimize pollution in the city center. Since many cities faced similar challenges, Frankfurt in 1891 serves as an origin point for our story. This chapter aims to show how European cities appropriated the new electricity systems and how electricity fared against the older, better-established gas system. I draw attention to the role of electricity not merely in providing energy but also in expanding the political power of municipal authorities and making cities more autonomous. The chapter illustrates how energy and transport networks "empowered cities" in the double sense: by providing sources of power for lighting, engines, and transportation, and by giving them new political powers to shape their social, economic, and spatial development.

Gas—The First Energy Network

Nonnetworked energy systems persisted well into the late nineteenth century: paraffin lamps lit most urban households, and coal or wood was used for heating and cooking. The only urban "energy system" possessing a network infrastructure at that time was gas. Following a technology developed in England in the early nineteenth century, gas was generated principally by coking coal (baking it in an enclosed oven), and in some instances also wood, and supplying this gas through a network of pipes, using a gasometer to store any excess. By 1890 extensive gas systems served most areas in practically all large European cities; gas was used for illuminating urban streets and shops, cultural facilities, restaurants, public buildings, and wealthy private households. Although gas pipes ran below most urban streets, many poor households could not afford the high installation and operating costs. A trade union survey of 85 Dresden working-class households revealed that none of them had gas or electric lighting as late as 1903 (Gransche and Wiegand 1981).

Gas networks across Europe were creations of private initiative. Cities typically granted concessions to gas companies in exchange for cheaper rates for street lighting and sometimes for a municipal share in the profits. Paquier and Williot (2005) identify four types of gas-using countries. In Britain and France, private companies took the lead and also initiated the circulation of technology and entrepreneurial forms across Europe. In Germany, Switzerland, and the Scandinavian countries, medium-sized and smaller towns established gas networks after the 1860s, while in the Iberian countries, Italy, and the Austro-Hungarian Empire, gas remained limited to larger urban centers. In the Balkan states, the first gasworks appeared only after the middle of the nineteenth century, and diffusion was restricted mostly to capital cities until the First World War.

From the outset the gas industry embodied a contradiction typical for urban network technologies. The industry was mostly financed and run by private entrepreneurs or share-holding companies, and it flourished in an age that glorified laissez-faire capitalism. However, with their high installation costs discouraging multiple competing systems, network technologies were usually organized as monopolies. Indeed, most often the heavy investment costs for installing the gas-generating plant and the network of pipes prohibited any real competition within a given territory. Moreover, since gas companies depended on municipal concessions to use city streets for laying their pipes, and since these concessions were usually limited in time, the gas companies only rarely accepted competitors on their grounds.

This de facto monopoly in nearly all major European cities exposed private gas companies to public criticism. Exasperated with poor service and high costs, a strong reform movement, emerging in the 1870s in several countries, proposed to transfer gas and water companies to municipal ownership. This "municipal socialism" was particularly strong in Britain, Germany, Switzerland, and the Scandinavian countries. It constituted an important part of the emergence of a new paradigm in the attitude of municipal authorities, departing from a traditional laissez-faire position and gradually adopting a more interventionist stance toward town planning and economic activities (Kühl 2002; Giuntini et al. 2004; Hård and Stippak, this volume; Krabbe 1990). City councils and mayors—most famously, Birmingham's mayor Joseph Chamberlain—came to discover that technology could serve as a means to increase their power.

As the dominant urban energy network, gas lighting helped transform daily life in European cities. By giving an enhanced sense of security and order to nighttime streets, gas lighting cultivated street life and sociability, and enabled shops and department stores to exhibit fashionable goods irrespective of the time of day or season (see Capuzzo, this volume). Gas also made possible round-the-clock work in factories and late-night entertainment (Schivelbusch 1983). Profits from municipal gasworks even funded general municipal budgets, further expanding cities' effective powers.

Notwithstanding these significant benefits, a growing hostility against gas welled up in the 1880s when gas lighting caused dramatic fires in theaters and public buildings. In 1881 a catastrophic fire in the Vienna Opera killed almost 400 spectators. To be trapped in a theater fire became a nightmare of the urban middle classes. Besides these significant dangers, gas lighting had serious health disadvantages. The open gas flame consumed immense amounts of oxygen, produced smelly and poisonous emissions, and uncomfortably heated up large public spaces. Soot from burning gas stained wallpaper, curtains, paintings, dresses, and furniture. Electrical engineering companies eagerly exploited these nuisances and soon launched public campaigns that hailed electricity as a fairylike pure and clean light. Their publicity campaigns against gas helped usher in the new "age of electricity" (Binder 1999).

Electricity and the Reform Agenda

As Frankfurt's mayor had proclaimed in 1891, electricity seemed to solve several urgent social problems. Since electric motors were flexible and could be made relatively small, they were considered particularly promising for artisans and small businesses, which were otherwise marginalized by big industrial corporations. It now seemed possible to reverse the century-long trend toward industrial concentration that had been driven by the steam engine's operating economies of scale (Wengenroth 1989). Since electricity was initially laid out in cities, it was particularly promising for urban industry. Urban transportation was yet another promising vista. The horse-drawn streetcar, set up in most capitals and large European cities by 1890, had reached the practical limits of expansion in the face of rapid suburbanization. Steam locomotives generated major environmental problems, especially within cities, and posed their own serious fire and safety hazards. With the direct-current electric motor and the trolley system, developed by Frank Sprague in the United States in the 1880s and marketed by the enterprising AEG in Europe, the electric streetcar promised an attractive technological alternative (McKay 1976; Teaford 1984, 234–240).

Although urban elites were initially reluctant to permit the installation of overhead electric wires in the streets, by the turn of the century the electric streetcar was introduced in many European cities and quickly became a stirring symbol of urbanity, mobility, and modernity. A major reason that municipal councils overcame their initial reservations was the anticipated decentralizing potential of the electric streetcar. The "housing question" loomed large for many city officials and members of the urban elites who were represented in city councils and social-reform societies. They were convinced that the population density in inner cities was unacceptably high and that new housing in less polluted environments had to be created to avoid either social revolution or a general degradation of the working classes (Zimmermann 1997). The electric tramway seemed to offer a perfect solution. It would open up the urban periphery for decent working-class housing on affordable land and still make it possible for workers to reach their urban workplaces (Ladd 1990). In the mid-1890s, driven also by innovations in the chemical and electrical industries, many European city officials and technology boosters embraced electricity as a panacea to a wide range of urban problems.

Despite its evident promises, electricity was still tangled up in a confusing variety of technological systems and economic arrangements. The mature direct-current (DC) system was perfectly capable of driving heavy motors and illuminating city centers, with the advantage that surplus DC electricity could be stored in batteries and later uti-

Figure 8.1
The construction of a transformer shaft in the German city of Hannover. Unlike American cities, European cities preferred sinking electric shafts beneath city streets—avoiding the profusion of wires running overhead. Source: Wuttke 1904, II: 139.

lized at times of peak demand. However, to keep transmission losses down, DC generating stations had to be placed close to the areas of maximum consumption, usually the city center. Having a steam-powered DC station in the middle of town posed major environmental problems because of the smoke and noise, the transport of coal and waste, and the provision of cooling water, in addition to the expensive land costs. Given the prospect of future expansion, it became simply unacceptable to have major DC electricity plants in the city center. Urban planning schemes aimed to relocate such polluting industries to the urban fringes (D. Schott 2002). In fact, the problem of urban pollution had prompted Frankfurt to stage the 1891 exhibition in the first place.

Alternating-current (AC) technology provided an attractive solution. AC power could be generated at peripheral locations with easy access to fuel sources or at accessible waterfalls, and then could be transmitted to the city center at high voltage with only slight losses, as Frankfurt again had demonstrated in 1891. But since AC electricity could not be stored, the generating capacity of the power station had to equal the system's highest possible demand. The first AC systems thus tended to be overdimensioned, which reduced profits. A further problem was that initially no AC motor on the market could be easily applied to industrial workshops. For streetcars, AC had to be converted to DC, leading to further losses. The polyphase system, using two or three wires carrying alternating current, had its spectacular appearance at the Frankfurt Exhibition along with the groundbreaking long-distance transmission. Polyphase motors were well suited to heavy-duty industrial purposes. The polyphase system thus opened new horizons for exploiting energy resources that were situated far from the consumption centers.

Reports from leading experts indicate that none of the three systems—DC, AC, or polyphase—was deemed superior in the early 1890s. A sound choice could be made only in relation to a specific local context, taking into consideration the relative share of local users and the anticipated growth of power consumption. Local administrations were confronted with a bewildering array of technical alternatives. They also had to wrestle with a number of nontechnical issues: Who should build and run the electric utilities? Which electric applications were most important? What should be electricity's relation to existing gas networks? Unable to handle the daunting complexity, cities frequently consulted independent experts. Professors of electrical engineering, such as Erasmus Kittler, acted as experts and thus contributed to the circulation of knowledge and experience between European cities (D. Schott 1999a; König 1986). Similarly, directors of established power stations gave advice on new projects, designed projects, made calculations, or assessed projects submitted by engineering companies.

These city officials and independent experts never acted in isolation. Local decisions about electrification were often entangled with a range of urban planning issues, such as municipalization and electrification of streetcars. By projecting a certain technological path, experts in effect also charted a specific course of economic and spatial development for the city. The consequences of their technology choices were often far-reaching and long-lasting. An early decision for a DC system could, for example, hamper a city's later industrial development or spatial extension. Since these choices involved heavy capital investments, early developments were locked in for considerable lengths of time. In this way, decisions about energy systems resulted in emergent urban structures.

Contemporary urban actors were aware of the momentous character of their decisions. Cities like Frankfurt, where local governments valued environmental quality, strongly favored the AC system (Steen 1991, 676). In Mannheim on the Rhine, which underwent rapid industrialization in the 1890s, a polyphase system arose within the spatial contexts of a new industrial port. The desire to provide the industrial port with municipal power, and the fact that the port itself offered a suitable location for a power plant, determined the choice of both system and location (D. Schott 1999c, 379–389).

By the turn of the century, polyphase technology was chosen most often because it provided great flexibility in locating power generation and fulfilling the widest range of energy demands. Increasingly, the polyphase system powered new urban electric stations or extensions of existing networks. Still, the significant capital already invested in DC or single-phase AC technology prevented any quick changeover. In many cities, ranging from metropolises like Paris to smaller municipalities like Darmstadt, different technological systems and even different voltages coexisted over several decades (Fernandez 1997). Only after the Second World War were fairly uniform technological standards established across Europe.

Electricity and Gas: Competition or Complement?

Urban electrification developed while gas still dominated public illumination as well as private illumination for commercial premises and high-income households. Remarkably, the challenge of electricity also brought about major technological innovations in the "old" gas technology. Gas revived as a serious competitor with the Auer incandescent light (or "Welsbach mantle"), which radically reduced gas consumption and did away with the troublesome open gas flame (Braun 1980). Gas lighting thus gained a new competitive edge over electricity in the early 1890s, as it cost significantly less. To compensate for lower gas consumption due to the very economical Auer light, gasworks also introduced new rate structures to expand the market for heat and power. As a consequence, gas consumption for heat and motor power around 1900 actually exceeded that for illumination in many German cities. Gas companies fought back against the electricity campaigns with their own energetic campaigns promoting gas for cooking and water heating. They opened showrooms parading new gas appliances, staged popular lectures on gas cooking, and offered to rent new gas appliances on attractive terms. By introducing slot meters, where a certain quantity of gas was paid for by coin, the companies made gas usage more popular among working-class households (Goodall 1999).

Besides their obvious self-interested economic motives, gas companies' activities also had a health dimension. The findings of bacteriology and the preoccupation with tuberculosis as the main urban killer drew increasing attention to domestic personal hygiene. Bathrooms became part of standard housing design for the middle class, and relatively cheap and efficient gas-fired water boilers thus found a rapidly growing market (D. Schott 2006). Public health reformers also campaigned against coal fires, claiming that coal-fired stoves caused massive domestic pollution through ashes and bad air, and connecting these problems with tuberculosis and respiratory diseases. Using gas for cooking and heating was thus presented as a clean and hygienic alternative.

The most effective incentive to stimulate gas consumption was the differentiation of rates for lighting, cooking, heating, and motor usage. Gas consumption expanded considerably after the rates for cooking gas had been reduced: private gas flames in Germany more than doubled in the 1890s (from 4.1 to 8.5 million), while total private consumption increased nearly as much (from 405 to 721 million cubic meters). Nevertheless, as late as 1914 only about half of urban households in selected German cities used gas (D. Schott 2005, 497–502). This massive expansion of gas usage forced many cities to construct new gasworks. Technological innovation also permitted better utilization of the by-products of gas generation. The sale of such valuable products—especially coke, tar, benzole, pitch, and ammonia—helped pay for the gasworks themselves. The marketing of these by-products also forced municipal gasworks to integrate better with the local and regional chemical and industrial economy; consequently, directors of gasworks acquired considerable knowledge of relevant industries near their cities.

Gas continued to light the large majority of street lamps in European cities, since the cost of gas remained low. Only certain "representative" streets and squares were singled out for electric illumination. When Munich decided in 1894 to introduce electric street lighting, it cost the city four times as much as the traditional gas lamps (D. Schott 1998, 224). Expensive electric street lighting was not merely a demonstration of civic pride; it was also an investment in traffic safety, which had become a major problem in cities due to the rapid diffusion of the new electric tramway (see Buiter, this volume). Many felt that the streets were safer because the new electric lamps were much brighter than gas ones. Nevertheless, German cities did not reach the same density of electric street lighting seen in U.S. cities such as New York, Chicago, and Philadelphia (Teaford 1984, 230).

While competing on some markets, electricity and gas had thus established a relationship of complementarity around the turn of the century, especially in cities where both utilities were jointly owned and where rate policies could support it. Usually, electricity reigned supreme for prestigious private and public illumination and also served public transportation, while gas catered to the emerging heat and cooking markets, as well as the lighting of lower- and middle-income households.

Electricity—Social Reform or Luxury Technology?

The optimism that urban reformers had so fervently expressed in 1891 about the democratizing and decentralizing potential of electricity remained a utopian dream for some decades. In actual practice, electricity remained a luxury technology that reached only a small group of affluent customers. A decade later, no more than 10 percent of private households in German cities had been wired. Wider diffusion was hampered by high costs, especially for installations. Furthermore, the rather selective layout of electric networks distinctly favored higher-income neighborhoods. As long as lighting dominated demand, electric utilities were unable to substantially lower the prices they charged consumers, although they tried to diversify their customer structure. The general strategy was to attract load at off-peak periods, in particular during the day, when little lighting was required. Municipal and private power stations therefore tried to attract motor load from artisan and industrial customers and, especially, traction load from electric streetcars (see Buiter, this volume). Consequently, streetcar lines were rapidly built in many European cities during the 1890s and 1900s. This diversification strategy, supported by differential pricing for power rates and traction, eventually gave municipal electric utilities a more even load curve.

Clearly enough, urban electrification enjoyed some successes, but questions persist about whether electricity was a democratizing economic force. Electricity did not, in fact, become the savior of depressed craftsmen, because most of them simply could not afford it. Where electricity actually did enter workshops and small industries, it replaced gas motors or small steam engines and supplied power for a range of tools and applications. Only a few trades developed into major users—such as wood or metal turners, bakers, and butchers—and these tended to have a significant but intermittent need for mechanical power.

How much did electricity really help to solve the housing problems of overcrowded cities? There can be little doubt that the electric streetcar reinforced the trend toward suburbanization, although it did not by itself actually produce suburbs in most cases. If we examine the social composition of streetcar riders, we find a surprisingly varied picture. In many cities, the more affluent middle classes dominated streetcar ridership for many years, so of course they gained a significantly wider choice of residential locations and leisure destinations. Whether streetcars became a vehicle of housing reform depended on political rather than technical choices. Only where town councils took determined steps—introducing cheap workers' fares, constructing lines that connected working-class housing districts with factory districts, and running cars at the times when workers actually needed them—did the streetcar become the "common man's vehicle" (McKay 1976).

What the streetcar clearly did accomplish was the opening up of the urban periphery for leisure activities. Even before 1914 many streetcar lines had their terminal stations at or slightly beyond the edge of the built-up city, close to popular woods or leisure grounds. For this reason, Sunday was usually the day when tramway lines had their peak load, particularly during the summer. On these days one could find riders on the tram from all reasonably prosperous walks of life. In cities where streetcars were run by the municipality, the new means of transportation was used to create leisure grounds, stadiums, zoos, and industrial estates accessible to the wider urban public.

Figure 8.2
An elevated urban railroad in the German city of Elberfeld. Relatively affluent riders typically dominated electric transit for years—unless cities took determined steps to run early morning trains for workers with cheap fares. Source: Wuttke 1904, II: 193.

Urban Electrification in Britain and France

Although a pioneer in gas, Britain was something of a latecomer to electricity. As with gas, private entrepreneurs initially took the lead in electricity, but after 1895 British municipal utilities became increasingly prominent. Already by 1900 electricity works under local government ownership outnumbered private ones by five to two, and continuing until nationalization in 1946 a clear two-thirds majority of power stations were owned by local councils (Millward 2000, 326).

The relatively weak regulatory regime in Britain from 1840 to 1870 had supported numerous private undertakings in gas and water. In gas, where private entrepreneurship was quite entrenched, public ownership never surpassed 40 percent. Because of widespread discontent with poor services and high monopoly prices, municipalities took over significant parts of the gas industry beginning in the 1860s and with this background soon became engaged in electricity and streetcars. The 1870 Tramway Act and 1882 Electric Lighting Act gave municipalities the power to purchase a private company after 21 years. These acts provided local councils a formal entry into local transport activities when the private concessions expired (Armstrong 2000). The resulting municipalization of streetcar companies in the 1890s encouraged cities to set up local power stations.

The strong role of local government in electric utilities is not difficult to explain. Cities experienced significant difficulties in regulating the industry by means other than direct municipal ownership, and local politicians often desired to widen the tax base and to use the municipal utility's profits to finance local expenditures. Profits from utilities were sufficient in many towns to finance their entire expenditures for public health and police (Millward 2000, 334).

Cities in France differed from those in Germany and Britain both in the pace of development and the local balance of power. Whereas in the 1880s French cities had numerous initiatives to set up power stations and construct electrical networks, the picture changed considerably after 1890 (Kühl 1997; Caron 1997). With private gas companies vigorously defending their monopolies on illumination, there was no wave of municipalization of gasworks in France comparable to that in Germany and Britain. French gas companies furthermore opposed most attempts by other private entrepreneurs or the municipality to set up electric networks that would compete with gas illumination, or they insisted on carrying out these projects by themselves.

The French legal system, above all the Conseil d'Etat, vigorously upheld the monopoly rights of the established companies and discouraged municipalities from granting entrepreneurs the right to use public streets to construct new networks (Fernandez 1997, 115ff.). The distinct structure and function of French municipal administration also worked against municipalizing utilities. In France, positions in municipal administrations often served as launch pads for careers on the national level, which meant that bright and energetic urban leaders continually had their eyes on the national political stage. In German cities, by contrast, the well-paid and prestigious office of mayor was a highly attractive career for a middle-class professional. Delivering a significant local benefit such as municipalization, then, meant more for a German mayor than for his French counterpart. The French public was also less ready to accept municipal ownership of utilities than was the German or British.

The situation in France changed only somewhat after a 1906 law recognized electricity as a "collective service." The law gave utilities the right to construct delivery lines across private property. Accordingly, gas companies could no longer insist on their monopoly without showing readiness to engage in this new service. The law also meant that in principle municipalities had the right to organize the distribution of electricity on their territory, although few actually did so. Most towns attempted to ensure that electricity provision would be universal and prices reasonable by working out detailed "books of charges" for the concession companies. In many cities a mosaic of technical systems developed, as authorities gave different concessions in different parts of the city; this system sometimes resulted in one company buying up the shares of others, effectively recreating a monopoly-like situation. Frequently, as in Paris, electric concessions did not extend to motor power, so small islands of electricity generation for workshops and small factories persisted among the surrounding districts of electric lighting systems (Fernandez 1997, 120).

From Local to Regional Power Supply, 1905–1920

Cities were the decisive actors as well as the most important playing field during the first two decades of electrification. With their concentrations of potential customers, they were obviously the most promising markets. Furthermore, municipalities were deeply involved in shaping and influencing the course of electrification, as owners of streets and granters of concessions, on both municipal and private property. These varied activities can be seen as the direct municipal appropriation of the supposedly universal electrical system. After 1905, however, the roster of actors and scale of activities changed dramatically, with the main arena in electrification passing from an urban to a regional scale and thus threatening to disempower cities. Especially in Germany, the individual states (*Länder*) became active in energy issues. At the same time, larger utilities, some of them combining public and private ownership, developed new models of regional electrification. Rather than producing electricity on urban territory, they set up highly efficient power stations on or near the energy resource (most often water or brown coal), and aimed at creating larger regional networks in order to bring electricity to smaller towns and the countryside.

Regional systems also effectively tapped "white coal," or hydropower. An early example is the hydropower station of Rheinfelden on the Rhine, which was financed by Swiss capital and set up in a small German border town in 1898. Within a few years its cheap waterpower had given birth to a thriving industrial district and a regional power system, much as the waterpower from Niagara Falls had done near Lake Erie. Because its network crossed national borders, it made up the nucleus of a European power compound system (Herzig 1992b). The large-scale exploitation of hydropower by private companies eventually ushered in a policy of state electrification based on the systematic use of hydropower, especially in Scandinavia, Switzerland, and those German states that were devoid of coal but potentially rich in waterpower (Stier 1999; Thue 1995; Gugerli 1996).

More prominent on the whole, however, were large regional companies such as the Rheinisch-Westfälisches Elektrizitätswerk (RWE). Established in 1898, the RWE increasingly set up networks of power stations and high-voltage distribution lines in the heavy-industrial districts of the Ruhr. Under the leadership of industrial tycoons Hugo Stinnes and August Thyssen, the RWE developed a strategy of integration. The idea was to create a flexible technological mix by combining anthracite-fired thermal stations in or near towns, brown-coal-fired thermal stations on the coalfields, and waterpower stations in the adjacent hills, with the hope that the resulting regional system would meet daily and seasonal patterns in power demand (Todd 1999; Pohl 1992). The RWE expanded rapidly by pursuing a rigorous low-price policy, using the largest generators available to reduce coal consumption and pioneering the use of steam turbines. Stinnes, a typical "system-builder" in Thomas Hughes's terms (1983), aimed to turn the whole industrial basin from Aachen to Cologne, including the lower part of the river Lippe, into one gigantic energy and transport utility. To counter rising fears among city councils and state authorities about its monopoly behavior, the RWE actively cooperated with municipal and district authorities, offering them shares and seats on the board of directors. In the 1920s the RWE linked hydropower stations in the Austrian Alps with brown-coal-fired thermal stations in the Rhineland via a 220-kV transmission line. The RWE thus created an influential model of regional system-building (Gilson 1999) in which cities played only a marginal role.

Power struggles over urban hinterlands further weakened German cities' once strong position, even before World War I. Demographic changes in the late nineteenth and early twentieth centuries had accelerated suburbanization in the vicinity of larger towns and had transformed the economic and social character of nearby villages and small towns. As a result, energy networking had also entered their agenda. In search of new markets, gas companies and smaller electrical engineering companies actively sought concessions for gasworks and power stations in small towns and even in villages of as few as 3,000 inhabitants. Such companies often found themselves at odds with larger cities that were anxious to retain influence over their economic hinterlands. In the past, network technologies had proven highly attractive for residents of suburban villages when they negotiated their terms of incorporation. Frequently, incorporation contracts set clear timeframes for introducing running water, gas, and electricity (D. Schott 1999b). Because of the prohibitive costs involved in building extensive power networks or gas pipes in less densely populated and less industrialized hinterlands, many cities chose to form mixed public-private companies to develop regional energy networks and to modernize public regional transport (Buchhaupt 1994).

Regions in France also gained the upper hand over cities in the years before World War I. Private companies erected plants close to waterfalls, coalfields, or a coal-importing port to generate electricity on a large scale and at a lower price than local coal-fired power stations could achieve. This strategy made good sense for private companies, since power stations on municipal territory ran the risk of being municipalized once the concession expired. Thus, where waterpower lent itself to easy exploitation, regional networks developed at an early stage; in general they were more common in the mountainous south of France than in the north (Bouneau 1987). A pioneering regional company was the Société Méridionale de Transport de Force, which operated in the southwest of France.

In Scandinavia, waterpower also hastened the transition from city-centered to regional electrification. In Finland, urban electrification usually began with thermal power stations, but local governments turned to hydropower plants due to growing demand after 1910 and the coal crisis during World War I. In 1919 the city of Helsinki became a shareholder in the South Finland Power Company, organized by the forest industry. Already by 1923 more than 50 percent of the capital's electricity came from hydropower, and by the 1930s the share of hydropower rose to over 90 percent, when Helsinki also bought power from the state-owned Imatra power plant. Thus, the transition from the urban to the regional phase of electrification was mainly performed through a change from thermal generation to waterpower. In terms of actors, Finland presents a special case, because the forest and wood-processing industries possessed large waterpower resources and were highly dependent on power. Finnish industry successfully refused to subject itself to a national electricity plan and undercut several emergent national efforts by preemptively investing in its own rival system of hydropower plants and high-voltage transmission lines (Myllyntaus 1991). In this instance, established industrial structures remained in the driver's seat.

Geography, particularly in terms of the availability and distribution of waterpower, played a major role in the shape of electrification. In Denmark, which had very little available waterpower, electrification depended on thermal power stations close to bigger cities, and power companies gradually extended their networks across the surrounding regions. Rural cooperatives played a significant role in rural electrification, for the Danish state was not particularly interventionist. In Norway, where waterpower was abundant and could be tapped close to the centers of consumption around Oslo, regional electrification was often driven by local authorities that might form consortia to finance capital-intensive projects. In Sweden, on the other hand, where significant waterpower resources were situated far from the large cities, the state played an important role quite early in developing waterpower resources and constructing long-distance transmission lines as backbones of a state-controlled transmission grid (Kaijser and Hedin 1995).

Toward a National Grid

World War I introduced powerful incentives for electrification across Europe. The need to save oil products for military uses, especially in the oil-poor Central Powers, made it difficult for private households and other consumers to obtain paraffin for lighting. Gasworks and electric utilities suddenly faced an enormous upsurge in demand for lighting, and they were more than happy to comply as long as they could acquire the necessary fuel and construction materials for extending their networks. In Germany the number of small consumers of electricity effectively doubled during the war (Herzig 1992a, 133). The shift in industrial production toward armaments furthermore required large capacities for power generation to satisfy the demand for aluminum, munitions, and nitrogen. Toward the end of the war, serious shortages in coal and copper supplies temporarily stopped the expansion of power stations. Power cuts and imbalances, partly due to the lack of linkages between different regional networks, intensified the debate about the future organization of electricity provision. The electrical industry in fact called for state intervention, demanding that a national grid of high-voltage transmission lines be superimposed on the existing mosaic of local and regional utilities to ensure more efficient generation and distribution of electricity (Gilson 1998).

Coal shortages during the war also enhanced the attractiveness of waterpower for such favored regions as southern Germany, southern France, Switzerland, Italy, and Scandinavia. Higher costs of coal and the enormous increase in electricity demand made numerous waterpower projects economically feasible. Regional German states now established state-owned power companies, set up hydropower stations, and established distribution networks in the yet-unwired provincial regions. Companies such as Badenwerk, Bayernwerk, and Preussen-Elektra became powerful actors that competed with private and mixed regional companies (Stier 1999).

For cities, these developments spelled *dis*empowerment. They could neither exert the spatial control necessary for such large-scale regional projects nor raise the required capital. Their only solution was to try to compete with regional companies by entering into partnerships for the construction of state-of-the-art power stations serving multiple interests. In Mannheim the large power station built in 1923 downwind from the city served as a safeguard for local supply as well as a node for several regional utilities that needed to balance their seasonal requirements due to uneven waterpower (Hochreiter 1994). Mannheim's power station was constructed to utilize waste heat in a local district-heating network, an early instance of cogeneration. Local and regional utilities were often linked up with each other by means of multilayered supply contracts, in the absence of any formal state program. In the end, however, while many German cities still generated at least part of their own electric power, the bulk came to be provided by large power stations operated by regional utilities.

While relieving environmental problems, this out-sourcing led to cities becoming dependent on other companies, especially in pricing. Profits from electricity and gas acquired an even greater significance for German cities in the interwar years. The national tax reform of 1920 had stripped municipalities of other regular sources of income (Reulecke 1982; Krabbe 1989), and they therefore tended to cling even more obstinately to their utility revenues. Beginning in the late 1920s, cities faced a fierce front of powerful adversaries. The director of the German Central Bank, Hjalmar Schacht, attacked the practices of local authorities to overprice gas and power, and business interests and regional utilities challenged the economy and efficiency of municipal power generation. During the Great Depression, cities were forced to lower

Figure 8.3
The Mannheim harbor. Interaction with the Rhine, and rivals upstream and down, shaped Mannheim's industrial development and electrification (see chapter 2). Source: Wuttke 1904, II: 236.

Gas and Electricity in the Urban Environment

their rates in support of the national government's deflationary policies. At the same time prices were driven even lower by regional utilities facing massive overcapacities due to their oversize power stations.

The Nazi government made its own interventions in the power economy in 1935 by issuing the Energy Economy Law, drafted by the now Minister of Economic Affairs, Hjalmar Schacht. This law supported the forced rearmament policy and introduced far-reaching state regulation of electricity and gas industry just short of nationalization. Municipal energy policies thus became subordinated to the National Socialist goal of economic rearmament (Hellige 1986; Löwer 1992). By the late 1930s German cities had almost entirely lost their independent role in shaping the evolution of energy systems. New legal frameworks—which remained intact even after local democracy was reestablished in Western Germany after 1945—took away city councils' ability to control the operations of their municipal utilities.

In the United Kingdom power networks changed from urban to national control during the interwar years. Profits declined for most local urban utilities, as their networks stretched farther afield where customers were less dense. In turn, lower utility profits weakened municipal finances and thereby hampered towns and cities from resisting externally imposed organizational changes. The type of mixed enterprise exemplified by the RWE did not emerge in Britain; instead, the action shifted to the national level. In the context of coal shortages and debates about economic reconstruction at the end of World War I, the Williamson Committee, appointed by the Board of Trade to "consider the Question of Electric Power," proposed establishing district boards to coordinate the transmission of electricity between urban areas (Hannah 1979). A few years later the Weir Committee went even further and proposed a Central Electricity Board that would construct a national grid, close down small stations, and standardize the patchwork of different frequencies. This solution simply incorporated local utilities into a national system. In the decade between 1926 and 1935 a massive £20 million was invested annually in the electricity industry, and by 1933 a national grid of high-voltage lines interconnecting the regional networks had been completed. The result was a much wider usage of electricity; the number of consumers in fact rose roughly tenfold in this period (Millward 2000, 343). Nevertheless, the patchwork of 581 local electricity companies (in 1938) jealously defending their local monopolies was increasingly seen as counterproductive in terms of national economic efficiency.

Reports from several government committees in the 1930s and 1940s gradually established a consensus beyond the Labour Party that not only gas and electricity but also other industries, such as transportation, aviation, and coal, should be put under national control. The rationale was to realize economies of scale and make basic services available at uniform prices across the whole country. In 1948 a nationalization act removed ownership and control of gas and electricity from the hands of local governments, and instituted central authorities, organized in 12 regional boards. By the late 1940s British cities no longer had any autonomous or semi-independent role to play in the management and development of their energy networks (Hannah 1979). Although it might sound unusually radical, this policy could be found elsewhere in Europe—and not only in the East. Similar nationalizations also occurred in France and Italy in the immediate postwar period.

The All-Electric Home

While cities had sharply reduced room for maneuver, given the rise of regional public-private utilities and state-owned utilities in the interwar period, many cities still owned the network within their territory and thus could shape the process of diffusion to private households. Electrification of the home gained an almost messianic appeal. Although the initial fascination with electricity of the 1880s and 1890s had faded, a new range of electrical household appliances—such as the refrigerator, the vacuum cleaner, and stove, and particularly the extremely popular electric iron—had widened household uses of electricity far beyond lighting (Orland 1990). Furthermore, the social and cultural shocks of World War I transformed household work. Especially in the defeated countries of Central Europe, the middle classes had suffered a massive financial drain due to inflation and redistributive government policies. Few middle-class households were still able to employ maids and servants (Schlegel-Matties 1995). What is more, industrial rationalization and structural changes in the growth of services opened a wider range of respectable job opportunities for young women from middle-class families.

Both of these processes increased the profile of electrical appliances as mechanical substitutes for domestic servants. They would—at least according to the propaganda of the electrical industry—allow middle-class women to run their households without outside help while preserving their beauty and physique. Furthermore, the electric fridge would cut food waste and make daily shopping unnecessary, electric irons would prevent glowing coal particles from spoiling the laundry, and getting rid of the coal stove would banish coal and ashes and improve general domestic hygiene. Housing reformers, public health activists, advocates of women's emancipation, and marketing strategists of electrical engineering companies and utilities thus formed a powerful coalition to spread the "electric gospel."

By comparison, how did domestic electrification progress in interwar Britain, where the middle classes had not experienced such a massive erosion of their financial standing? A major reason for the relatively slow pace in British home electrification, if the case of Manchester is representative, was the reputation of electricity as a "rich man's energy" (Luckin 1990, 53). In 1924 only one in seven of Manchester households used electricity, predominantly those in the well-to-do suburbs. The slow but steady reduction in the "flat rate" and the increasing range of electrical appliances expanded this share of households to five out of seven by 1937. The most attractive tariff was available only to owners or long-term tenants of larger houses, however, and not to shorter-term renters. Because of this disparity, the Manchester Electricity Department faced heavy criticism in the 1930s. Laissez-faire critics challenged municipal economic activity on ideological grounds, while Labour Party councilors attacked the social exclusivity of electricity. The Conservative council majority reacted by introducing prepayment slot meters for electricity, which had been so successful for gas. The electricity department in some cases even paid for the wiring. Local measures such as these, backed up by the development of the national grid, almost doubled electricity consumption in British cities between 1933 and 1940 (Luckin 1990, 63).

The mass diffusion of electricity did not yet endanger the economic position of gas. The gas department of Manchester modernized its marketing and pricing strategies, and by the early 1930s offered a wide range of gas appliances for hire or sale. Protests erupted when the city of Manchester forced residents of new municipal housing estates to use electricity, and tenants found it difficult to control their consumption with the new technology. Once again, efforts by determined modernizers ran into stern resistance by citizens. Despite the overblown rhetoric of the electricity lobby, gas kept its lead over electricity in terms of customers and consumption through the late 1930s.

The All-Electric House, a major activity of the Electrical Association of Women (EAW), reflects the deep cultural significance of the promise of electricity. Established in 1924 when the national-grid debate was just getting under way, the EAW promoted the electrification of the home and pleaded for a greater voice for women. Above all, it embraced a progressive reform agenda of improving homes, making cities cleaner, and dispersing the population over a wider area. Linked by personal and financial bonds with the electrical industry, the EAW nevertheless maintained its own profile in highlighting women's specific expectations. It published a journal, *Electrical Age for Women*, established 30 branches by 1932, and sought to influence the way in which houses would be electrified by publishing a Women's National Specification.

The high point of EAW activities was the construction of an All-Electric House in Bristol in 1935. Supported by the attractive power rates of the expansion-minded municipality, this house demonstrated the comprehensive scope of domestic electrification within a wider reform agenda. The house featured all kinds of electric appliances and incorporated the very latest in domestic kitchen design. In its outward appearance, the house presented a striking contrast to most contemporary suburban houses, using an uncompromising modernist architectural language of cubic proportions, flat roofs, and large windows. The house had no traditional heating system whatsoever; even the fireplace was electric.

The All-Electric House generated much public interest. In one month, 20,000 people visited the house, while the utility and building press, as well as women's magazines, wrote generally admiring articles about it. However, cost calculations clearly showed that not everyone could afford all that electricity. The house had been constructed for

£1,000, a sum that only well-off professional families could afford. Its running costs in electricity were projected at a stiff £30 per year. And even these costs owed much to Bristol's cheap combined tariff (Reece and Roberts 1998). Other branches of the EAW did not succeed in replicating the Bristol example. Electricity was not yet the universal and cheap source of energy for housework that its promoters claimed. Although two-thirds of British homes were wired by 1939, the dominant use continued to be for lighting. Overall, the house remained "a machine heated by coal, coke and gas and powered by women" (Hannah 1979, 208).

Similar coalitions of pro-electricity forces, including women's associations, were at work in German cities. Whereas the interests of utilities and engineering companies need little explaining, the obsession of housing reformers with electricity is quite astonishing. In Frankfurt, one of the vanguard cities in housing reform in the 1920s and early 1930s, the famous "Frankfurt kitchen," designed by Grete Schütte-Lihotzky, was installed in more than 10,000 municipal apartments. The design of this kitchen translated principles of scientific management and Taylorist motion studies to the spatial organization of domestic work. The kitchen, reduced to a mere working space, was sized for just one person. With the goal of freeing housewives from unnecessary drudgery, the design was meant to minimize physical labor by placing all essential requisites within easy reach and equipping the whole kitchen with electrical appliances.

However, these supposedly scientific spatial arrangements ran counter to traditional patterns in working-class households. Women found themselves isolated in a merely functional space, unable to watch their children or to share the kitchen work with friends and relatives. And the all-electric kitchen in many of the new Frankfurt flats was far from ideal. Many tenants complained that cooking and heating water with electricity took much longer and cost more than with coal or gas. The new electric appliances were not always quite up to the task, but cultural adaptation and appropriation of the new technology were also necessary. Many tenants resisted this enforced modernization of kitchen work and energy technology, and some tenants even fought sustained battles with the municipal housing departments to reintroduce gas or coal stoves (Heßler 2001).

Overall, household electrification advanced quite rapidly in German cities. In Berlin, 55 percent of households were connected in 1928, almost double the share from 1925, and in other cities the percentage could be considerably higher (Orland 1990, 93–102). Nevertheless, household consumption of electricity in the interwar period still focused on lighting and small appliances like the electric iron. Electrical stoves, washing machines, clothes driers, and refrigerators found their way into private households only slowly. By 1939 only 800,000 stoves were connected, despite vigorous campaigning. And less than 1 percent of households could boast of a refrigerator in 1937. Mass diffusion of these appliances in American style set in only during the mid-1950s (Zängl 1989, 245; Hellmann 1990, 154; D. Schott 2006).

Empowering and Disempowering Europe's Cities

Residents' ambivalence about the Frankfurt kitchen might indicate more fundamental problems in the relationship of European cities with their energy networks. The installation of gas and electricity networks clearly empowered European cities in the late nineteenth century to illuminate city streets, to propel public transportation, and to power industry. In emphasizing a political reading of "empowering," however, one sees a swelling and receding wave. When cities municipalized utilities and developed visions of comprehensive town planning in the late nineteenth century, they progressively acquired a wider spectrum of instruments to intervene in urban developments and even to direct them. An interventionist stance can clearly be observed in German, Swiss, Scandinavian, and British cities. At a time when national governments expressed no interest in energy issues, city governments across Europe discovered how electrification could be used strategically to support a range of spatial and social policies.

Yet it was not technology per se, in terms of its automatic and intrinsic qualities, that empowered cities. How far the potentials of the technology could be utilized, and to what effect, depended on the institutional setting in which energy (and transportation) policies became embedded and how they were defined. For example, municipal intervention was motivated by a strong desire to reform industrial capitalism in order to stave off revolution and to mitigate physical degeneration by progressive housing, mass transportation, and urban planning (cf. Hård and Stippak, this volume). A municipal streetcar could support a middle-class leisure city as well as promote industrialization and workers' mobility, depending on political preferences and path decisions. Municipal ownership certainly seemed to give cities more freedom of maneuver to pursue their long-term planning goals.

This heyday of "empowered cities," roughly from 1890 to 1910 in the case of Germany, was followed by slow but steady disempowerment, in which technological, economic, and political factors interacted in distinct ways. Large-scale power stations were no longer compatible with the urban framework, owing to pollution, financial considerations, and their resource base. The expanded use of waterpower and brown coal brought private entrepreneurs and the regional states on the stage. Political competition for tax resources further limited the scope of municipal energy policy. Since the two world wars had strongly highlighted the strategic potential of energy, nationalization of the energy industry or corporative regulation under state control was regarded as a necessary step to control and direct local energy policies. Only in exceptional cases could cities continue to exert strong direct influence over municipal energy and supply to their citizens.

On a more general level of operating principles and structures of energy systems, one may ask what was lost in this process of disempowerment. After all, the path taken by European cities during these decades was surely not the only one imaginable, nor the one that cities must follow today. Some historians have argued that technological alternatives, such as decentralized energy systems that make better use of waste heat through cogeneration, might have been feasible in the 1920s and 1930s. Recall that Mannheim used "waste" heat to generate electricity as early as the 1920s. If such schemes were feasible elsewhere, greater local or municipal activities during these decades could have significantly altered the mix of energy sources. In light of today's energy-related problems and the quest for the sustainable city, it might be high time to rediscover and realize such alternatives, to reempower European cities through a sustainable local energy policy (Hewett 2001).

9 In the Shadow of the Factory: Steel Towns in Postwar Eastern Europe

Dagmara Jajeśniak-Quast

Socialist cities in Poland, Czechoslovakia, and the German Democratic Republic (GDR) created a distinctive "type" of European urban development. Springing up behind the Iron Curtain, these cities were subordinated to the priority that socialist industrialization placed on heavy industry, in particular iron and steel. Although such cities have been intensively studied, and we have learned a great deal about their social history and architecture, we know too little about their character as industrial cities.[1] The specific relationship between the factory and the city, while obvious, is nearly unknown. This is a significant omission since factories strikingly dominated many aspects of city life during early socialist industrialization (see figure 11.3).

This chapter sets socialist cities into a broad European context by discussing their origins and development as well as their similarities with and differences from other industrial cities in Europe. It highlights the circulation of ideas, technology, and people between West and East as well as within the Eastern European countries. In addition, this chapter discusses the political effort to differentiate socialist cities from capitalist cities in the West. Finally, the chapter outlines the paradoxical use of nationalism in legitimizing the new socialist cities.

Modern European cities depended in many ways on technologies that circulated internationally, as this volume emphasizes. Socialist cities that developed in Eastern Europe in the 1950s were no different. What is distinctive about them is that they depended to an extreme extent on heavy industry for their location and employment. For ideological reasons, they were shaped by preexisting Soviet urban planning models that explicitly displaced capitalist models (Ludwig 2000, 42). They are also a recognizable "type." The attentive visitor can even today see many similarities between three leading examples: Nowa Huta in Poland, Eisenhüttenstadt in the former GDR, and Kunčice in the Czech Republic. Most striking are the similarities in their city centers from the 1950s, when construction behind the Iron Curtain was defined by the elaborately decorated Stalinist architectural style and its embrace of "national tradition" (Pehnt 2005, 287). All the same, socialist cities and socialist urban planning were often based on modern, Western planning ideas. In fact, many architects from the United

States and Western Europe worked on socialist urban planning in the 1930s. Modernist architecture and urban planning were, somewhat ironically, nearly perfected under strongly national influences within socialism.

For some time, scholars have explored the nationalist legitimation of the supposedly internationalist communist system.[2] This nationalist appropriation also applies to socialist architecture and urban planning (Arvidsson and Blomqvist 1987). The three socialist cities demonstrate the appropriation of Western ideas and concepts within specific historical, national, and local aspects of architecture and culture. For example, Nówa Huta contains Renaissance elements, so often found in Polish architecture. While Soviet architecture rather grandly directed that "the characteristics of the 'golden age' would be transferred from the classical past into the concept of the future" (May 1999, 98), it was left to Polish planners to select the particular golden age, which in Poland was the sixteenth century. Similarly, Eisenhüttenstadt followed the 16 general principles that the GDR government adopted on July 27, 1950, including the overall Soviet dictum that "Architecture must be democratic in its contents and national in its form. In doing this, architecture applies the experience of the people embodied in the progressive traditions of the past." Yet again it was local architects, this time German, who identified which specific traditions to deem progressive and how these might be considered democratic and national. In the early years of the GDR, classicism was associated with the French Revolution and the contemporary progressive tendencies in the German bourgeoisie. Of course, these choices hardly stand above history; similar classicism can be found in public buildings in the United States and in Europe's authoritarian regimes of the 1930s (Ludwig 2000, 44).

The circulation of ideas about city development relied on the substantial mobility of architects and urban planners themselves (Misa, this volume). Especially during the global economic crisis of the 1930s, numerous prominent architects and planners from the United States and Western Europe took active roles in Soviet industrialization and urbanization. In addition, there were study tours by experts to Western Europe and the United States. Contemporary trade journals and books reported extensively on these experiences (Bodenschatz et al. 2005, 47). Twenty years afterward, in the 1950s, the Soviets in turn brought these models and ideas to Eastern and Central Europe.

Socialist cities in the 1950s repeated the universal experience that industrialization structured cities in distinct ways. First of all, in a direct echo of nineteenth-century company towns, each socialist city depended on a single industrial company and indeed was established as a residential community for that company's iron and steel workers. The choice of location resulted from favorable conditions for the forced development of heavy industry. As a result, a special city-plant relationship developed. For two decades or longer, the socialist cities and their factories were inseparable.

Second, the deliberate ideology of socialist transformation and central economic planning marked these cities. Socialist cities, even more than other European cities, reflected the large-scale political changes in Europe. It was consequential that Eisenhüttenstadt, Nowa Huta, and Kunčice represented the first socialist cities of the GDR, the People's Republic of Poland, and the Czechoslovak Socialist Republic (ČSSR). Explicitly political decisions shaped these socialist cities much more than did the simple expansion of population or the building of infrastructure. Their city centers epitomized the influence of socialist ideology. Quite strikingly, there was no traditional market square, owing to its historic association with capitalism. Instead, a new center was designed for state-sanctioned political rallies, military parades, May Day demonstrations, and other forms of collective life. Clearly exercises in prestige, the new cities expressed socialist power (Aleksandrowicz 1999). Socialist cities founded after the Second World War were exemplary in this regard. During the 1950s sizable investments flowed into them, while far less was spent on older urban centers, despite their own great need of reconstruction from wartime damages. In 1955 and 1956, more than two-thirds of the 18,000 residential units built in the metropolis of Kraków went to the relatively small industrial district of Nowa Huta (Chwalba 2004, 69). Ideology also directly shaped the inhabitants, who were to form the "new man," recruited from the class of peasants and workers, and separated accordingly from the feudal and reactionary past. The factories and the cities were to become visible symbols and material manifestations of these changes in society.

One expression of ideology in these new cities were the socialist work competitions that took place on the construction sites. Thousands of young people took part in paramilitary youth organizations in this "race for socialist deeds." The cities themselves followed ideals broadly similar to classic modernistic urban planning in Western Europe, with the basic goal of providing comfortable and healthy housing. The houses were built at a generous distance from one another to guarantee each resident a maximum of light as well as garden space. Yet "healthy" socialist cities were intended to sharply contrast with the "unhealthy" capitalist workers' slums. Planned for 5,000 to 6,000 people, individual residential areas within these cities featured a comprehensive infrastructure with such facilities as a department store, Culture House, schools, kindergartens, cinema, library, and a club house. Each house was even to be equipped with loudspeakers and a telephone.

The new cities embodied socialist industrialization on the symbolic level, too. In the 1950s they often bore the names of communist leaders; thus, the newly built district of Ostrava became Stalingrad. At the end of 1951 the new plant in Ostrava was named for Klement Gottwald, president and first secretary of the Communist Party of the ČSSR. On May 8, 1953, the new steel plant in the GDR was named after Stalin, just a few months after he died, as was its newly built city, Stalinstadt. The plant regained its old name (Ironworks Combine East) in 1961, when the town of Stalinstadt along with two adjoining districts was renamed Eisenhüttenstadt.[3] In Nowa Huta the steel plant in 1954 became the Lenin Steel Mill on the thirtieth anniversary of Lenin's death. The mill retained this name until 1989, when the workers' council renamed it the Tadeusz Sendzimir Steel Mill after a prominent Polish inventor-metallurgist who lived and worked in the United States from 1939 and who (as discussed below) was personally involved in providing a circulation conduit to Poland for American industrial technology in the midst of the Cold War (Choma and Gil 1999, 22, 47).

Finally, all these high-profile, state-directed investments were called "new steel mills." The adjective "new" was intended to represent the new epoch to the outside world. Archive records for the Ironworks Combine East (EKO), in Stalinstadt/Eisenhüttenstadt, label it explicitly as an antithesis to the "old" Maxhütte in Unterwellenborn, Thuringia.[4] These names formed a signpost, sometimes literally, to distinguish the "new" socialist factory complexes from the "old" capitalist ones.

Overall, a complex jumble of difference and similarity can be seen in comparing the socialist cities with other European cities. Socialist cities can look back on an urban history of at most fifty to seventy years, a mere blink in the eye compared with Europe's long urban history. Central planners certainly meant to differentiate socialist cities from Western, capitalist ones. Nevertheless, and in spite of the Iron Curtain, influences and technical solutions from Western Europe and the United States certainly found expression there. Initial plans for the socialist iron and steel mills explicitly drew on Western technologies and sometimes even on Western raw materials. In many cases, experienced engineers from older, prewar cadres acted in accordance with Western-style economic premises rather than socialist ideology. The political situation in Europe after the crises of 1947–1948 was a watershed. As detailed below, the worsening East-West conflict resulted in planners' relocating these plants and cities. The original openness to Western technical know-how and raw materials also changed, with socialist cities looking instead to the Soviet Union. Yet the city and mill designs retained some of their Western features even after these political changes. By the early 1960s the liberalization of political and economic relationships within Europe provided a broader base for East-West cooperation in the field of technology.

Location Decisions for the New Iron and Steel Cities

Stalinstadt

In the Soviet Occupation Zone that subsequently became East Germany, the consequences of the Second World War governed the location for the iron and steel industry. Compared with Poland or Czechoslovakia, the eastern part of divided Germany had an especially pressing need for a new iron and steel mill, for most of the iron and steel industry was either in the western occupation zones or in the eastern territories lost to Poland. The only steel mills in the city of Brandenburg and in Hennigsdorf, and the Maxhütte in Unterwellenborn, had been destroyed, dismantled, or had insufficient capacity to satisfy the urgent demands for iron and steel. In August 1949, the Blast Furnace and Low-Shaft Furnace Commission introduced the idea for a new iron and steel mill for eastern Germany. The early plans indicated that the projected East German plant would import iron ore from Sweden, the traditional source of this key raw material also commonly used across Poland and Czechoslovakia. Coal would come mainly from West Germany and Czechoslovakia. For this reason, the commission suggested a location close to the raw materials—near a Baltic port or West German coal supplies.

However, the increasingly severe political tensions dramatically altered these plans. In the aftermath of the political crises of 1947–1948, West Germany, especially after it was officially established as a separate state, could no longer be viewed as a reliable source of vital raw materials. The small town of Fürstenberg on the river Oder, near the newly created German-Polish border and on the Oder-Spree canal, was selected as the site for the new steel complex and associated city. Officially, "three essential principles of socialist location planning" favored this location. These principles stressed close proximity to raw material sources, favorable transport routes, and installation of industrial centers in farming areas to remove the "essential difference" between town and country.[5] With the tightening of the Western embargo in 1950, raw materials for the new iron and steel mills came exclusively from the East. The changes in raw materials, with ore to come from the Soviet Union and coke from Poland, now favored a factory location close to the German-Polish border. The third official argument—reducing differences between urban and rural areas—was a standard goal of socialist industrialization.

The location on the German-Polish border was significant also as a way of constructively absorbing people expelled from Germany's former eastern territories (Niethammer et al. 1990; Schwartz 1999). After the war, a large number of expellees settled in the region along the new German-Polish border in the hope of returning home (Jajeśniak-Quast and Stoktosa 2000; Jajeśniak-Quast 2001a). They created a worrisome presence along the border. Its new location in this very region made Eisenhüttenstadt into a symbol of the friendship with Poland and the USSR. New jobs and homes were to be created as quickly as possible to avoid a concentration of dissatisfied people who might threaten the new socialist system. The expellees—described by the state's official language rules as "repatriates"—were to form the workforce in the GDR's first socialist town.

Military considerations were also important in eastern Germany, as everywhere else during the Cold War. The changed political situation after 1948 significantly accelerated investment in heavy industry. The new iron and steel mill also had to be built as far from the West as possible. The final decisions in favor of Fürstenberg on the Oder and the location for the EKO residential facilities were adopted in 1950 by the German

Socialist Unity Party (SED) and the Council of Ministers. In the beginning, the prefabricated facilities south of the works consisted mainly of buildings from the former "repatriates" camp in Frankfurt-Gronfelde. On August 18, 1950, Industrial Minister Fritz Selbmann broke ground for the new works. Six months later, the foundation stone was laid for the EKO residences and work was started on the first residential block. Early in 1953 the Council of Ministers of the GDR designated the much expanded EKO dormitory facilities as an independent city.[6]

Nowa Huta

In 1945 work began on the largest postwar Polish heavy industrial investment. Discussions had started about developing the Polish iron and steel industry shortly after the end of combat. As early as July 1945, the Commission for the Development and Rationalization of the Iron and Steel Industry drafted a proposal for constructing a new, modern steel mill in the traditional industrial region in Dzierżno, Upper Silesia, on the Gleiwitz Canal (Kanał Gliwicki). Zygmunt Wiadery, the engineer, drew on prewar German plans for an iron and steel mill in Laband (today Łabędy) that would produce one million tons of steel annually.

In the first three-year plan for industrial development following World War II, Poland assigned the new factory complex a "bridging function" between West and East. Polish industry would retain significant contacts with the West through the new iron and steel industry project. Its technical design was intended to make the new steel mill independent of any one type of iron ore, to allow imports from Sweden, Brazil, Morocco, or the Soviet Union. The original plans also called for importing technical know-how and purchasing rolling mills from the United States. The steel mill would employ 5,000, with new housing built for 2,500 workers. The commission even awarded the project's contract to the Freyn Engineering Company from Chicago. Construction costs were estimated at $200 million, with a loan of $23 million to be secured from the World Bank (Salwiński 1999, 82–84).

The events of 1947–1948 intervened forcefully. Plans for the new steel mill changed drastically after Poland, on orders from Moscow, rejected the Marshall Plan and signed the Polish-Soviet Trade Treaty in January 1948. The trade treaty provided for Soviet assistance in Poland's industrial transformation, which would now include specifically Soviet technical plans and a corresponding industrial investment policy.[7] Needless to say, Poland would no longer serve as an East-West bridge. The American embargo of 1948 sharpened the divide by preventing so-called strategic goods from being exported to Eastern Europe. The embargo prevented the new venture's rolling-mill equipment from leaving the United States, even though it was already paid for (Kaliński and Landau 1998, 246; Loreth 1999, 171–172).

Just as with Eisenhüttenstadt, the emerging Cold War tensions altered the location for the Polish steel mill and its city.[8] In January 1949 a delegation of Soviet experts visited Poland. Together with Polish colleagues, the group of Soviet engineers, architects, and urban planners evaluated eleven locations in southern Poland between the Oder and Dunajec rivers. The delegation specifically rejected the original location on the Gleiwitz Canal, ostensibly because it was unsuitable for the new, higher capacity steel mill. On February 21, 1949, the delegation visited an area just east of Kraków. Soviet experts judged this region as the most suitable for the new steel mill (Salwiński 1999, 91), and within days, the Minister of Industry and Trade signed off on the new location.[9] The Kraków region had much to recommend it, but most Polish experts wanted the mill placed much farther from the center of Kraków. They noted several liabilities with the location chosen by the Soviet experts. First, the site near Kraków did not create jobs where they were most needed. Second, the site's exceptionally fertile soil had for generations provided Kraków with high-quality vegetables, and it was clearly a waste to have prime agricultural land covered by an industrial plant. Finally, the decision would force the relocation of Kraków airport.

The rapid decision in favor of the substitute location near Kraków was something remarkable. Whereas the preparatory work for the original location on the Gleiwitz Canal had lasted two years, the decision to build a huge metal-working combine near Kraków took scarcely a month. The first geological tests were done only after the location decision was already made. Furthermore, construction work at the site began while the project was still on the Soviet planners' drawing boards. The city plan was not confirmed until March 1951, when construction was already well along. This extreme haste resulted partly from the sweeping political and economic upheavals, but scholars agree that ideological factors were decisive (Czubiński 1992, 218; Tokarski 1999, 138; Pawlitta 1979, 100; Bogdanowski 1996, 14; Chomątowski 1999, 50; Delorme 1995, 35, 69). A key point was that the arrival of massive numbers of young industrial workers was designed to alter the social structure of Kraków; considerations of this nature were discussed openly. The six-year plan stated that "a strong proletarian pillar was to be guaranteed for the capital through the development of industry in Warsaw and its environs."[10]

Kunčice

As in Poland, prewar plans were influential for the new iron and steel mill in Kunčice, the largest single investment in the Czechoslovak five-year plan. However, Czechoslovakia actually adhered to its prewar plans in choosing Kunčice, while increasing the scale of the enterprise owing to the imperatives of socialist industrialization. Prewar plans had called for expanding the hundred-year-old iron works in Vítkovice (Vítkovické Železární). Bordering on Vítkovice to the south and including the open space on the opposite bank of the river Ostravice, Kunčice's excellent location was also close to the coking plant at Ostrava. However, the threat of war in 1938 ended the plans of the then-owners of the Vítkovice ironworks, the Rotschild-Guttmann family. Instead, operations were partially modernized and some facilities turned to armaments production. During World War II, the Vítkovice ironworks—as part of the Hermann Göring Works—mainly produced ammunition and components for the V-2 rockets (Kána et al. 1966, 149).

On October 20, 1942, construction of additional installations began in the district of Kunčice. At the end of the war, construction employed over 1,000 workers. October 20 would have been the unproblematic birthday of the new Kunčice plant, except for unforeseen postwar events. Immediately after the end of fighting around Ostrava, some 12,000 German prisoners were detained in the new plant's buildings and shops.[11] Then, as a result of the treaty of March 31, 1945, between Czechoslovakia and the Soviet Union, the newly constructed plant was designated as war booty and transported wholesale to the Soviet Union.

A decree by President Edvard Beneš on March 23, 1946, nationalized the remaining buildings, and the remnants of the Kunčice plant were incorporated into the Vítkovice ironworks, which itself was nationalized as the fourteenth plant.[12] Soon, the Central Office of Czechoslovakian Iron and Steel Mills directed the Vítkovice ironworks to prepare construction plans for a new plant with an annual capacity of 1.5 million tons of crude steel. When the Ministry of National Defense identified the Vítkovice ironworks and the planned mill as having special significance for national defense, armaments production gave the plant a special position in the Czech economy and a priority for investment, similar to the Polish case.

The First Socialist Cities, Then and Now

Like the steel plants, the first socialist cities in Eastern Europe—Eisenhüttenstadt, Nowa Huta, and Kunčice—began as showpieces of the new socialist society. They were to provide for all the needs of workers, including social and cultural institutions such as municipal common facilities, the house of the party and its subsidiary organizations, and the all-important central square. Obviously, in each of the three cases, the heavy industrial plant provided the essential reason for building the city (Apolinarski 2005, 323). All the same, planning for each city expressed slight differences. A church was originally planned for Eisenhüttenstadt,[13] although it was not built until after 1989. Nowa Huta struggled for and won a church as early as 1961. In Nowa Huta, as in Eisenhüttenstadt, the main street was to connect the city center with the plant and blast furnaces at the other end (see figure 9.1).

Socialist ideology came to govern many aspects of urban design. "Small urban planning units" subdivided each of the three cities into residential complexes with homes for about 4,000 to 6,000 people and "stores for everyday needs and for services, as well as institutions for the education and health of our children."[14] In the beginning, be-

Figure 9.1
Stalinstadt's main street ran straight to the steelworks on the horizon. Source: Fricke/ Dokumentationszentrum Alltagskultur der DDR (Eisenhüttenstadt).

fore the central planning directives really took effect, houses were built with bricks and with relatively generous fixtures and fittings. Each home had a large kitchen and a bathroom with a window, and rooms even had parquet wooden flooring. Socialist transformation ended such "imperialist" construction methods, as was stated in the case of Eisenhüttenstadt:

> The imperialist philosophy in the field of architecture is known as functionalism, i.e., we have to start from the function of a house and from the needs. Building therefore takes place from the inside to the outside, and the unimportant enclosure of the cooker, the bed and the WC is created. Beauty and dignity are unnecessary. The construction is cheap and profits are high. This period of decay is now behind us!... Ahead of us in a happy future lies a new epoch, and we have every reason to believe that the greatness of our era will also be expressed in the architecture of our new town.[15]

This ideal of a new epoch was pursued at the beginning of the 1950s. But in reality, for some buildings, no expense was spared. Workers referred to the lavish administrative center for the new Polish iron and steel mill as the Polish "Vatican." Its construction costs per square meter (at 553 zlotys) exceeded those of the Warsaw Philharmonic Hall including all of its lavish decorative materials! Even the landmark Warsaw Palace of Culture cost a third less per square meter. At the time, the average investment cost for office buildings in Poland was around half the cost of the Polish "Vatican."[16]

Figure 9.2
Small urban units in Nowa Huta in the 1950s. Source: Muzeum Historyczne Miasta Krakowa, Oddział Nowa Huta.

As the iron and steel plants expanded, however, the dictates of ideology gave way to the pressing practical demands for new housing. At the end of the 1950s, Eisenhüttenstadt turned to large prefabricated slabs that were a substantially cheaper method for construction. By the end of the 1960s the prefabricated slab had become standard. Today, this *Plattenbaut* housing, consisting almost completely of prefabricated buildings made of concrete slabs, is one of the most challenging legacies of socialist industrialization. The stark difference between the "old" and the "new" houses can still be seen in all three cities.

The Soviet Union used the prefabricated slab method extensively. Curiously, the technology of constructing houses from prefabricated semifinished products had been developed in France, Sweden, and Finland in the early 1950s (Zblewski 2000, 170). Initially Nowa Huta held back, with both the director of the industrial association for constructing the plant and the director of city construction opposing the use of prefabricated or semifinished products (Żerebecki 1955, 8). However, the enormous expansion of the iron and steel mills in the 1970s generated a huge growth in the city's

Figure 9.3
The first prefabricated building in Eisenhüttenstadt in the 1960s. Source: Wolfgang Timme/ Dokumentationszentrum Alltagskultur der DDR (Eisenhüttenstadt).

workforce and created severe housing shortages. To deal with the problem, Nowa Huta turned to relatively quick construction methods using prefabricated slabs. Standardized construction could supply new housing quickly and cheaply in the required volume. These Western solutions were still adapted to local needs. For instance, Nowa Huta built "Swedish blocks" without basements.

The situation in Ostrava was similar. It had been an industrial city since the end of the nineteenth century, but the drive for socialist industrialization brought a huge wave of new workers from all over Czechoslovakia. Obviously they needed residential areas. In the 1950s, new housing projects developed next to traditional workers' housing. As in Nowa Huta, beginning in the 1960s both traditional and prefabricated construction methods expanded the housing stock.

Even today the prefabricated buildings are in demand in Nowa Huta, since Poland still lacks affordable housing. In Ostrava most of the population lives in the new prefabricated housing projects. However, the population today is declining in the older workers' settlements (Jiřík 1993, 711). Eastern German cities such as Eisenhüttenstadt have many vacant homes as a result of the exodus from the city and from the *Platte—*

housing projects built of prefabricated concrete slabs. The exodus that afflicts Eisenhüttenstadt might also be the future of Nowa Huta or Kunčice, for the massive turn toward prefabricated building construction in the 1970s resulted in inexpensive housing that now desperately needs costly renovations. The required funds are simply not available because most housing cooperatives are deeply in debt. In Nowa Huta's largest cooperatives, up to 30 percent of the residents—including many former workers from the iron and steel mill—have not paid rent for years or are unable to service the loans they took out to purchase their apartments. In a vicious circle, a large concentration of residents depends on a single company and carries the heavy burden of transformation to a market economy. This is a major sociopolitical challenge, not only for the cities themselves but also for the surrounding regions.

Cities and Plant, Residents and Workers

The new cities were entirely subordinate to the steel plants, especially in the 1950s. The iron and steel plants most obviously dictated the pace of urban life through the rhythm of their shift work. Moreover, in accordance with the character of a socialist city, all municipal utilities were state-owned and closely associated with the plant. The steel combines owned not just the industrial plant and all ancillary operations for supplying the workforce with food and consumer goods; they also owned the power plants, the water supplies, and the greater part of the urban housing and social institutions. In Eisenhüttenstadt the iron and steel mill owned the city heating plant, the large bakery, the meat combine, a trucking company, a motor vehicle repair shop, and an artisanal service provider. The city itself was built almost exclusively with building materials produced in the EKO plant; for example, lightweight pumice slag brick for walls, ceilings, and roofing tiles came from the plant's discarded blast furnace slag.[17] Even the children participated in the special relationship between city and plant. Worker brigades sponsored nearly all school classes as well as made possible many school festivals and trips.[18] In all three cities, the steel plants provided crucial support for many major municipal projects including stadiums, Culture Houses, and schools.

The plants also determined the size of the cities. As the plants grew, so did the cities. Most people in the newly created cities were also employed, directly or indirectly, by the plants. Nearly every family had a direct relationship to the iron and steel mill, as the example of Nowa Huta shows. Compared with the other two locations, the population of Nowa Huta and the workforce of the new iron and steel mill grew mostly from migration during the period of socialist industrialization. Even though it initially lacked a traditional working class, Nowa Huta developed into a typical workers' city with relatively small differences between classes and occupational groups. Another typical feature was the high proportion of men in the town, especially in the 1950s

Figure 9.4
A shop in Nowa Huta in the 1960s. Source: Muzeum Historyczne Miasta Krakowa, Oddział Nowa Huta.

(see table 9.1). Nowa Huta had many workers' hotels, occupied mainly by male construction workers.

With a large population of young adults, the early socialist cities bore considerable social costs. Torn from familiar environments, young workers frequently experienced difficulties coping with the strain of living in an unfamiliar place and working in often-hazardous conditions. Alcoholism and crime became major problems in all three cities. Time and again, young people formed diverse subcultures that were not regarded favorably by the party and decision makers. In Nowa Huta in the 1950s, the young *Bikiniarze* wore bright, conspicuous clothing, rejected political and social standards of the socialist system, and identified with Western and American mass culture. This rebellious attitude manifested itself in music and fashion (Kuroń 1989, 44; Lebow 2002, 2005).

Over time, municipal administrators of the socialist cities gained a measure of independence and began making many municipal decisions. Thus, the relationship

Table 9.1
Population and Number of Workers in Nowa Huta, 1950–1970

	Population of Nowa Huta (Town or District)			Number of Workers in the Plant		
Year	Men	Women	Total	Men	Women	Total
1950	10,926	7,920	18,846	567	166	733
1955	44,741	37,371	82,112	13,205	2,884	16,089
1960	54,400	46,690	101,090	16,709	2,276	18,985
1965	65,605	56,605	122,210	23,321	3,479	26,800
1970	85,115	76,304	161,419	26,864	4,660	31,524

Sources: Borkowski (1976, 22) and data from the Personnel Department of Huta im. Tadeusza Sendzimira S.A.

Figure 9.5
A market in Nowa Huta in the 1950s. Source: Muzeum Historyczne Miasta Krakowa, Oddział Nowa Huta.

between city and plant solidified: the city's development path moved from great dependence in the creation phase to a degree of municipal autonomy in later years. Again, the three cities experienced striking differences. Where the new municipality became a district of a larger city, as in Kraków or Ostrava, it depended far less on its steel plant than did Eisenhüttenstadt, where the complete dependence of city and steel plant persists even today.

Even though the plants continue to drive the demographic situation, differences remain between the three cities. As they experience reductions in the workforce, the cities and their districts shrink as well. Although Kraków still has a positive demographic balance, Eisenhüttenstadt—once Germany's youngest town—is rapidly aging. In 1960, its average age was 27.5. Now, Eisenhüttenstadt suffers alarmingly from outward migration, especially of the young. In the fifteen years since 1989, it lost almost 15,000 residents. Unlike Nowa Huta and Kunčice, Eisenhüttenstadt was built on a greenfield site and can fall back only on the town of Fürstenberg (Oder), which is rich in history but very small. Today, because of the lack of jobs, Eisenhüttenstadt with its 38,000 inhabitants cannot absorb the population as can larger cities like Ostrava with 315,000 citizens or Kraków with 750,000 residents, which incorporated Nowa Huta in 1951. In those cities, inhabitants can take up work in other industrial locations, services, or universities, whereas in Eisenhüttenstadt, residents face a grim choice of out migration or social decline. Around 18 percent of the homes in Eisenhüttenstadt are vacant, a situation entirely inconceivable in Kraków or Ostrava, where owning a home remains a goal, especially for young people.

Despite these severe challenges, these three steel towns even today are not completely deindustrialized, in sharp contrast to many districts in Western Europe or even the United States, where fully half of the entire country's steel industry jobs vanished in the 1980s. In hard-hit Eisenhüttenstadt, the steel mill employs less than one-third of its peak (1989) workforce, but even this is hardly the complete mill shutdown that devastated many of the mill towns in the Pittsburgh region. While Eisenhüttenstadt lost 25 percent of its population in fifteen years, the worst-hit Pittsburgh mill town (Braddock) lost nearly 40 percent of its population in just ten years. Ironically enough, the capitalist global economy may well be the savior of Eisenhüttenstadt, Nowa Huta, and Kunčice. Today, after many twists and turns since 1989, all three locations are divisions of Arcelor Mittal, the world's biggest iron and steel producer. For the moment, continued prosperity for these steel communities depends on the continuing steel boom driven by industrialization in India and China.

The Circulation of Technology

During the much-vaunted socialist transformation, each country in effect forced its own industrial development. Industrialization drew on raw materials and capital goods of the individual members of the Council for Mutual Economic Assistance (COMECON). After the hardening of the Cold War in 1948, COMECON stressed the significant possibilities that might flow from close cooperation among the socialist planned economies.[19] The Iron Curtain made a socialist division of labor and integration an economic and political necessity in Eastern Europe as well as formed an alterative to integration into the West (Jajeśniak-Quast 2002). The USSR controlled the key sectors of iron, steel, and energy during the Cold War. In the case of Poland, this affected both iron ore and production facilities. Similar to the transition of today's market economy, the geographic focus of foreign economic relations changed rapidly in late 1947 and early 1948.

Poland's rejection of the Marshall Plan in 1948 ended cooperation with America in the construction of the new iron and steel mill. Like the Soviet Union in the 1930s, Poland depended on foreign technology to develop its heavy industry. Thus, the Polish Board of the Central Association of the Iron Industry contacted experts in the Soviet Union for help constructing the iron and steel mill. In place of American engineers, the Polish-Soviet Treaty on the Supply of Industrial Installations of January 1948 sent Soviet engineers from Gipromez in Moscow to design the mill.[20] Besides planning the new iron and steel mill, the Soviets would also supply equipment. This change in direction surprised and even overwhelmed Polish engineers. Even as they tried to familiarize themselves with Soviet metallurgical processes, they experienced many problems in acquiring the new technology. They grappled with such problems as radically overdimensioned Soviet blast furnaces, the conversion of existing furnaces to high-pressure technology, and the construction of blast furnaces without what they considered proper cooling boxes.[21]

Moreover, the Soviet technology circulating across Eastern Europe remained recognizably American in origin. The technology for Magnitogorsk, the most important iron and steel mill during Soviet industrialization in the 1930s, came from Indiana and Ohio. The engineering and contracting firm of Arthur G. McKee of Cleveland, Ohio, developed the plans for this pioneering Soviet installation, and it set the pattern for others that followed (Bacia 1994, 8; Sutton 1968, 1971; Kirstein 1984). German architects associated with Ernst May from Frankfurt designed the city of Magnitogorsk. Indeed, many U.S. and German specialists, engineers, and architects became enthused with socialist industrialization in the 1930s (Bodenschatz and Post 2003, 29). The West was plagued with depression, while Soviet industrialization was in full swing (Scott 1942; Schattenberg 2002, 253; Lewin 1985, 57 ff; Davies 1997, 70; Becher 1959, 72, 105). In 1930, 4,500 foreign workers and experts—including 600 architects from Germany and 1,000 specialists of various sorts from the United States—were working in the Soviet Union. Toward the end of the first Five Year Plan, the population of foreign experts and their family members increased to around 35,000 (Šarapov 1976, 1113).

Along with at least 30,000 forced laborers—including "kulaks" opposed to collectivization and other victims of Stalinism—voluntary workers from all over the world, including the young German communist Erich Honecker, worked in Magnitogorsk.

Although coming originally from the United States via the Soviet Union with a twenty-year delay, the technology for Nowa Huta had experienced modifications. About 800 Gipromez employees prepared the project and brought with them twenty years of experience in building and developing the Zaporozstal steel mill. In sharp contrast to industries in the United States, the new steel mill was conceived from the start as a state-owned enterprise. And, significantly, the original Soviet ideal was adjusted to Polish conditions. At Magnitogorsk the Soviets had shipped in workers from over 1,000 kilometers away (Kotkin 1992, 1995; Schlögel 1995; Kirstein 1984). Since that system of forced labor was simply not possible twenty years later in the GDR, Poland, or Czechoslovakia, the smaller scale and local conditions had to be taken into account. In contrast to the model Soviet cities of the 1930s, built from scratch in the middle of nowhere, nearly all the socialist cities of the 1950s were closely linked to such existing cities as Kraków, Ostrava, Frankfurt an der Oder, or Berlin.

The three Eastern European locations depended to different extents on Soviet technology. In Eisenhüttenstadt and Kunčice, the influence was much less than in Nowa Huta, especially in the 1950s. Each site depended more or less equally on raw material supplies from the Soviet Union. In deciding where to locate the plant in the GDR, planned deliveries of raw materials played a decisive role. On the other hand, especially when compared with Poland, Soviet technology was less important for the Eisenhüttenstadt/EKO plant, since there was a long tradition of iron and steel working in Germany. For instance, in the early 1960s on a site visit, the Soviet chairman of the COMECON Standing Committee on the Black Metallurgical Industry objected to the blast furnaces in Eisenhüttenstadt being of purely German design. He attempted to convince the EKO managers to install a supposedly "more solid and more efficient" Soviet furnace.[22]

The fact that some mills in the Soviet Union were off limits to foreign visitors made relations with the USSR rather difficult. For example, in the mid-1960s, the EKO tried to evaluate whether the rolling-mill experiences of the Iljitsch Works in Sdanow could be transferred to its facilities, but its study was stymied because foreigners could not enter the Soviet plant.[23] Polish specialists faced similar problems and depended much more heavily on Soviet technology than did their German colleagues. In fall 1949, Polish managers from Nowa Huta paid their one allowed visit to the Zaporozstal Soviet iron and steel mill, which served as a model for the Polish mill.[24]

In the Czech case the new investments, which were originally discussed only for the development and modernization of the existing old mills in Ostrava, were redefined and extended after the socialist turning point in foreign economic relations. In

Kunčice, as in the GDR, the influence of Soviet technology was initially much less than in Poland. This mill drew above all on the experiences of the neighboring Vitkovic ironworks, and therefore on domestic Czech technology.

Although the Cold War and the embargo abruptly interrupted East-West contacts, the three steel plants repeatedly sought cooperation with Western partners. The embargo actually hurt Western companies that depended on the interchange, especially causing financial losses for certain companies in Great Britain and the United States. All the same, in some cases deliveries of capital goods did take place. Until August 1950, when Churchill severely criticized the exporting of equipment to supply Polish and Soviet heavy industry, British companies made many offers to Polish and Soviet trade representatives in London. Even at the end of 1950, the British concern Craven Brothers successfully interceded with the British government in favor of increasing exports to Eastern Europe.

British cabinet members fiercely debated severing the trade agreement with Poland and restricting trade with Eastern Europe. Not all parliamentarians supported this decision. Unlike Churchill, Prime Minister Clement Attlee favored further deliveries to the East in the framework of existing contracts.[25] Cold War propaganda campaigns began in the East and the West (Körner 2002). The government in London held trade unionists who had "returned from Warsaw with the appropriate instructions" responsible for the strike wave in the United Kingdom during September 1950.[26] However, the GDR directed its propaganda most heavily against the United States (cf. Buckow 2003): "The US interventionists have divided Germany in order to make West Germany into the assembly zone for their planned third world war against the Soviet Union, the peoples' democracies and the German Democratic Republic. They maintain spies and saboteurs in the German Democratic Republic to destroy the peaceful development work of the workers."[27] The workers at EKO and the whole population of the town were mobilized through these propaganda campaigns. Following the example of the EKO shop stewards, people sent countless letters to the president of the United States of America to protest the U.S. Supreme Court's decision to ban the Communist Party.[28] The tone of Cold War propaganda was therefore similar in both parts of the continent, despite serious differences between the systems.

The strident verbal confrontations made trade and cooperation with the West increasingly difficult. Many enterprises and former business partners in both economic systems attempted without much success to improve the situation. Only in the 1960s did the situation change gradually, allowing equipment to be installed at Nowa Huta from both Austria and Japan. It even became possible to buy equipment that was on the West's embargo list. For example, personal contacts with the Polish diaspora (the eleven million Americans of Polish origin) enabled deliveries from the United States. After liberalization began in Poland in 1956, two Polish-American industrialists met with the deputy prime minister of Poland for the first time to discuss delivery of a galvanizing plant for the mill in Nowa Huta. Together with Armco Steel, the Sendzimir Company, founded in 1956 in Waterbury, Connecticut, by Polish emigrants, delivered the plant in 1961 with financial support from the Import-Export Bank in New York (Jajeśniak-Quast and Lenczowska 2000).[29] The project succeeded in part because during the confrontation between the systems, the United States provided strategic support for Poland—believed to be the "weakest link" in the socialist bloc—including direct diplomatic negotiations between the two countries. The project also succeeded because Tadeusz Sendzimir was very well known and much appreciated in the industrial field. Sendzimir's motives were not just economic; he felt a duty to help his homeland. As another mark of his reputation, the Nowa Huta mill took his name in 1989, as noted above.

In the 1960s the GDR also began negotiating with Western companies to gain solutions for engineering problems at steel mills. Formally neutral states, such as Austria, played a key role in this countercurrent circulation. In summer 1961 the GDR carried out an intensive exchange on steel mill engineering with Austria. Delegations from the EKO as well as the responsible ministries and state agencies spent a good deal of time in Linz and Donawitz inspecting the pioneering oxygen converter steel plant owned by Vereinigte Österreichische Eisen- und Stahlwerken (VÖESt).[30] Their goal was to benefit from a decade of experience in Linz and Donawitz with the basic oxygen process and to acquire Austrian know-how for the planned new steel mill in Eisenhüttenstadt.[31] Intriguingly enough, the basic oxygen furnace, known for a time in the West as the "L-D process," came in subsequent decades to be the world's predominant steel-making process, accounting for fully 63 percent of the world's steel output in 2003.

This Cold War circulation of technical knowledge was, of course, not independent from the Soviet Union. Along with the GDR, Poland, Czechoslovakia, and Romania, the USSR also negotiated with Austria. At the end of the 1950s, particularly after Khrushchev's visit to Linz, negotiations with VÖESt were taking place in the Soviet Union as well, focused on planning two complete steel mills for the Soviet complex at Lipezk. Represented on the Soviet side by the Moscow State Planning Commission, the discussions concerned problems that were of significance for all COMECON countries in the field of West-to-East trade and technology transfer. Moscow's consistent aim was to avoid payments in convertible currency. In addition, the difficult problem of license rights and license fees was never really resolved.

Neutral Austria was not the only prospective supplier of Western technology for the socialist states. In 1966 the GDR delegation at the Leipzig Trade Fair negotiated with other Western European companies for new technology to rebuild parts of the EKO. Two years later the GDR's export trade agency began negotiating with Japan to supply technology for texture-rolled strip production for the VEB Strip Steel Combine.[32] The Federal Republic of Germany played an increasingly important part in deliveries to the GDR. Close inter-German contacts continued between the EKO and the West

German August Thyssen steel mill in Duisburg. In 1954 Thyssen had received a contract within the framework of the mill's "United German Effort" (*Gesamtdeutsche Arbeit*).[33] These contacts played an important role later. For example, steel from EKO was often rolled in the West, because EKO did not have a hot-rolling mill. The GDR frequently used its special trade status between the two German states in both supplier and purchaser relationships, especially after the mid-1960s. The Federal Republic facilitated inter-German trade by providing direct benefits to West German companies and by exempting goods sent to the GDR from the valued-added tax.[34]

This window of opportunity gave the GDR a special position within COMECON, not only in the supply of Western technology but also in selling its own products in the West. These benefits helped create the GDR's relatively high productivity and the high quality of its industrial products, in comparison with those of other socialist states. Countries within COMECON quickly recognized the possibility of using the GDR as a bridge to the West (much as Poland had once been designated) and also of profiting from the higher productivity in the GDR. At the end of the 1960s the GDR was already ahead of Czechoslovakia and even ahead of the USSR in Poland's ranked list of cooperative relationships.[35] In this way East-West contacts between the systems continued throughout the Cold War. Thus, although it is not often appreciated, in a real sense the 1960s marked the beginning of the famed "return" to Europe (this term usually refers to the efforts especially prominent after 1989 among intellectual and political leaders from the former Eastern Europe who sought to restore their countries to their proper place in the center of Europe) and an intensified circulation of ideas, technology and people.

Conclusion

Eisenhüttenstadt, Nowa Huta, and Kunčice are prototypical industrial cities. They were created as out-and-out industrial settlements just like the Hungarian cities that Pál Germuska discusses in his chapter. However, regional differences existed and persisted, as can be seen most obviously in Eisenhüttenstadt. At the same time, these are undoubtedly European cities. The first socialist cities clearly drew on Western European urban developments despite their comparatively brief histories and the self-conscious efforts to distinguish them from capitalist cities. Despite the socialist legitimation of these cities through connections to national historical traditions, the socialist cities did not grow organically from interactions with the environment and population. They differ from other European cities in the following aspects.

1. Even to the present day, these cities continue to depend on a single industrial plant. Especially in the 1950s, the plants governed entirely urban affairs. They also took over many municipal functions, such as providing food and leisure activities.

The plants also determined the size of the cities, because many inhabitants were employees or former employees of the plant. As the plants grew and shrank, so did the cities. Today, with the reduction in the size of the workforce, the cities and districts are decreasing. Outward migration, especially of the young, most severely affects Eisenhüttenstadt. The socialist cities were once young but have grown old.

2. Much more than other European cities, socialist cities during socialist industrialization were characterized by migration from rural areas. More and more people were required for expanding production in the new industrial plants, and these employees had to be recruited mainly from agriculture. This phenomenon is evident above all in Nowa Huta. In Kunčice, migration from the less developed Slovakian part of the republic played an important role, whereas Eisenhüttenstadt drew on people expelled from the former German territories in Eastern Europe after the Second World War.

3. Socialist industrialization and economic planning marked the foundation of the socialist cities. The degree of dependence on ideology and the enormous politicization of the process of urbanization are distinctive, perhaps even unique in these cities. Thus socialist cities represent the political changes in Europe to a much greater extent than do other European cities. Eisenhüttenstadt, Nowa Huta, and Kunčice became the first socialist cities in their respective countries.

In spite of the Iron Curtain, Western European and U.S. technology remained important in these cities. In all cases, autarchy or decoupling from the West was not the original objective in the early search for a location. The changed political situation in Europe, intensified by the 1947–1948 crises, altered the selection of locations for plants and cities. Even the origin of the technical know-how changed from the West to the East. Surprisingly, the technical solutions remained partially Western, even after the change of direction. Socialist transformation promoted the development of industry under its own models and ideas, but before long, cooperation developed across the Iron Curtain. The situation changed gradually in the 1960s, when Western equipment from the Federal Republic of Germany, Austria, the United States, or Japan was installed in Nowa Huta, Eisenhüttenstadt, and Kunčice. These relationships with the West for transferring know-how did not develop completely independently of the Soviet Union, which exercised a controlling role, especially in questions of payments and license fees for the Eastern bloc.

Moreover, the history of Soviet industrialization in the 1930s leads back to the United States and to Germany. The technology for Magnitogorsk, the most important iron and steel combine during Soviet industrialization in the 1930s, came from the United States, especially Indiana and Ohio. German and French architects planned the city of Magnitogorsk. Twenty years later, Magnitogorsk's technology and its urban planning circulated, especially to Poland and Hungary. As demonstrated in this chapter, the circulation of ideas, technology, and people modified the relationship between

Soviet technologies, cities, and ideals and the socialist cities that grew up in their shadow.

Notes

1. Arbeitsgruppe Stadtgeschichte 1999; Breit 1980; Gayko 1999; Jajeśniak-Quast 2001b; Karnasiewicz 2003; Knauer-Romani 2000; Lebow 2001; Lozac'h 1999; Ludwig 2000; Marek 2005; May 1999; Richter et al. 1997; Skarbowski 1971; Terlecki et al. 2002; Towarzystwo Miłośników Historii 1999.

2. See Agursky 1987; Connor 1984; Rigby and Fehér 1982; Seton-Watson 1964; Zaremba 2001; Zwick 1983.

3. EKO Stahl company archive (henceforth referred to as UA EKO), Volkseigener Betrieb Bandstahlkombinat "Hermann Matern" (henceforth, VEB BKE), file no. A 75, Zeittafel des VEB Bandstahlkombinat Eisenhüttenstadt Stammwerk EKO, 13.

4. Ibid., 16.

5. UA EKO, VEB BKE, file no. A 648, District Committee "10 Years of Stalinstadt," ed. Center for Press Information: "Stalinstadt: The First Socialist Town in Germany: Material on the the First Ten Years of the Ironworks Combine 'J. V. Stalin' and Stalinstadt," 129; and the same in UA EKO, VEB BKE, file no. A 969, p. 127.

6. UA EKO, nn 4–5; Tägliche Rundschau, November 2, 1950, in EKO Stahl GmbH, 2000, 48; UA EKO, VEB BKE, file no. A 645, 1.

7. Archiwum Akt Nowych w Warszawie (Archive of the New Files in Warsaw; henceforth, AAN), Ministerstwo Przemysłu i Handlu II (Ministry of Industry and Trade II; henceforth, MPiH II), "Protokół nr 1 z posiedzenia w dniu 7 maja 1948 r. w sprawie projektowania zakładów hutniczych w Polsce, sporządzony przez Ambasadę R. P. w Moskwie" (Minutes of the meeting on May 7, 1948, on the subject of planning the iron and steel mills in Poland, made by the Polish embassy in Moscow), Sign. 8/15, 3.

8. AAN, MpiH II, "Notatka z posiedzenia w Ministerstwie Przemysłu Metalurgicznego ZSRR w dniu 10.12.1948" (Note from the meeting in the Ministry for the Metallurgical Industry of the USSR, dated December 10, 1948), Sign. 8/15, 166.

9. AAN, Państwowa Komisja Planowania Gospodarczego (PKPG; State Commission for Economic Planning), Sekretariat Dyrektora Generalnego (Secretariat of the Director-General), Założenia organizacyjne, sprawozdanie 1949–1952 (Organization principles, reports 1949–1952), Sign. 435.

10. Archiwum Państwowe w Krakowie (State Archive in Kraków; henceforth, APKr), Wojewódzka Komisja Planowania Gospodarczego Inwentarz, nr 31 (Woiwodschafts Commission for Economic Planning, Inventory, no. 31; henceforth, WKPG), Sign. WKPG 32, "Ustawa z dnia 21 lipca 1950 r. o 6-letnim planie rozwoju gospodarczego i budowy podstaw socjalizmu na lata 1950–1955" (Law dated July 21, 1950, on the six-year plan for economic development and construction on the basis

of socialism in the years 1950–1955), in Dziennik Ustaw Reczpospolitej Polskiej (Statute Book for the Republic of Poland), Warsaw, 30 August 1950, Nr. 37, Pos. 334, 445f.

11. Company archiv Nová Hut Ostrava—Kuničice (henceforth, PA NH), Edmund Grygar: "Kronika Nové Huti Klementa Gottwalda Ostrava-Kunčice, I. Díl: 1952–1958" (Chronicle of the New Clement Gottwald Iron and Steel Works, Ostrava-Kunčice, Vol. I: 1952–1958), 15.

12. Ibid., 16ff.

13. UA EKO, VEB BKE, file no. A 255, 5.

14. UA EKO, nn 5, 128.

15. Ibid., 213a.

16. AAN, Państwowa Komisja Planowania Gospodarczego (PKPG; State Commission for Economic Planning), Gabinet Zastępcy Przewodniczącego (Cabinet of the Deputy Chairman), "Huta im. Lenina, Organizacja, koszty i nakłady budowy" (Lenin Iron and Steel Works, organization, costs and funds for construction), Sign. 437.

17. UA EKO, VEB BKE, file no. A 108, Cyrulik, Alois: Stalinstadt, ein großes Eisenhüttenkombinat, Translation from the Polish plant newspaper *The Voice of Nowa Huta*, 4 October 1958, no. 56, 244.

18. UA EKO, nn 5, 129.

19. AAN, MPiH II, Departament Planowania (Planning Department), Wydział Planowania Długofalowego (Department of Longterm Planning), "Tezy do 6-letniego planu rozwoju przemysłu w latach 1950–1955" (Propositions for the six-year plan for the development of industry in the years 1950–1955), Sign. 2385, 4.

20. AAN, n 12.

21. UA EKO, VEB BKE, file no. A 1598, 10f.

22. UA EKO, VEB BKE, file no. A 579, memorandum on evaluating the consultation with COMECOM's Standing Committee on the Black Metallurgical Industry, 5 May 1961, 153f.

23. UA EKO, VEB BKE, file no. A 1057, "Report from the VVB Iron Ore-Pig Iron on the Introduction of the New Economic System of Planning and Management and the Status of the Introduction of Modern Technologies," 93.

24. AAN, n 12.

25. AAN, PKPG, Gabinet Przewodniczącego (Cabinet of the Chairman), "Sytuacja gospodarcza Polski w ocenie zagranicy" (Poland's economic situation as seen from abroad), Raporty polityczne, notatki, sprawozdania (Political reports, notes, communiqués), Sign. 156, 10–11.

26. Ibid., 21.

27. UA EKO, VEB BKE, file no. A 450, government of the German Democratic Republic, Coordination and Control Office for the Work of Administrative Bodies, "Lesson 1, 1953, for the

Representatives of District and County Councils: Conclusions from the XIXth Party Conference of the CP-SU for the Work of State Bodies in the GDR, 162/2."

28. UA EKO, VEB BKE, file no. A 771, letter from the general meeting of the EKO shop stewards on 24 November 1961, 124 and 129.

29. Interview with Michael Sendzimir, Kraków, January 2000.

30. UA EKO, VEB BKE, file no. A 498, "Technical Results of the Visit to the Vereinigte Österreichische Eisen- und Stahlwerken in Linz," 158.

31. Ibid., 160.

32. UA EKO, file no. A 1044, 49f.

33. UA EKO, VEB BKE, file no. A 73, 7.

34. UA EKO, VEB BKE, file no. A 319, "Information on Price Policies as against West Germany and the Independent Political Unit of West Berlin of 28 November 1968," 111, 111/2.

35. Ibid., 123.

10 Science in the City: European Traditions and American Models

Martina Heßler

For many centuries, cities have been regarded as major hubs for the generation and diffusion of new ideas and innovations. The conception of the city as the home of science and technology is very old, and has been expressed by such classical thinkers as Socrates and Ibn Khaldūn as well as the Renaissance utopians Thomas More, Tommaso Campanella, and Francis Bacon (Basalla 1984). In Europe, universities were founded in urban areas from the twelfth century on, and cities enabled the exchange of knowledge and material goods during the commercial expansion that followed. In the early modern period, city-based academies, salons, clubs, and coffeehouses played a critical role in intellectual life, with formal institutions in Paris and London making especially important contributions to scientific knowledge. The research university expanded in the nineteenth century, especially in German cities, and attracted worldwide attention. Vienna was a center for intellectual and scientific life around 1900 (Janik and Toulmin 1973). More recently, urban areas have emerged that typically feature scientific institutions, associations, high-tech companies, research groups, and universities. In short, science and urban life seem to be inseparably connected.

Even today, cities exert disproportionate influence on intellectual life, with their vibrant networks of universities, museums, and publishers of all sorts. Politicians and representatives of the sciences are still convinced of the importance of cities. Thus, the Foundational Association of German Science (Stifterverband für die Deutsche Wissenschaft) recently advertised a contest for a "city of science" to encourage German cities to support their creative capabilities and to tap into their scientific potentials.

Simultaneously, contemporary observers perceive a decrease in the importance of cities for the development of science, for today's information and communication technologies enable scientists to conduct research with fewer restrictions on physical location. The massive expansion of the Internet implies that the centuries-old hegemony of cities over intellectual life, including scientific and technical research, is no longer certain. Some scholars even talk of the "electronic requiem of cities" (H. Böhme 2000, 13). Furthermore, we can observe a trend, over the course of the twentieth century, of science institutions and universities being established not in the center of the city but on its outskirts. These new suburbs, whether they are strictly speaking science cities or technopoles, emerged at the edge of old cities and metropolises. Silicon Valley and Bangalore are paradigms for this suburban science.

Research on science cities and high-tech regions in their varied and myriad forms has boomed in the last thirty years. Every local politician, so it seems, wants a recipe for creating his or her own Silicon Valley. One important point in the scholarly literature involves distinguishing between science cities and what are called "technology parks," "technopolises," or "technopoles." Science cities, such as Cambridge, Massachusetts, or Cambridge, England, are said to traditionally aim at "scientific excellence through the synergy they [are] supposed to generate in their secluded scientific milieux," whereas technology parks seem to reflect more or less deliberate attempts to promote technological innovation and industrial production (Castells and Hall 1994, 8, 10). It seems helpful to use the term "science city" for cities with a high number of science institutions, academies, and universities that aim to "reach a higher level of scientific excellence," while reserving the terms "technology park," "technopolis," and "technopole" for cities that aim specifically at integrating science and industry in order to enhance economic prosperity.

To illustrate the research on science cities and high-tech regions, two books are worth referring to. Manuel Castells and Peter Hall's *Technopoles of the World* (1994) and Raymond W. Smilor, George Kozmetsky, and David V. Gibson's *Creating the Technopolis* (1988) examine these phenomena from a worldwide perspective. Summing up the findings of these scholars and others working in this area, three observations may be made. First, nineteenth-century science cities such as Göttingen, Oxford, and Philadelphia already existed before they were so designated; in contrast, many nation-states in the twentieth century built new "science cities" from scratch. One famous example is the Soviet city of Akademgorodok in Siberia (Josephson 1997). Second, there have been several, partially overlapping objectives: the creation of new science cities was an effort to reduce the agglomeration of scientific institutions and to assist the decentralization of scientific research. Often, they were regarded as tools for regional development, and, maybe most importantly, they were generally conceived as instruments for national scientific developments. Such developments were "considered a positive aim in its own right, in the hope that better scientific research will progressively percolate through the entire economy and the whole social fabric" (Castells and Hall 1994, 39). Finally, after the 1970s, science cities have tended to evolve into technopoles.

A comparison of international science cities in the second half of the twentieth century makes very clear that their transformation into technopoles constitutes a trend that can be observed all over the world. In the first decades after World War II, science cities were founded to carry out pure research. Sites such as Los Alamos or Akademgorodok gathered scientists at a secluded place, detached from the world to enable them to pursue their basic research. However, from the 1970s on, the model of a science city lost its importance, and instead technopoles emerged all over the world. Sophia-Antipolis in France, erected in the early 1970s between Nice and Cannes, represents this change. The idea behind Sophia-Antipolis was to create a site for a mixture of innovative technologies, housing, and living activities (Castells and Hall 1994, 85ff.). Many technology parks and technopoles followed the same path.

It is striking how similar the development of these science cities and technopoles appears. Worldwide influential concepts of organizing scientific research and concepts of urban planning determined their development. Focusing on one case study—the German science city of Garching in southern Germany—this chapter will scrutinize the guiding visions of urban planning and the modes of scientific research, as well as their interplay. It will also pay particular attention to the question of how internationally circulating concepts were modified by local traditions and local contexts. How did actors appropriate these ideas? National and regional factors play a role in the adoption of these concepts, as well. Scholars have not yet systematically asked whether European science cities differ significantly from American and Asian ones, or whether any special features can be discerned that mirror European traditions. To investigate these issues, this chapter analyzes the emergence of Garching as a part of the science city of Munich, and considers wider German, European, and even international developments. It discusses the ideas and concepts that drove the rapid changes in Garching, especially in the areas of urban planning and scientific organization. Thereby, it pays special attention to the question of how European and U.S. traditions and ideals influenced Garching's development. I maintain that it is possible to observe a partial transition from a European to an American tradition in the second half of the twentieth century. Before the 1970s, the model of organizing science cities could be described in analogy to a European Renaissance model, Bacon's New Atlantis, characterized by "separated functions." After the 1970s, however, the success of Silicon Valley as a role model indicated the emergence of rival ideals of integration and interconnection. Still, this American model did not simply push aside an older European tradition. Rather, two other European traditions came to serve as sources of inspiration for urban planners, politicians, and scientists: on the one hand, the idealized Greek polis, and, on the other hand, the well-integrated, prefunctionalist European city of the nineteenth century.

The Emergence of "Municon Valley"

While Munich had long been known as a bourgeois city and royal capital, its image changed during the course of the nineteenth century to that of a city of art and science, with lively artist communities and a university (Zimmermann 2000). In the 1950s Munich replaced Berlin as Germany's electropolis, and a couple of decades later began its development as a technopolis (Sternberg 1998). Noting that Munich boomed economically, Castells and Hall (1994, 173) fittingly call it an "upstart city." In the 1980s, the greater Munich area went on to become one of the most successful high-tech regions in Germany. City officials enthusiastically adopted the label "Municon Valley," coined by a journalist during this decade. The concept is closely connected to Munich's dominant position in the German—and European—electronics industry. However, its development into a technopolis was not restricted only to microelectronics. Today, Munich's outskirts host various centers for different technologies. For example, we find nuclear physics in Garching, microelectronics in Neuperlach, and biotechnology in Martinsried.

Each of these suburban sites reflects a different model of urban planning and another mode of organizing scientific research. In fact, one could say that Munich is split into different science cities. By focusing on Garching, Munich's first science city, this chapter traces the consequences that the changing ideas of scientific organization and urban development had on a local level, while at the same time discussing these changes against the backdrop of broader trends. A key turning point occurred in the 1970s, when efforts were made to transform the science city of Garching into a technopole. Garching thus partly mirrors changes that can be discerned in many other places.

Simultaneously, Garching is unique because it was not the result of a master plan, as were almost all science cities or technopoles that emerged after the Second World War. During the last half century, it was gradually transformed from a small Bavarian village into a science city. As this change took place, conflicts arose from a clash of different social groups. Garching's heterogeneous population set it apart from science cities that were built from scratch, such as Los Alamos, Akademgorodok, or the Japanese city of Tsukubu, which had more homogeneous populations. Life in Garching, located some ten miles northeast of Munich, was slow and tranquil before the arrival of science institutions, and nothing distinguished it from other Bavarian villages. People made their living mainly from agriculture, and village inhabitants all knew each other. An early twentieth-century observer described Garching as a village with cows on the street, small cottages, and dung heaps (Deutinger 1999, 225ff.). The second half of the twentieth century brought dramatic changes: the village expanded rapidly, and the number of inhabitants, workplaces, and buildings grew enormously. According to its officials, the town had evolved into a "center of science and technology of international significance."[1]

The starting point for Garching's new developments was the erection in 1957 of the first German nuclear research reactor, soon dubbed the "atomic egg" (*Atom-Ei*) on account of its shape. Many research institutes and departments of the Munich universities settled around the atomic egg. Garching was notable for its exceptionally open attitude toward housing scientific institutions. Led by Mayor Josef Amon, the community welcomed the building of both the nuclear reactor and additional research

Figure 10.1
Garching was the home of Germany's first nuclear research reactor. Erected in 1957, the reactor building quickly became known as the atomic egg (*Atom-Ei*) on account of its distinctive shape. Source: Deutsches Historisches Museum Berlin.

institutes. Only in the years since 1990, when Garching was officially designated as a town, have its inhabitants become somewhat more critical toward science and technology. In 1997, following a dramatic expansion in the students and employees there, Garching was officially named a university town.

Although specific factors and local actors played crucial roles in the transition of Garching into a science city, this chapter attempts to relate this particular case to wider European traditions and features that shaped its development. First, it considers the general process of the suburbanization of science. Second, it critically examines the ruling concepts of urban planning and the guiding ideals of scientific organization in the 1950s and 1960s. To support my claim that a decisive change took place in the 1970s, this chapter analyzes at some length the underlying concepts and guiding ideals of urban planning and of the mode of knowledge production from 1970 until today.

The Suburbanization of Science

Before the twentieth century, academies of science, universities, and institutions of higher learning were usually located within cities (Burke 2002). The reigning idea was that the main activities of a university should be located within one building at a central place in the city. In only a few select cases in Europe, such as in Tübingen, were the university and an agglomeration of science institutions located outside of the city (Paletschek 1997, 45). Plans to build a "university city" outside Berlin in 1910 prompted worries that such a transfer of "modern American institutions" was not appropriate for German circumstances (Nägelke 2003, 19). Nevertheless, the "suburbanization of science" spread across Europe in the 1960s, when the campus—a typical U.S. institution—became the model for building new universities and science cities as independent units outside city centers. Typically, campuses have remained close to the city and constituted an experimental field on the outskirts of a truly modern city. This concept has become the guiding ideal in most European countries (Muthesius 2000), mainly due to the lack of space within the cities, along with the universities' desire for their research institutes to be close together.

Europeans did not simply copy American ideals, but rather appropriated the American model in different ways. Under some circumstances, American ideals actually converged with European traditions. In Great Britain, for example, the concept of the campus merged with the model of the college, while in Germany the influence of Humboldt's grand nineteenth-century tradition was still evident. Indeed, German education policymakers in the 1960s revived the idea that the university should represent an integral whole. However, with the increasing number of students and the differentiation of the sciences, it became ever more difficult to realize this ideal within cities. In this situation, the American model seemed to offer an appropriate solution: a campus university "on the green meadow" at the outskirts of the city was seen to be compatible with the Humboldtian idea of a unified whole.

These kinds of arguments were also mobilized in the case of Munich. In 1963 the administration of the Munich University of Technology (TUM) made public its plans to move some of its institutes to Garching, and even to develop Garching as a new main location. The university administration invoked markedly Humboldtian language in stressing the need to concentrate research institutes and departments in one location: "The university is a complex whole. It is not separable. All departments are more or less connected to each other.... Communication between various research areas is absolutely necessary."[2] Even though the idea of the spatial proximity of scientists is a very old one, it converged with the model of a campus when universities became too huge to be located within cities. When the American campus model replaced the century-old European tradition of placing universities, academies, and science institutions in city centers, the outcome was a new model: scientific sites at the edge of the city.

Moreover, the trend to locate universities and science institutions in outlying areas of the city corresponded with the trend toward the suburbanization of housing, industry, services, shopping, and leisure. Suburbanization occurred in the United States and Europe at different times. The process started in the United States over a hundred years ago, and initiated a debate in Europe on whether to follow the model of the United States—the "prototype of a suburban nation" (Sewing 2002, 29). Still, suburbanization did not decisively structure urban development in Europe until the 1950s. Generally, inhabitants first began to move out of the city, with employment, services, and eventually leisure following, more or less in that order.

Recently, European and American cities have also witnessed a phenomenon that Joel Garreau (1991) calls "edge cities." These are new urban agglomerations characterized by having more jobs than bedrooms, by having most of their employment in the information industry, and by having a full complement of shopping malls, retail spaces, and leisure facilities. Despite their dispersed location, people perceive them to be one place. Although Garreau located edge cities first in America, similar developments also appeared in Europe. In contrast to the United States, however, suburban university areas or science cities usually do not include a full-fledged infrastructure of services and leisure. At times, these developments led to a heated debate within Europe. While some criticized a strong American economic and cultural influence, others stressed the structural similarities of European and American societies and claimed that these underlying similarities necessarily lead to similar urban forms (Müller and Rohr-Zänker 2001). Of course, such a claim raises some questions: Which historical traditions are effective, which concepts influence the development of cities of sciences, and, in particular, do national or even regional appropriation processes lead to differences?

Garching in the Period of "Separated Functions"

Considering that Garching was a small village where little had changed for centuries, contemporaries in the 1970s observed that the townscape had rapidly undergone a radical transformation. Garching had become a town with high-rise buildings and residential districts, while the old cottages had almost vanished.[3] It had also lost its agrarian character: the number of farms fell from 63 to 36 in the decades between 1953 and 1974, and in 1988 its last farm closed down. As the town's agricultural nature had come to an end, industry and science moved into the old stables. As a community chronicle noted: "The research area of Garching is one of the largest and most important research centers in Europe. This has entirely determined the development of the town of Garching."[4]

When the nuclear research reactor was built in Garching in 1957, it provided a starting point for relocating numerous research institutions and university departments there, including the School of Physics of the Munich University of Technology

(1956–1957), the Max Planck Institute (MPI) for plasma physics (1960), the MPI for extraterrestrial physics (1964), and the institute for radiochemistry (1964). In 1966 the Walther Meißner Institute for Low Temperature Research—a section of the Bavarian Academy of Sciences—moved to Garching, and in 1967 it was followed by the accelerator laboratory (a joint initiative of the TUM and the Munich Ludwig Maximilian University) and the TUM departments of chemistry and biology.

This tremendous influx of scientific institutes altered the social structure and everyday life of Garching. The importance of traditional habits and institutions, such as churches, associations, schools, and bars, diminished, and their integrating function in daily life correspondingly weakened. The power and influence of the Catholic church similarly petered out. Residents no longer knew each other as well as they used to, and life became increasingly anonymous.

Garching's evolution from a small Bavarian village into a science city was influenced by a variety of factors: the technology policy of the Bavarian government, the favorable attitude of inhabitants and local politicians toward the settlement of science, and geographical characteristics. Most strikingly, the morphology of Garching was decisively shaped by concepts of urban planning that were influential across Europe at the time. The modernist principle of the physical separation of functions, formulated by the International Congress of Modernist Architects (CIAM) in the Athens Charter of 1933, was at the height of its influence (see Misa, this volume). In Garching, this principle was transformed to cover the functions of industry, habitation, and research. Local planners and politicians regarded these to be the community's main functions, and they accordingly located them in spatially separated parts of town.[5] Functionalist urban planning split Garching into different areas corresponding to its different functions, and planners even took special care to create and maintain distinct boundaries between the areas. Local politicians, under the sway of the ideal of separated functions, did not for a moment perceive this separation as a problem. They followed the dominant model of urban planning, believing it was the best way to organize urban space.

Even today, the main "functions" remain in three distinct areas, which lie at a certain distance from each other and are only tenuously connected by public transportation. Garching has evolved into a typical functionally segmented "suburb" with a highly innovative research area at its edge. The various social groups of inhabitants and their different functions—such as living, science, and industry—are effectively segregated. In 1959 a new Max Planck residential quarter was built specially for scientists,[6] making manifest the separation of the scientists from the other inhabitants of the village. Unsurprisingly, the lifestyles and habits of scientists and their families differed from those of older residents. Likewise, the research institutes were not integrated into the city. Today, the industrial zone is located in Garching Hochbrück. Residential areas are concentrated in the city center, while scientists carry out their research in the science area.

Figure 10.2
The "atomic egg" in Garching iconography. Garching expressed its identity as a science city by incorporating the city's iconic nuclear reactor into its official town crest in 1967. Source: *Die Stadt Garching* (Garching bei München: Stadtverwaltung Garching), cover.

As the village was socially transformed and spatially restructured, the community of Garching, led by its local politicians, started to regard itself as a science city. At the beginning of the 1960s, the district council named streets after such famous German physicists as Albert Einstein and Max Planck, and the local high school became the Werner Heisenberg Gymnasium. In 1967 the community held a weeklong celebration of the reactor's 10-year anniversary. It is worth noting that this idea came from the community of Garching itself, rather than from the scientific community. In the same year, the atomic egg was included in the local coat of arms. Clearly, Garching had acquired a new identity as a "science city."

Nevertheless, Garching as a whole was not characterized by science. A sharp divide existed between, on the one hand, the symbolic politics and discourses that stressed the inseparable connection of the town of Garching and science and, on the other hand, everyday life in Garching, which was hardly related to science. Idiomatic expressions invented by long-time inhabitants reveal this separation. Scientists were labeled "those over there," and inhabitants used the term "science city" only for the research area, not for the town as a whole. A local newspaper stated that Garching's inhabitants did not pay much attention to the research area, while scientists were not interested in the events of the village.[7]

Thus, the research area evolved as an independent site at the edge of Garching. I would argue that this gulf corresponded with contemporary theories of urban planning, namely the separation of functions, fully in line with the practices of urban planners across Europe. For a time, the prevailing concepts of organizing scientific research reinforced these urban planning ideas. During the decade when Garching first took form, both urban planning and scientific planning were based on the ideal of separated spheres. While urban planners kept such functions as residing, work, consumption, and leisure separate from each other, scientific research was also organized in a similar "mode of separation." Exactly what this mode embraced will be analyzed in the following section. Suffice it to say here that the spatial separation of the research area apart from the town of Garching, as well as its evolution into a "scientific ghetto," can be read as a logical outcome of the similarities in urban planning and in organizing scientific research.

Garching—A Secluded Space for Scientists

If we look for scientific traditions, concepts, and ideals that guided the evolution of modern science in Europe, several roads lead back to the Renaissance utopian Francis Bacon. Yet Bacon has been interpreted quite differently at different times. Whereas writers in the eighteenth and nineteenth centuries saw him as a founder of modern science and the inductive method, more recent writers have emphasized his ideas on the inseparable connection of scientific, technological, and human progress and his concept of science as a decisive resource for humanity and for the well-being of nations (Krohn 1987, 12). This difference is, of course, no coincidence, as interpretations of classical thinkers always mirror contemporary interests and concerns. When in the following I attempt to show that it is possible to discern certain analogies between Bacon's 1624 utopian fable *The New Atlantis* and Garching's research area, I am referring to the latter-day interpretations of Bacon, which suggest that science guarantees human and social progress. As many scholars have already emphasized, this ideal has shaped our concepts of modern science. Philosopher Ernst Bloch (1985, 207) described Solomon's House—the central research location in Bacon's utopian island Bensalem—as a university of technology. And as Paul Josephson shows in his *New Atlantis Revisited* (1997), Bacon's utopia even served as an explicit role model for Akademgorodok.

It is not difficult to find passages in Bacon's pamphlet that anticipate twentieth-century science cities. For example, the aim of Solomon's House was to gain "knowledge of causes and the secret motions of things; and the enlargement of the bounds of human empire, to the effecting of all things possible" (Bacon 1624/2001, 205). The main objective was to bring about scientific progress to improve society's well-being and wealth. Bacon's model suggests that the search for knowledge and truth automatically improve our lot: advances in the sciences will bring technical progress, economic prosperity, and higher standards of living. In this idea, we see a unity of truth and utility. Moreover, Bacon conceived of long-term projects financed by the "state" to keep science free from political and religious constraints (Felt et al. 1995, 35).

By reading Garching through the lens of Bacon's *New Atlantis*, I wish to call attention to certain parallels between this Bavarian town and the distant utopian island. Although I do not claim that the science city of Garching was a conscious reinvention of

Solomon's House, I suggest that certain elements of Bacon's utopia, as commentators interpreted them in the twentieth century, reappear in this Munich suburb. Unlike in the case of Akademgorodok, Garching's politicians and scientists did not explicitly refer to Bacon or consciously claim to recreate Solomon's House. Still, it can be argued that some of Bacon's ideas have become, so to speak, a part of our cultural sediment, and as such they play a role even though actors do not discuss them explicitly.

Thus, Garching—and probably most science cities in Europe and in the United States that were founded in the 1950s and 1960s—can readily be compared, at least metaphorically, to Solomon's House. The parallels between Garching and Bacon's ideal city-state can be found on at least three levels, concerning (a) scientists' role in society, (b) their seclusion from everyday life and political and economic constraints, and (c) their orientation toward pure science in service of the nation and humanity.

The Role of Scientists and Science in Society
When one of the "fathers" of Solomon's House—that is, a scientist—made a rare visit to the city, a ceremonial procession was held. Marking the leading role of science, the father led the procession, and he was in turn followed by representatives of the administration, politics, and industry (Bacon 1624/2001, 204). Like Bacon's scientists, who were expected to serve humanity by solving its fundamental problems, scientists in Europe during the 1950s and 1960s enjoyed unusually high prestige. In no small measure it was commonly believed that science would bring well-being and economic prosperity for society (Kreibich 1986, 24).

The treatment of scientists in Garching mirrored these Europe-wide trends. Several world-famous scientists—including Werner Heisenberg and Hans Maier-Leibnitz, who both pursued research in Garching—were honored with local decorations. Moreover, local politicians believed that science, and, in particular nuclear energy, would not only solve human problems but also transform Garching into a wealthy and modern town. The local priest even called nuclear energy a "gift of God."[8] In the 1950s and 1960s the peaceful uses of nuclear energy were typically accompanied by an enthusiastic discourse and unrealistically inflated expectations. This almost religiously colored discourse played a crucial role in Garching.

Obviously, the notion that scientists are benefactors of society corresponds to the central ideas of Bacon's *New Atlantis*. As Daniel Bell (1973) noted, Francis Bacon's Bensalem in *The New Atlantis* is different from Plato's utopian state because Bacon sees the king not as a philosopher, but as a scientist and researcher. Likewise, the central building in Bensalem is not a church but a research institute, Solomon's House, built by the "state" to create knowledge to serve humanity. Bacon describes an institution that cooperated and interacted with the government, but that was regarded as an independent unit (G. Böhme 1993, 12). Similarly, as I discuss below, Garching's research area was also perceived as an independent unit at the edge of the small village—

independent from Garching as well as from everyday life and from questions of technical application and the commercialization of pure science.

The Research Area as Secluded Space
From the outset, the Garching research area was a certain distance from the center of town, and it was also perceived as a place independent of everyday life. It was a separate site with only scientific institutions. Work times and leisure times were also sharply separated: in the research area, there were no residential homes, no cafés, no shopping area, and no restaurants. Scientists were supposed to pursue their research without any disturbance by daily affairs or disruptions by political and economic constraints. Although this spatial separation was certainly an outcome of the concept of separated functions, it can also be regarded as the manifestation of a clear preference for the social isolation of scientists. A certain distance from day-to-day conflicts and the short-term needs of society was intentional and deliberate. Castells and Hall (1994, 39) claim that the model of the secluded scientist, carrying out his research free from mundane material concerns, arose in the Middle Ages: "To build a community of researchers and scholars, isolated from the rest of the society ... goes back to the medieval tradition of monasteries as islands of culture and civilization in the midst of an ocean of barbarism." At the same time, we may add, the practice of sequestered researchers in Garching obviously corresponds with the ideals of seclusion expressed by Bacon's description of Solomon's House.

The Relation between Pure and Applied Science
Reinforcing the separation of scientists from everyday life and social and political constraints, Garching also embodied the prevailing emphasis on pure or basic research. Of course, it is difficult to strictly separate pure and applied research; the boundaries between them are by no means clear-cut. Industrial research laboratories often carry out pure research of the same quality as that performed at the best universities, and universities increasingly cooperate with industry. In the science-based industries, such as pharmaceuticals and microelectronics, it is difficult or even impossible to classify a given investigation one way or another.

Even if the separation of pure from applied science is a historical invention, the idea of separated spheres was extremely influential during the first two decades after the Second World War. The Garching institutes, their tasks, and the identities of their scientists underline the concentration on basic research. Scientists in Garching—both in the university departments and at the Max Planck institutes—pursued their research undisturbed by industrial necessities (Stumm 1999, 217ff.). Susan Boenke (1991, 252) emphasizes the "academic concept of science" within the MPI for Plasma Physics, labeling it a "retreat for basic research." This orientation did not mirror, as one might think, Humboldt's ideal that researchers were to carry out research as an end in itself. Rather, the dominating idea was Bacon's "knowledge of causes and the secret motions of things" that he thought would lead automatically to useful applications. The Institute for Plasma Physics was not concerned with immediate applications but had a longer-term goal: to develop a fusion reactor. Its task was designated as "aim-oriented pure science" as well as "gaining fundamental and general knowledge" (Eckert and Osietzki 1989, 125).

In the beginning, Garching's scientific community embodied only Bacon's concept of the unworldly and remote scientists who work undisturbed and bring "light" to society. In this model, scientific progress can occur only if scientists can conduct numerous experiments that do not have to result in immediate usability, let alone in direct applications. Bacon emphasized the value of such nondirected "light-bringing experiments" for scientific progress. In 1970, however, a new phase began in Garching when the Max Planck Society founded an institute there for the industrial application of research results.[9] Its main task was to review scientific and technological developments of the Max Planck institutes from the perspective of possible industrial applications and then conduct a market assessment to find interested companies.[10] As a government report put it, "When there is a possibility to turn the results of pure research into industrial application *without disturbing basic research*, this should be done consistently."[11]

In Solomon's House, different scientists were concerned with different tasks. Three of the "brothers," designated as "light bringers," were occupied with gaining a deeper understanding of nature. Three other brothers, called "benefactors," were concerned with evaluating the outcomes and deciding which results could serve for application and practice (Bacon 1624/2001, 213f.). Bacon regarded pure science as a means to guarantee human progress, wealth, and prosperity, and believed that gaining knowledge and discovering truth were not ends in themselves. It is obvious that Bacon aimed at the utilization of science, as many scholars have already emphasized (Lyotard 1994).

Bacon's ideal that fundamental research leads more or less automatically to economic prosperity and wealth resembles some developments that can be observed in the United States after the Second World War. According to Vannevar Bush's influential vision, set out in his report *Science—the Endless Frontier* (1945), pure science was the foundation for technical innovation, social welfare, and economic prosperity. Because pure science was regarded as the pacesetter for technical progress and economic prosperity, Bush believed it should be concerned only with truth and knowledge. In particular, it had to be independent of the short-term interests of politics and the economy. Bush claimed that governments should accept new responsibilities for promoting science. Consequently, basic research was financed and promoted by the national governments, in the United States as well as in European countries (Weingart 2001, 79,

177). This idea also determined the development of Garching's research area into a site where scientists concentrated on pure science and pursued their research remote from everyday life.

Like other influential ideas that circulated internationally, Bacon's program was appropriated into diverse national contexts. For example, Paul Josephson (1997, xvi) shows that Akademgorodok was both the product of Baconian ideals and the logical outcome of decades of Soviet political, economic, and ideological developments. Likewise, we also observe a particular process of national appropriation of the Baconian model in the United States. Already in the title of his report, Bush put forth a concept familiar to all Americans—the idea of the "frontier." The development of basic science corresponded with the conquest of new frontiers: "It is in keeping with the American tradition—one which has made the United States great—that new frontiers shall be made accessible for development by all American citizens" (Bush 1945, 6). In Garching, local and regional politicians saw the embrace of science as a powerful means to transform an agrarian town into a modern city. Simultaneously, Garching combined a new, emerging modern identity with local traditions, as symbolized by its coat of arms, which displayed not only the atomic egg but icons representing the village's traditional values: a wagon wheel and typical regional trees.

Garching in the Phase of Integration

Although it grew steadily, Garching did not develop into a lively, science-dominated town during the phase of separated functions. This is not terribly surprising; a general kind of lifelessness was the rule rather than the exception for new towns built according to these planning principles. Beginning in the early 1960s, critics blasted functional planning for bringing about the "death of cities" (J. Jacobs 1961). These critics called for a revitalization of the urban fabric, and emphasized the need for increased density and the blurring of functions. Their role model was what they regarded to be the traditional "European city," a place characterized by heterogeneity, high density, and the integration of functions.

At the same time that urban planners began to embrace the ideal of a dynamic and lively city with a traditional urban character, ideas about how to organize scientific research went through fundamental changes, as well. In the decades around 1970, many science cities were transformed into technopoles when they retreated from their earlier concentration on pure research and instead adopted efforts to stimulate application-oriented research. One main objective was to locate industry, science, and technology in the same area in order to guarantee economic prosperity. Thus, Garching was not unique in this respect but mirrors a worldwide trend. As mentioned earlier, Sophia-Antipolis was founded in France, Japan created a net of new technopoles, and today high-tech clusters and technopoles mushroom at the periphery of cities all over the world. These changes were accompanied by efforts to revitalize another European urban model, that of the Greek polis. The founders of technopoles were particularly prone to refer to the Greek polis, which they perceived as a "city of wisdom and light." Since the 1970s, the watchwords for science have been interconnection, integration, and the overcoming of separation of function. A pertinent example is the Multifunction Polis in Australia, founded at the end of the twentieth century. The concept of this city was extraordinarily ambitious in many respects: it was supposed to promote future-orientated high-tech industries; it was to evolve into an international pivotal point in the Pacific region; and at the same time it should—as is highly typical of newly founded cities—present itself as an "ideal city" in which a new lifestyle should emerge (Castells and Hall 1994, 217). Here, as in the case of Garching, changes in the organization of science and in urban planning went hand in hand.

Following these new ideals, the community of Garching tried to integrate the separated functions of living, industry, and research. To stimulate urban activities, the administration in the late 1970s created a community center (*Bürgerhaus*) as a "lively focal point," the goal of which was to "overcome the bad features of functionalism."[12] Local politicians instituted a "citizens' week" (*Bürgerwoche*), held annually for twelve days. Each of these efforts was intended to stimulate social and cultural life within the town and to increase contacts between different social groups. The new idea was that a healthy town should be a place where people live, work, meet, and communicate, where an urban atmosphere exists, and where public life is encouraged.[13] In this new age Garching adopted such goals as the "revival of urban life" and cultivating public life. Integration, interaction, and exchange became the main priorities. These ideals implied an orientation toward urban features that were typical for the European city of the nineteenth century: compactness, density, and proximity.

During these years, community members began to see not merely the scientists themselves but also their research as strangers in town life. Now perceived as a problem, the separated research area was described in terms of isolation and even labeled a "ghetto." In the late 1970s Garching's politicians made serious efforts to reduce the separation between the town and the science area. The aim was "to connect the alien element, the university, more closely to the center of Garching."[14] Complaints about lifelessness were increasingly vocalized, and the town was even given satirical nicknames such as Akademgorod and Novigarchinsk (Wengenroth 1993, 297). Obviously, these metaphors, alluding to huge, planned Soviet cities, were meant to invoke lifelessness, artificiality, and isolation.

Garching's efforts to create an integrated town were epitomized in a more or less desperate plan in 1977 "to construct a walking route between the town and the research center."[15] A more recent proposal from the early 2000s envisioned a bridge connecting the research area with Garching town, with its various shops, sports facilities, and meeting spaces.[16] Despite these efforts by the officials, few scientists live in Garching, and even fewer play a role in the social and cultural life of the town. Interaction and exchange between inhabitants, scientists, and local politicians are rare. The evident difficulties in overcoming these entrenched patterns hint at the long-lasting resilience of the urban fabric. Separated functions in materialized form are not easy to overcome.

Bavarian research politicians and university administrators also understood that the times were changing. While in the 1950s no one had called into question the secluded nature of science, in the 1980s the representatives of the Munich University of Technology explicitly criticized the "isolation" of the Garching scientists.[17] To address this problem, the university administration, in cooperation with the Bavarian government, endorsed the building of new facilities for shopping, sports, and housing. Nothing happened quickly, however. Not until recently did these plans materialize, as bookstores and copy shops, bars and restaurants, shops and apartments, medical services, and even a kindergarten began to appear.

The university administration's recommendation to integrate a multitude of functions into the research area obviously constitutes a strong effort to create an urban environment. From about 1980 onward, the administrators have been claiming that research can thrive only in a diverse and stimulating environment, while describing the "splendid isolation" of scientists as risky because it contributes to removing science from society. The ideal of the seclusion of scientists has been transformed into an ideal of integrating science into society as well as of integrating science and technology. And, interestingly, a central element of this new ideal is the concept that traditional urban structures may best contribute to such integration.

Thereby, the awareness of the importance of spatial proximity and "communication spaces" has dramatically increased. New "communication spaces" were created through new architectural forms. The buildings of the MPI for Astrophysics (1979) and the neighboring European Southern Observatory (1981) encourage lively, spontaneous communication between scientists. In both buildings an open cafeteria also serves as a reading room. Researchers can meet there and discuss scientific problems, and the nearby small conference rooms invite spontaneous debates.[18] Moreover, the departments of astrophysics and extraterrestrial physics are structurally connected. Even though the image of scientists holding spontaneous debates may be somewhat stylized, it is clear that these buildings embody an awareness of the synergetic effects of face-to-face communication and informal and spontaneous contacts.

At the same time, attempts were made to bring together basic and applied research in one location. In 1980 the University of Technology heavily criticized the spatial separation of engineering science from basic research: "The status quo is extraordinarily unsatisfying, because the necessary scientific exchange between basic research [in Garching] and the technical departments [in Munich] is very difficult."[19] In 1978 the faculties of chemistry and biology settled in Garching, and the faculty of mechanical engineering followed in the 1990s. Moreover, many applied science and application-oriented institutes were founded, including the Walter Schottky Institute for Semiconductor Technology and the Bavarian Center for Applied Energy Research. Recently, the faculties of computer science and mathematics moved to Garching, in immediate proximity to the faculties of physics, chemistry, and engineering, in order to create a "center for mathematical-scientific research and software technology." At the end of the 1990s the chancellor of the University of Technology asserted that an "entrepreneurial culture" (*Gründerkultur*) would evolve in Garching. To this end, the university and the town of Garching recently built an engineering and entrepreneurial center to encourage the establishment of new business companies. In addition, General Electric has decided to locate a research center in Garching. Without a doubt, Garching has started a transformation from a city of science into a technopole. Whether this change will be successful, especially economically, is still an open question.

Returning to Bacon, one could say that the description of science cities as analogous to Solomon's House has been called into question. Scholars, politicians, and researchers have begun to admit that pure science does not automatically lead to useful applications and technical innovations (Szöllösi-Janze 1999, 43ff.). Instead, they now praise an alternative model that embraces goal-directed research and the direct utility of science. Whereas Bacon believed that researchers had to understand nature prior to developing uses for science, the new modes of organizing scientific research focus on economic utilization, especially the possibility of applying and commercializing science from the very outset. This concept did not suddenly appear around 1980, but existed for many decades. Since the end of the nineteenth century, efforts have in fact been made to integrate science and industry and also to integrate them spatially. For example, Trafford Park in Manchester, established in 1894, organized an entire geographic region with the explicit goal of integrating science and technical applications (Kargon et al. 1992). Moreover, companies have employed numerous scientists and, in particular, founded research laboratories. However, only in the last decades of the twentieth century did this alternative, utility-driven model evolve into a dominant ideal for organizing science cities.

The orientation toward directly useful applications and commercial technologies is typical in the post-1980 period, and it is in no way unique to Garching. The role model is obvious: Silicon Valley. After the Second World War, Santa Clara County, a once-agricultural area south of San Francisco, was transformed into the most famous agglomeration of high-technology industry. This development can be traced back into the 1930s, when Frederick Terman, professor of electrical engineering at Stanford University, initiated close ties with regional companies. At the beginning of the 1950s, Terman built the Stanford Industrial Park, where companies were located close to the university. In the following years, as many new companies grew, Silicon Valley became extraordinary successful. It was characterized by the spatial proximity of companies, research institutes, and universities (most notably, Stanford and Berkeley). In particular, it became famous for a certain culture characterized by close ties between companies and universities, as well as by informal relationships between companies. The term "Silicon Valley" became a metaphor for a milieu where new companies were continually being founded, where the intense circulation of knowledge and a high density of contacts were the norm, where work life and private life were interconnected, and where there was a culture of cooperation and competition (Saxenian 1994).

Many governments, in Europe and elsewhere, tried to recreate Silicon Valley in their own countries: "It became the dominating model for organizing science areas; legions of cities and towns have endeavored to become the next Silicon Valley" (Kargon et al. 1992, 352). The name "Municon Valley," which Munich officials enthusiastically reproduce in the public discourse, hints at that tradition. In fact, Munich and Bavarian politicians went on a pilgrimage to the United States to become better acquainted with the California model. Thus, a U.S. model has seemingly taken the place of the European Renaissance model of Francis Bacon.

However, founders of other European technopoles also explicitly refer to a third tradition: a stylized model of the city, or the Greek polis. For example, the French Sophia-Antipolis was initially conceived as a new "city of science and wisdom" (Wakeman 2003). The term "technopolis" itself hints at a Greek tradition. Referring to a Japanese project, Raymond Smilor and colleagues (1988, xiii) use this word to characterize the close connection of "technology and economic development in a new type of city-state." "Technopolis" refers, as they stress, to the tradition of the antique polis as a "city of wisdom and light."

In the ideal-typical model of the polis, the city thus became a metaphor for urban life, humane cohabitation, and public life, as well as for exchange, communication, and the development of art and science. The city is regarded as a crucial element of European culture that guarantees the further development of democracy, art, philosophy, and science. This reference to the Greek polis is observable not only in debates about the organization of scientific research, but also in debates about the city at large and whether the "European city" should serve as a model for urban planning today (Hassenpflug 2000, 132). Thus, at the beginning of the twenty-first century, both science and urban planning are still discussing an ancient model and a genuine European tradition. Again, science and the city are inseparably connected.

Notes

1. Gemeinde Garching b. München, "Sitzungsvorlage (Dokumentation) zum Antrag auf Stadterhebung," (henceforth, "Antrag auf Stadterhebung") 1990, p. 22, Archives of the Community of Garching (henceforth referred to as GRS or—for the period after Garching was designated a town—SRS).

2. Memorandum on the relocation of the Munich University of Technology from Munich to Garching, published by Dept. 4 (Property) of the Central Administration of the Munich University of Technology, March 1980, p. 20 (henceforth, Memorandum).

3. *Süddeutsche Zeitung*, 11 December 1974.

4. "Antrag auf Stadterhebung," p. 10.

5. GRS, 20 July 1973.

6. GRS, 19 June 1959.

7. *Abendzeitung*, 24 June 1982, p. 14.

8. *40 Jahre Atom-Ei Garching*, 1997 (Munich: University of Technology), p. 47.

9. In German this institute was called Garching Instrumente Gesellschaft zur industriellen Nutzung von Forschungsergebnissen mbH. I would like to thank Michael Eckert for making available to me an interview he conducted with Dr. Meusel (27 October 1987) as well as source material on this topic.

10. *Mitteilungen der MPG zur Förderung der Wissenschaften*, Göttingen, Issue 6, 1970, p. 396.

11. Ibid.; my italics.

12. Stadtverwaltung Garching b. München, ed., *Stadt Garching: Stadtführer* (Garching: City Administration, 1990), p. 32; *Münchner Stadtanzeiger*, 17 August 1982, p. 6.

13. Attachment 1 of the minutes of the GRS, 18 July 1975.

14. Attachment 3 of the minutes of the GRS, 19 May 1978.

15. GRS, 13 May 1977.

16. Interview with the mayor of Garching, Helmut Karl, 31 January 2002.

17. Memorandum, p. 13.

18. *Süddeutsche Zeitung*, 6 May 1981.

19. Memorandum, p. 12.

IV Planning and Power

11 Between Theory and Practice: Planning Socialist Cities in Hungary

Pál Germuska

"Its mother was mining and its father was the planned economy": this parable neatly describes the family tree of Tatabánya, one of the Hungarian socialist cities (Faludi 2002, 51). In the postwar decades more than 1,200 new industrial towns were established in the Soviet Union and Eastern and Central Europe as a result of socialist industrialization. The attributes and features of these distinct cities were debated right from their beginning in the 1930s with the groundbreaking for Magnitogorsk. Most scholars writing on the Sovietization of Eastern and Central Europe stress the discontinuities rather than continuities of these cities with those of the interwar period. Although the Eastern European leaders tried to adopt the Soviet model, it proved impossible to do so exactly: what appeared to be a "transfer" of Soviet ideology, models, and practices was in reality a complex process of circulation and appropriation. Moreover, the socialist city as a type was rooted in prevailing European theories of construction and city planning, even though these were significantly reshaped by Stalinist ideology.

The historical literature on socialist cities stakes out a number of positions. Although early studies ascribed different characteristics to the capitals, the rebuilt cities, and the new towns, they labeled all of them as socialist cities (French and Hamilton 1979; D. Smith 1996). Some scholars emphasize that the industrial socialist cities were newly established (Merlin 2000; Szirmai 1988), drawing an analogy to the postwar British and French "new towns." Socialist cities also can be interpreted as a "myth" (S. Horváth 2004). Recently, following the emerging consensus that socialist cities are best seen as a functional concept (Szelényi 1983; Węcławowicz 1992; Aleksandrowicz 1999; Szczepański 1993), I have analyzed a roster of eleven Hungarian socialist cities selected for their (a) prominent position in redistribution systems; (b) specialized industrial function; (c) predominantly industrial workforce; and (d) uniform city planning. The empirical basis for this chapter is archival research on these eleven cities: Ajka, Dunaújváros, Kazincbarcika, Komló, Oroszlány, Ózd, Salgótarján, Százhalombatta, Tatabánya, Tiszaújváros, and Várpalota (Germuska 2004, 19–53).

The main aim of this chapter is to show how Hungarian socialist cities drew upon European connections and an international circulation of ideas and experts, as well as to investigate the complex ways in which influential planning theories were appropriated in these cities. In *The Socialist City*, French and Hamilton (1979, 2) drew attention to these very processes: "If anyone is to doubt the importance of the international flow of ideas, then bear in mind that many a Czechoslovak, Polish, Romanian, Soviet, or Yugoslavian planning office has substantial translated materials, plans, and maps of various 'Western,' but in particular British, new-town designs and strategies." This chapter first examines how city planning ideas emerging in Western Europe during the interwar years influenced Soviet architectural theory and practice. It then discusses in detail the Soviet planning directives that framed the building of Hungarian socialist cities in the 1950s. Next, I analyze how these city planning principles were modified by local appropriations. Finally, the chapter compares the emerging spatial structure of the Hungarian socialist cities listed above, and evaluates the circulation of knowledge, models, and practices between the Soviet Union and Hungary. This chapter complements that of Dagmara Jajeśniak-Quast by including material from Hungary, a fourth socialist country, and dealing with industrial cities built up around coal mining rather than steel production.

Theories of City Planning and Soviet Prototypes

The idea of a socialist city fits organically into the long historical lineup of utopias or ideal cities. In its most ambitious form, the socialist city aimed to create simultaneously an ideal workers' city and a new society. In the Stalinist era, however, the vision degenerated into a rigid system of dogmatic planning. Even this system later disintegrated with the weakening of the socialist system. By the 1970s the idea of a socialist city was little more than a set of routinized solutions in city planning and architecture.

The creation and structuring of space played a special role in socialist cities, for explicitly ideological considerations determined their spatial order. The traditional ("bourgeois") city had developed in accordance with economic rules, with its town center dominated by mercantile activities. By contrast, Aleksandrowicz (1999, 4) explains, "The aim of the socialist city planners was in turn to give the centre of the new city an ideological character. That is, instead of the marketplace, the centre of the town was here to be organized by political institutions and by official ideology." György Enyedi (1989, 21–24) argues that the theory of socialist architecture and spatial structuring arising in Eastern Europe after World War II had three sources: the modernist city architecture of the interwar era, communist ideology, and the actual Soviet construction practices of the 1930s and 1940s. The most influential element in the modernist tradition was a resolute belief in the omnipotence of planning. Modernist architecture, as a kind of holistic ideology, was believed to be capable of redeeming the world and bettering society. Communist ideology amplified the modernists' belief in progress and planning. It offered grand principles—in the end never realized—such as "make the difference between village and city disappear" or "equal quality of living in all parts of town." Actual Soviet architectural practices represent a particular interpretation of modernist architecture. Let us look at these three factors in more detail.

The Roots of Modern City Planning

Ebenezer Howard's *Garden Cities of Tomorrow* (1902) was a key starting point of modern city planning. In launching the idea of the suburb, Howard fundamentally changed the conception of planned urban space. Around the same time, Tony Garnier introduced the *cité industrielle* (industrial city), which he imagined as a new kind of township, conceived as an independent unit with its own industrial plant, sited near a series of houses with gardens. Along these lines, modernist architects in Germany were commissioned to design housing estates in the 1920s, including Ernst May in Frankfurt, Martin Wagner in Berlin, Fritz Schumacher in Hamburg, and Bruno Taut in Magdeburg. These housing estates sharply departed from the traditional city structure: buildings were arranged relatively freely in open, landscaped areas, and streets were given new functions. Also on the basis of May's plans, a series of satellite cities (*Trabantenstädte*) were established around Frankfurt, separated by a wide green belt (P. Hall 1992, 31–38, 55). Although Howard's garden city movement was well known in Hungary, its direct influence was not pronounced. There was only one large project in Kispest (now the nineteenth district of Budapest), where between 1908 and 1912 a new settlement was built with 8,000 flats for the *salariat* (the so-called black-coated workers with a high-school education but no university diploma). Funded by the government, this colony was even named after the prime minister, Sándor Wekerle. In time, due to the influence of the de Stijl circle and the Bauhaus school, functionalist ideas began to shape the design of Hungarian buildings and urban structures. Form became more and more related to practicality, with the guiding principle of removing all superfluous "fringes, tassels," and ornaments (Meggyesi 1985, 40, 46–50).

The Swiss-born theorist, architect, and writer Le Corbusier was perhaps the most influential modern urbanist. He designed skyscrapers and buildings supported by pillars, introduced roof gardens and open floor plans, and designed windows in strips as well as facades free of ornamentation. Of particular note for Hungary, Le Corbusier developed the urban planning concept of the "linear city" and emphasized the strict separation of functions into distinct urban spaces. Cities would be made up of three distinct zones: historical city centers for administration, commerce, and culture; linear industrial townships constructed along rail lines; and agricultural and recreation areas enclosed by transportation routes.

Le Corbusier was the central figure in the Congrès Internationaux d'Architecture Moderne (CIAM), founded in 1928, and the author of the famed Athens Charter (see

Misa, this volume). The Charter was not so much the starting point as the summary of many different modernist architectural trends and fashions. The document, a catechism-like version Le Corbusier published in 1943, defined a number of basic principles that architects and city planners worldwide would later transform into inescapable norms. These rules included that workplaces should be concentrated in industrial districts separated by a green belt from residential areas, as well as that the ideal city should be divided into work, residential, and recreational zones (Meggyesi 1985, 55–65). This particular concept of "zoning" had a remarkable impact after World War II. You can see Le Corbusier's direct influence right across Europe, including in Stockholm, Frankfurt, Belgrade, the so-called new towns in England, and many Soviet cities (as an example, see figure 12.4).

In Hungary a generation of young architects had taken notice of the new architectural trends. In 1929 they founded the Hungarian section of CIRPAC (Comité Internationale pour la Résolution des Problèmes de l'Architecture Contemporaine), the executive branch of CIAM. At first, these young architects designed only private villas for the wealthy and surprisingly few public buildings or apartment blocks. Only in the mid-1930s did modern architecture gain sufficient acceptance in Hungary that public buildings could be constructed in this style. Modernist city-planning theories remained marginal throughout the decade. There were only two larger towns, Debrecen and Szeged, where anything like a uniform development plan guided construction. As another sign of the resistance to functionalist planning, the city planning department established at the Technical University of Budapest in 1929 was closed down after only five years (Pamer 2001, 68, 167, 219–220).

Communist Theory and Soviet Practices

With the end of World War II, a new era in Hungarian architecture began. Following the communist takeover in 1948 and the first five-year national economic plan in 1950, Hungarian architects were assigned a task of no small ideological importance: to give a socialist face to already existing townships and to new cities under construction. The dilemma, however, was that there was no conclusive definition of a "socialist city." Even the voluminous writings left behind by Marx, Engels, and Lenin offered no practical guidance on city planning. The works of Stalin also did not give the urbanists much help, apart from some very general remarks. For example, in a January 1934 speech at the 17th Congress of the Soviet Communist (Bolshevik) Party, Stalin briefly discussed the difference between capitalist and communist cities:

> The look of our big cities and industrial centers has changed. In bourgeois countries a prevailing trait of big cities are the alleys, the so-called workers' quarters at the outskirts of town...dark, damp, shaky apartments mainly in cellars, with the mostly poverty-stricken populace living there,

swarming in the dirt, cursing their fate. In the Soviet Union, the revolution made these alleys disappear. Beautiful newly built, clean workers' quarters were erected instead, which in a number of cases look better than the central districts of town. (Stalin 1950, 544)

With little detailed theoretical guidance, Hungarian architects and planners paid close attention to the actual practices of Soviet city planning and urban architecture. This practice was itself the outcome of lengthy debates. In the 1920s several Soviet avant-garde architects had developed diverse ideas for eliminating the defects of "capitalist cities," but their proposals turned out to be unsuitable to the political and social demands that accompanied rapid industrialization under the five-year plans. The official expectation that new cities would be built rapidly was best met by the construction of huge, self-contained "superblocks." Soviet practices thus favored uniform "residential integrated factories" that could be easily replicated (Meggyesi 1985, 50–55).

Foreign influences were especially pronounced in the early Soviet years. Many young Soviet architects were too radical for the highly regimented tasks of state planning, while members of the older generation were effectively banished from their profession for political reasons. In this situation the Soviet government chose a solution well-tested by the czarist regime: they invited Western European and American designers to Russia to realize the grand program for industrialization and construction. Many of these architects and planners were out of work during the Great Depression, while others, more politically minded, were eager for a chance to implement their social and architectural ideas in the first socialist state of the world.

One of the first foreign experts was Johannes B. van Loghem from the Netherlands. Arriving in the Soviet Union in 1925, van Loghem worked on designing five workers' towns for the Khuzbas autonomous industrial colony in Siberia. His plans informed the construction of more than one thousand apartments and numerous public buildings in the following two years. After the start of the first five-year plan, the Soviets contracted with several U.S. engineering firms to construct new industrial establishments. Among them were the Freyn Engineering Company from Chicago; McKee and Co. from Cleveland; and a company from Detroit led by the industrial architect Albert Kahn. Starting with U.S. Steel's model factory in Gary, Indiana, McKee and Co. designed the showcase iron and steel factory in Magnitogorsk in 1930. A design competition for the accompanying township selected the famous German functionalist architect Ernst May (Kotkin 1995, 43; Nevzgodine 2002, 45).

In accordance with Le Corbusier's principles, May imagined Magnitogorsk, the first Soviet socialist city, as a linear city. He planned a series of uniform superblocks parallel to the iron factory, separating the residential and industrial areas by a green belt. However, by the time May and his colleagues finished these elaborate plans, the construction of the factory and the residential buildings had advanced significantly, and there was no room left for the green area (Kotkin 1995, 110–114). Contrary to May's plans, the huge iron factory completely dominated the landscape and the city, and the "temporary" buildings were never dismantled. Instead, these barracks became integral parts of a disorganized city that stretched out in an amoeba-like fashion.

Meanwhile, Stalinist ideology asserted itself in Soviet architecture. A view emerged that a "socialist" city was thoroughly planned and should serve as a "social condenser." In short, a socialist city would be an environment designed to produce a new type of human. A comparatively free and multicolored era of city planning and architecture came to an end in 1931 when the party's Central Committee stated that the primary aim of a socialist city was to serve the socialist economy. A year later party leaders officially condemned modern functionalist architecture as Western "cosmopolitanism," dissolved several different architectural groups, and mandated membership in the official Association of Soviet Architects.

The rejection of Western ideas made the task of foreigners working in the Soviet Union precarious, to say the least, and they faced constant suspicion and xenophobia. Ernst May left Magnitogorsk before his first superblock could be built: he was removed from the construction site in November 1932 and was soon expelled from the Soviet Union. Maybe this was just as well, for May missed hearing firsthand the commissar Grigory Ordiokonidze, during a 1933 visit, state that Magnitogorsk was "a direct insult to Socialism" (Kotkin 1995, 119–120).

All buildings designed under the second five-year plan that started in 1933 followed a new style called social realism. This was a fundamentally nonmodernist, neoclassic ideology that strived for monumentality and applied strict symmetrical forms (Meggyesi 1985, 69–71). The architectural and city planning solutions that would eventually grow into a rigid, unquestionable canon emerged from the large-scale construction projects of the 1930s and 1940s.

Despite the ideological attacks on Western functionalism, some of its principles continued to guide Soviet urbanism. For example, Soviet architects tried to assert the physical separation of industrial and residential zones and the equal distribution of services in the newly founded or remodeled cities. In fact, the separation of city functions became a "golden rule" of socialist city architecture (French and Hamilton 1979, 7–9). The superblock neighborhood system, as noted above, also became a doctrine in Soviet practice. In the 1935 development plan for Moscow, new residential districts were divided into areas from 9 to 15 hectares in size, each accommodating up to 7,500 people and including all basic facilities for the residents. Each superblock consisted of multistory residential buildings and was surrounded by larger streets. Kindergartens and schools were placed in the areas between the buildings, and shops were located on the ground floor facing the streets (Meggyesi 1985, 93, 96).

During and after World War II, the neighborhood units changed somewhat in the Soviet Union. The blocks were integrated into even larger units, called *rayons*, to increase the utilization of the service facilities. The *rayon*, which was also called a residential district, was surrounded by main streets and was planned for up to 30,000 inhabitants. Each *rayon* consisted of four neighborhood units (or *mikrorayons*), each with its own school and kindergarten. Curiously, the superblock system also had its parallel in the West. Already in the 1920s, Clarence Perry had designed neighborhood units as part of the New York Regional Plan. According to his notion, each unit should house about a thousand families, and these units should not only supply the basic facilities (shops, elementary schools, etc.) but also contribute to the formation of local identity and community. The idea had a particularly successful career especially after World War II. This space-structuring solution was also used in the construction of numerous other cities, notably in the British new towns (P. Hall 1992, 43).

Directives for Socialist Cities

The Soviet occupation of Eastern and Central Europe brought about a significant change in architectural theories and city planning in the region. After 1948 the Stalinist model became the sole officially approved building style, and Soviet practices in architecture also became the norm. As we have seen, long before the occupation, Western city-planning ideas and architectural conceptions had been appropriated to fit local conditions as well as to suit communist ideology. By the early 1950s the Soviets had turned into "colonizers" who "exported" social realism and the directives of socialist city planning as the Soviet style and ideology (Oppenheim 2001).

Architectural Stalinism came piecemeal to Hungary and distinctly later than to East Germany. Architects and designers in the GDR felt the first brunt of Stalinism when they were called to Moscow in April 1950 and were instructed to adopt Soviet planning guidelines (Jajeśniak-Quast, this volume). They learned in Moscow much that was familiar. Following functionalist zoning principles, city planning should provide people with appropriate spaces for work, housing, culture, and recreation. Under the Soviet-mandated vision, however, urban centers were defined as monumental places for officially sanctioned demonstrations, processions, and public celebrations. In compliance with Soviet practice, the guidelines laid down zoning as an absolute rule. Each district had its own center for the leading cultural and social institutions. Each housing complex was made up of several housing blocks, and car traffic was not allowed inside the area (Durth et al. 1998, 172–173).

Despite these formal constraints, Hungarian architects and planners were for some years able to appropriate them in their own way. Hungarian architects were not called to Moscow as the East Germans were in 1950, and had no centrally mandated directions about their designs. For example, during the design of Dunai Vasmű, a large iron factory, and of the adjoining city of Dunapentele, the first Hungarian socialist city, Hungarian experts had lengthy coordination talks with Moscow, but only the engineering problems concerning the huge metallurgy plant received detailed attention. In striking contrast, city planning and transportation networks were left largely to local (Hungarian) judgment.[1] Thus the chief architect, Tibor Weiner, drew the city plan of the new Dunapentele and designed its most important buildings without ideological constraints. Weiner knew Western and Soviet interwar architecture firsthand, and was able to design the party headquarters and the first apartment houses in a recognizably international modernist style. Weiner had grown up with the Bauhaus school, worked next to Hannes Meyer in Moscow in the early 1930s, and later spent some years in Paris with Grete Schütte-Lihotzky, a former collaborator with May (Oppenheim 2001).

Weiner was hardly alone in his international outlook; Western visits and experiences figured in the careers of many significant Hungarian architects of the 1950s. Károly Dávid, designer of the main building of Ferihegy Airport and of the Népstadion (People's Stadium) in Budapest, had worked in Le Corbusier's Paris office in 1932. Lajos Gádoros, designer of the MÉMOSZ Offices (the central building of the construction workers' trade union), the Ministry of the Interior, as well as the Hungarian pavilion of the 1958 World Exposition in Brussels, had studied at the Stuttgart Institute of Technology in 1929–1930 and at the State Academy of Architecture in Düsseldorf between 1930 and 1932.

While construction moved ahead at full speed on at least four other socialist cities besides Dunapentele (Komló, Várpalota, Kazincbarcika, and Tatabánya-Újváros), a debate raged among Hungarian architects about "the right" style to follow. In the 1951 volume of the professional journal *Építés-Építészet* (Building-Architecture), a number of architects criticized their own earlier buildings, as well as those of each other, in an effort to appropriate the Stalinist guidelines to the needs of practice. The First National Congress of Hungarian Architects handed down theoretical guidelines in October 1951, declaring that Hungarian architecture was, in accord with Stalinism, to be socialist in content and nationalist in form. This implied that architects ought to break away from "cosmopolitan modernism," find "progressive elements of the past," and create a suitable style that followed the example of the Soviet Union (Simon 2000).

Only the Country Planning Institute (TERINT), a subdivision of the National Planning Office, offered a precise definition of the ideal socialistic city. TERINT in 1951 aimed at turning all existing settlements into true socialist cities: "For a Socialist city, the presence of the working class is essential. The task of the cities in certain parts of the country is to take on a guiding role in politics and economic life, and that is unthinkable without the presence of the working class." The authors made a direct connection between economic policy and the new society, especially the new cities: "the industrialization of our cities is an essential condition of building Socialism."[2]

At about the same time a directive from the Department of Architecture and City Planning at the Ministry of Construction shaped the development plans of already existing and newly constructed cities. One of the numerous requirements of the directive stressed the almost sacred placement of newly built factories: "When placing industrial works we have to use the possibilities that a well-located and well-built factory means for the enrichment of the cityscape. Thus, it is desirable that new industrial works should not be established at the outskirts, but should be given a striking position in the cityscape, in agreement with their importance."[3]

This line of thinking soon recurred in Weiner's own description of Dunapentele, by this time renamed Sztálinváros (Stalin Town). As Weiner described the allegedly Stalin-inspired socialist principles that had guided him:

The Socialist city has no periphery or downtown. The democratic nature of socialist order manifests itself in the fact that all areas of the city are built in identical quality.... Socialist cities and socialist industrial establishments are two poles of an organic unit. Thus the center of town and the main entrance of the factory have to have a direct relation to each other.... Socialist cities have to cater for all manifestations of public life, both in their structural make-up and architectural solutions, regardless of whether they concern individual and family life or the whole of society. A direct consequence of this demand is the shape of routes, city squares, public buildings, and residential areas, all of which have to reflect the monumentality that different social manifestations demand in a Socialist society, from the construction of street areas to the architecture of buildings. (Weiner 1951)

In addition, Weiner stressed that party institutions, public administration, mass organizations, and institutions for public health and culture had to be placed in the center of the city, "according to the different functions of Socialist life." It is noteworthy that commercial buildings, presumably leftovers from the capitalist past, were completely missing from his list.

Weiner would soon be able to demonstrate his ideas to the Stalin-prize-winning Soviet architect G. M. Orlov. On a visit to Hungarian planning institutions, Orlov paid particular attention to the construction of Sztálinváros (formerly Dunapentele), assessing the results during a walk from building to building. He considered the apartment houses to be situated too far from each other, and criticized the iron factory's management building and the cinema. Surprisingly enough, the only structure to get Orlov's whole-hearted approval was the party office building, one of Weiner's assertively modernistic creations (Simon 2000).

The official evaluation of the general plan of Sztálinváros, drawn up by the National Planning Office late in 1952, criticized Weiner's architectural solutions, despite his effort to apply socialist principles. The National Planning Office suggested taking the incoming main road, the "procession route," straight to the main entrance of the iron factory. They wanted to locate the city's main square at the intersection of this road and the one leading from the railway station. This square should become the city center, to "be developed with a due high concentration of buildings, in order that it becomes the most striking, most important part of town." Here they wished to place the most important political, administrative, and cultural institutions, as well as the

Figure 11.1
The Vasmű út (Ironworks Street) in Sztálinváros, or Stalin Town (1952). Beyond the barracks is the first finished building of the House of the Communist Party, planned by Tibor Weiner. Source: Intercisa Museum, Dunaújváros.

requisite statue of Comrade Stalin. The housing areas were to be designed around the main road, which ran in a north-south direction so that the iron factory could be easily reached from all parts of the "city developed in a quadrangle shape."[4]

However, as events unfolded, this officially revised plan for Sztálinváros never received final approval. Indeed, the orderly bureaucratic central plans for all the socialist cities remained something of a fiction. Just as with Magnitogorsk, the planners simply could not keep pace with the actual construction of the houses and factories. Due to the magnitude of the projects, the chaotic circumstances of city construction, and the constant demographic changes, the official plans were continually in a provisional state. Even at the end of 1953 there were no approved development plans for Sztálinváros, Komló, Kazincbarcika, or Tatabánya.[5] Construction simply proceeded without them.

The political changes in 1953—not merely the death of Stalin, but also the workers' uprisings in East Berlin—had a great influence on Hungarian socialist cities. While the building of socialist cities had been an absolute labor of love for party leader Mátyás Rákosi, the new Hungarian prime minister, Imre Nagy, managed to slow down the breakneck development of heavy industry and to interrupt the grandiose city-building projects. With the loosening of Stalinist ideology, the Soviet city planning canon could no longer be enforced. The general development plan of Tatabánya and the plan of its main square prepared in 1954–1955 may be the last large-scale social realist city plan in Hungary (Weiner et al. 1959). Uncharacteristically lacking an impressive industrial factory, the project tried to conjure up monumentality from the administrative center at the main square. The impressive square was to be placed in the foreground of the Gerecse mountains and created by broadening the main road coming from the city of Oroszlány. The square, capable of accommodating up to 30,000 people, was to be surrounded by the buildings of the county and municipal council and of the party committee. A palace of culture stood at the other side. The housing blocks, organized into neighborhood units, would have been built in a symmetrical order starting from the two sides of the main square.

However, this Stalinist pipe dream never materialized either. Social realism had failed as the ruling architectural style, and despite the comeback of Mátyás Rákosi, Hungarian architecture returned to modernism and functionalism (Prakfalvi 2004, 474). Of the previous city-planning elements, only the zoning system and the principle of neighborhood units remained. In the end the general development plan of Tatabánya, approved in 1962, struggled to solve pressing problems of space organization and dropped the attempt to build an imposing main square. The 1962 plan set out the following guidelines (G. Horváth 1972, 201–202):

• a clear, large-scale city structure needs to be created instead of the disintegrating and unmethodical "structure" developed from a number of villages;
• the zonal division of the city should be done taking into consideration the interaction of the types of occupation and townships;
• in place of the existing eccentric and sporadic institutional and commercial network a new city center has to be created, while providing sub-centers for the individual residential area units;
• in compliance with the administrative and cultural tasks, a new municipal council building, party offices, a post office, department store, cultural center, library, procession square and commercial center have to be developed in the center of the city;
• a correlating green-belt system has to be created.

By the 1960s Hungarian city planners had freed themselves from the direct political-ideological grip of Stalin, and their ideas were influenced more by Western architectural fashions than by Soviet practices. However, in the 1970s modern Hungarian architecture became schematic, stereotypical, and repetitive; and people began to react unfavorably to the mass construction of housing estates made of prefabricated concrete slab units. While in Western Europe and other parts of the world a change of paradigm took place in this area, architects in socialist countries were not able to go beyond this concrete-slab style owing partly to ideological considerations, and partly to perceived economical reasons. Modernist mediocrity beset Hungarian socialist cities:

The housing estates were gray and bleak with prismatic buildings of equal height, with hole-like windows and undiscoverable entrances. The blocks drifted into a neutrally diffuse space. Streets, squares and orientation points were lacking. No more imagination was mirrored by public buildings either, no matter if they were constructed in a new or old environment. The plans were mostly based on the simple contrast between tall houses and one-storey buildings of large horizontal expansion (*constructions en galette*) and lacked architectural wisdom, while mirroring the low standards of the building industry in their construction. (Simon 2001, 347)

Precisely because of the prominent nature and political role of socialist cities, alternative architectural ideas—such as applying forms from Hungarian folk architecture or the appropriation of Finnish or Japanese models—had no chance of success there. The most extreme form of homogenizing city planning can be observed in the townships of Salgótarján (1968) and Ajka (1987). The other socialist cities became a sort of waxworks of modern architecture, where all kinds of Hungarian technologies of prefabricated buildings were represented. It is not accidental that the residents of Dunaújváros (formerly Sztálinváros), Tatabánya, Várpalota, Kazincbarcika, and Oroszlány today consider the neoclassicist social realist quarters, amazingly enough, to be the most pleasant: "behind the representative, hard, and expressive neoclassicist main roads [we find] friendly residential blocks of maximum three or four floors with closed, intimate courtyards" (Meggyesi 1985, 72).

The Spatial Structure of Socialist Cities

Along with the prevailing political regime, city planners have to deal with the geographical environment. The existing building stock and historically established urban structure also limit the ability to transform and rearrange a city. This was true with socialist cities, even though in the 1950s planners tried to disregard the existing conditions, convinced as they were of the omnipotence of planning. While discussing the role of economic geography in city planning, Rezső Ruisz stated: "Stalin ... taught us that geographical environment is not a deciding factor and the effect of social development should be better integrated into geographical analysis" (Ruisz 1954, 39).

Stalin notwithstanding, physical geography was an important consideration for most of the Hungarian socialist cities. Eight of the eleven cities were mining towns (Ajka, Kazincbarcika, Komló, Oroszlány, Ózd, Salgótarján, Tatabánya, and Várpalota), where the location of coal mines strongly shaped the urban plan. Oroszlány was the only town with few physical restrictions, while three other mining towns lay in such narrow valleys that the possibilities of expansion were very limited. In the case of Kazincbarcika, lying in the Sajó valley, changes in the water level had to be taken into account. Tatabánya was hemmed in by the Gerecse mountains and could develop only in a north-south direction.

The new industrial cities lacked a traditional built environment, and thus the urban fabric had to be created artificially. Yet of the eleven cities, only three (Dunaújváros, Leninváros, and Százhalombatta) had geographical features that allowed the artificial establishment of a complete city structure. Százhalombatta reached its limits after only 10 or 15 years, however. Jammed between the Danube, the industrial plant, and a nearby motorway, it was constrained into a ribbonlike expansion much like Tatabánya.

Naturally, coal mining towns are dependent on the amount of exploitable coal supplies. In Salgótarján, for example, planners saw a clear risk that the mines in the immediate vicinity would soon be depleted. If in this event the mining activity moved to a nearby valley, one writer admitted rather ominously, "We should then rather consider the establishment of a new city in the direction where mining will develop." Another consequence would have been long commutes to distant coal mines, if Salgótarján's residential areas had been extensively developed (Ruisz 1954, 29).

Tatabánya faced a similar problem in a more creative manner. The general development plan of the city, drawn up in 1954–1955, predicted that the nearby coal beds would be emptied in less than 40 years. As discussed in a long-range development plan, the coal bed lay outside of the city and stretched in the direction of the rival town of Oroszlány. Accordingly, the coal pits would move ever farther from Tatabánya and ever closer to Oroszlány. "Tatabánya will grow into a Socialist miners' town exactly during the 40 years while the coal in its vicinity runs out," went the sharp diagnosis. To deal with this problem, the planners suggested building other chemical industries to process local bauxite and coal, since this might better employ the city's working population.[6]

Despite these dire predictions, the development of Salgótarján and Tatabánya hardly slowed down even when their coal supplies diminished, because both cities were county seats in addition to being mining towns. Instead, regional and city politics solved the problem in the 1950s and 1960s by defining so-called mining housing centers (Kóródi and Márton 1968, 126). This policy directed the construction of housing estates for miners and tried to ensure acceptable urban living conditions. Thus, even if certain mines were exhausted, the workforce itself did not have to be relocated; the problem was redefined as "only" one of proper transportation. This centralized solution embodied a strong bias in favor of larger cities, since after this period no substantial investment was ever made in the numerous smaller mining villages.

All newly designed cities have to develop a relationship to already existing human settlements. In this regard, the socialist cities that had the easiest time were the "greenfield" projects, such as Dunaújváros, Leninváros (renamed Tiszaújváros in 1990), Oroszlány, and Százhalombatta. They had been built close to relatively small villages, and only these previously rural parts—later called "the old city"—had to be integrated into the new city. The situation was somewhat more complicated when a number of existing villages were joined together, or when cities were established by means of administrative integration. In the cases of Ajka, Kazincbarcika, Tatabánya, and Várpalota, the previous villages and settlements were from three to nine kilometers apart. Ózd and Salgótarján existed as separate administrative units before being joined into a single city, although the resulting districts remained fragmented.

All six towns mentioned above had significant industrial establishments. In Salgótarján there were the Rimamurány-Salgótarján Ironworks Ltd., the mining company, the glass factory, and the Hirsch factory; in Ajka, the mine, the power plant, the alumina factory, the glass factory, as well as the railway; in Kazincbarcika and Várpalota, the mining company. In the villages that preceded Tatabánya the Hungarian General Coal Mining Company built colonies, while in Ózd it was the Rimamurány-Salgótarján Ironworks. These settlements obviously served the needs of the companies for ready industrial labor, and at the same time the workers' housing could be considered as a kind

Figure 11.2
The "six-door houses" in Tatabánya built as a separate colony by the Hungarian General Coal Mining Company for its workers in the 1910s. Each long house contained six flats (hence the name "six-door"), each with one room and kitchen. The common outhouses and coal holes were on the buildings' ends. The last six-door houses were abandoned in the 1980s. Source: Museum of Tatabánya. Photographer: Ferenc Sebestyén.

of social welfare. The resulting lack of a recognizable city structure, however, hampered the spatial and cultural integration of different parts of the settlement. As Zoltán Szabó put it as early as 1938 in his famous sociography entitled *Cifra nyomorúság* (Genteel poverty): "today there are four Salgótarjáns.... the actual city center lies in-between the four company areas. It does not belong to any of them and is not important to any of them, just like the city of white-collar workers across the railway lines" (Szabó 1938/1986, 231–232).

The administrative unification of separate townships might be sought for economic or administrative reasons. Still, strange situations often developed. In the Tatabánya area, for example, the kitchen of one newly built house was in the village of Bánhida, while its other rooms were built on Felsőgalla's soil; and the village hall of Tatabánya-Bányatelep stood on the territory of neighboring Bánhida. No wonder that Zoltán Magyary and István Kiss wrote a whole book in 1939 arguing that the fast-developing neighboring industrial villages should be integrated with existing structures and turned into cities as soon as possible (Magyary and Kiss 1939). And it is no surprise that the reason for designating the townships of Tatabánya in 1947 and Ózd two years later as cities was the development of these townships into integrated city units. Even decades later the different parts of these cities had not melded together. Administrative changes since 1990 have recognized the futility of forcing integration of these once-independent districts. For example, Farkaslyuk was separated from Ózd, and Berente from Kazincbarcika.

The Appropriation of Soviet Planning

Sztálinváros (from 1961, Dunaújváros) was the first Hungarian city where Soviet-style urban planning was successfully executed to a significant degree. City quarters were separated by extensive parks in the 1950s, and although these were partly turned into construction land later, the city still remained pleasant and breezy. In Leninváros and Százhalombatta the huge industrial plants were well separated from the housing estates. Due to its expansive oil refinery and power plant, the industrial zone of Százhalombatta will soon reach its residential area, however. In Oroszlány the mining facilities were not situated in the town, and the power plant was also built at some distance.

In Kazincbarcika and Várpalota, the separation of the industrial and residential zones did not succeed completely. Due to the terrain, the Borsod Chemical Factory in Kazincbarcika was built in the area between the former townships of Sajókazinc and Berente. When the industrial works began to expand in the 1960s, it had to grow around the old village. In Várpalota a smaller industrial zone became wedged into the old central part of town. In opposition to the zoning principle, however, so-called stand-by housing estates were built next to the industrial works (for example, the Nitrogen Works of Pét and the power plant of Inota).

In the other socialist cities (Ajka, Komló, Ózd, Salgótarján, and Tatabánya), "the factory itself became the organic center," to use the words of Lewis Mumford (1985, 425). The metallurgical works, power plant, and mining facilities lay in the very center of Ózd and Komló, and covered the best soil in the valley. Factories, head frames, mountains of slag, flue dust deposits, waste rock piles, red mud depositories, and external mining areas alternated with village-like areas and residential colonies. Although in the early 1950s planners tried to separate residential areas from industrial zones, almost all open areas were quickly filled up when huge numbers of workers flooded into these cities. The demolition of older settlements had always been on the agenda, but even as late as the 1990s thousands of residents still lived in one-room apartments nearly a hundred years old that entirely lacked modern conveniences.

Due to their limited authority in the socialist years, municipal governments could not effectively use zoning to remove or restrict the industrial activities that created the most pollution. Not even the development plans were effective devices. In the 1962 general plan for Tatabánya, for example, the possibility of removing industrial works from the city center was not even proposed; the planners only allocated a separate zone for newly established factories and urged an increase in green landscape areas (G. Horváth 1972, 204). The "interests of the national economy" proved to be more important than the health of the population, and city councils were unable to win political battles with the national ministries. Although pollution reduction became an issue for socialist cities in the 1970s, the only factory shut down for this reason was Tatabánya's cement works, which was closed in 1984. Between 1945 and 1990, no genuine proposals were made to remove or relocate the polluting industries in socialist cities. In part, this passivity had economic reasons: after all, the industries usually provided the financial basis of the towns. The most polluting industries were removed from Budapest in 1959, but only to be relocated at other sites outside the capital.

The construction of socialist cities as homogeneous compositions was possible only in the cities built in a "greenfield" manner. But, as I have already shown, even Sztálinváros had no centrally approved urban plan. The long-term or general development plans of socialist cities—embodying the principles of city center creation, land use, and zoning—were commonly approved only after the fact in the 1960s. Moreover, due to the changes in development policies and an unexpectedly rapid population increase, these plans frequently had to be revised repeatedly. For example, Ajka's first development plan was drawn up in 1954, the next one in 1957, then another from 1963 to 1965. The latter was finally approved, but had to be changed again in 1975. This version, in turn, was approved in 1979, only to be revised again in 1985 (Giay 1990, 35–36).

In these newly built cities, industrial establishments were usually given a central, almost sacred position—a demand that reflected the ruling ideology of the early 1950s.

Figure 11.3
The metallurgical works of Ózd in 1972. The steel mill is situated in the center of the town, so the downtown was polluted by dust and smoke. In the 1970s a new rolling mill was built, but the technology had not changed radically. Source: Ózd Municipal Museum.

Certain factories in Ajka, Ózd, Salgótarján, and Tatabánya were so closely intertwined with the residential areas that nobody thought of emphasizing them further by architectural means. In some cities, however, this principle was not realized, as in Várpalota, where the Inota power plant and the aluminum smelter were too far away from the city to comply with this idea. In other cases, the industries were potentially too dangerous to be situated among residences. Since the chemical works of Kazincbarcika had originally been conceived as a gunpowder factory, it was out of the question to locate it in the downtown area. On the contrary, there was a debate about whether the safety zone surrounding the potentially explosive factory should be 500 or 1,500 meters wide. Similarly, in the cities belonging to the two great chemical factory complexes built in the 1960s and 1970s, Leninváros and Százhalombatta, the idea of integrating the dangerous factories into the cityscape never arose.

It would also have been senseless to turn the mines of Komló, Oroszlány, and Tatabánya into shrines, as work was done underground. Because it had technical tasks to fulfill, the mine's head frame also would have been virtually impossible to decorate. The only measure architects could take was to build a modern office for the mine's director, but even this building had to be erected outside the city center. Consequently, Dunaújváros is the only one of eleven Hungarian socialist cities where a close relationship between town and factory was created, in terms of both space and spectacle.

Two of the requirements mandated by Soviet planners were somewhat contradictory. The particularly emphatic formation of the socialist city's central square contravened the often-quoted principle of "no separate downtown and periphery." But, as already mentioned, the socialist city center took on a new ideological function. During the construction of socialist cities in Hungary, planners and party leaders aimed whenever possible at a monumental structuring of space. For example, Mátyás Rákosi himself scrutinized the plans for the main square of Sztálinváros, and the Political Committee of the Hungarian Workers' Party also discussed them.[7]

In the end, however, hardly any part of the grandiose projects conceived in the Stalinist era actually came to fruition. No main square and no monumental city center were built in Sztálinváros; only the city hall was completed, in addition to the party offices, erected between 1965 and 1967 after a fifteen-year delay. This delay was evidently connected to the lack of an approved general development plan even in 1967. In fact, similar to the case in Ajka cited earlier, by 1967 almost twenty various city plans had been drawn up (Szíj 1967/1981, 283). In Tatabánya, as well, only fragments of the main square's monumental building complex were realized. Interestingly enough, these buildings were integrated into the traditional structure, and placed on the axis of the monumental statue of Turul, the mythical eagle of ancient Hungary, which had been raised on the peak of Kő Mountain in 1896 in commemoration of the millennium. The palace of culture and the building complex of municipal, party, and administration offices were supposed to be designed in the style of the *cour d'honneur* of eighteenth-century palaces. The only part to be constructed, however, was the wing of the county council. A step-by-step development of the square continued into the 1990s, and today there is a sprawling car park in the middle of the square. One can sense this type of uncertain, temporary character in the center of most socialist cities.

Nor was the core Soviet planning principle of "equal quality of living in all parts of town" ever realized. A class system of apartments developed in Sztálinváros in the early 1950s. As Sándor Horváth (2000) describes it, the quality of one's apartment depended both on one's status at the workplace and on one's position in the hierarchy of professions. The best and biggest apartments naturally went to the managers and higher ranking employees of the ironworks, as well as to some favorably treated skilled laborers. Similarly, Viktória Szirmai (1988) observed that leading intellectuals and officials lived in the latest housing estates, the so-called elite staircases. Mihály Andor and Péter Hidy (1986, 22–23) also show that the spatial structure of Kazincbarcika in the mid-1980s directly reflected social stratification.

At the end of the 1950s, the once-central role of the monumental city center disappeared with the realization that the grand main squares were never to be built. Socialist cities instead became typified by the "superblocks" or neighborhood units. The result is the cityscape described in *The Geography of Kazincbarcika*: "On our way toward the interior we do not find the usual divisions as in other cities, where you gradually reach an enormous, densely populated center with a high concentration of buildings through a field of one-story family houses with gardens. [Here] the modern blocks of houses almost grow out of the plow lands" (Frisnyák 1979, 156).

City centers retained one vital function, intriguingly enough connected to consumption. A strictly hierarchical supply scheme organized consumption in the system of neighborhood units. As József Kóródi and György Kőszegfalvi (1971, 51) described the situation in Dunaújváros: "the way in which the system of residential areas were structured and organized reveals an effort to create an equally comprehensive communal and public supply network: the hierarchic, step-by-step method of institutions has to be able to satisfy daily and periodical needs." Much as cities were hierarchically ranked, city planners thought of matters within the city in terms of a hierarchical system of distribution and supply. Planners intended to assure that daily needs were met at the housing-block level and weekly needs at the district level, but to meet more serious needs, one still had to go to the city center.

Even though the debate about what to "do" with city centers was never conclusively settled, there was an evident change of policy around 1970. From then on, socialist city centers benefited from large-scale investment in new shops and public buildings. In Leninváros, a building boom in the 1970s brought the construction of many public buildings, as well as the central department store and several larger grocery stores, while the city hall was inaugurated in 1981. In Salgótarján, large-scale construction of the city center and a homogeneous main square was realized just before 1970. The layout of the center, however, followed the previous—social realist—structure. The square, which opens onto the main road at right angles, is surrounded on three sides by bulky masses of buildings, the department store Pécskő, point blocks, and a tall block of flats (Prakfalvi 2004, 495). The new educational, medical, cultural, and administrative institutions were established at this time, too. In Ajka the city center was redeveloped during the 1970s, complete with a new cultural center, office building, department store, shopping street, hotel, post office, and coach station. However, there was no money left for the construction of the district court, the museum, the cinema, and the library.

Yet, even with all these public institutions and offices, lively city centers rarely came into being. One reason for this may be that the function of these cities—with the exception of the two county seats—remained all the while industrial. Socialist cities entirely lacked wholesale and long-distance trade. Educational institutions conformed mostly to the needs of industry; of the industrial cities, only Salgótarján had an institution of higher education. Furthermore, these cities never became the cultural center of their regions; adequate restaurants and hotels were never built. The majority of socialist cities simply lacked the community institutions and public spaces that can create and sustain social relations and a vibrant city center.

The presence of unbuilt areas also contributed to the fractured and undeveloped nature of socialist cities. In a number of cities the housing estates, rural areas, and town districts are separated by some kilometers from each other, with nothing but weedy fields between them. The spatial disintegration was further increased in Ajka, Komló, Salgótarján, and Ózd by the annexation of their surrounding villages. A total of sixteen villages disappeared as independent units, but none was architecturally integrated with the "mother city"; in fact, some of them were as much as eight kilometers away. Of course, placing them under city jurisdiction made a certain sense, but the villages kept their previous character, and their inhabitants never identified with the "socialist city."

Conclusion

This chapter has pointed out the importance of circulating ideas and conceptions, as well as their national or local appropriation. At the beginning of our story, in the 1950s, the Soviet model imported from abroad appeared to organize apartment buildings into blocks and neighborhood units. Soon, however, Imre Perényi and his colleagues—modifying the principles of the Soviet architect Ia. P. Levchenko—developed the Hungarian version of *rayon* theory, and tried to organize groups of housing blocks into districts or quarters on a larger scale (Perényi 1952). From the second

Figure 11.4

In the 1970s and 1980s the city of Ajka developed in a new way, with more parks, a lakeside leisure area, and a reshaped city center. In 1987 Ajka won the most prestigious Hungarian urban studies award, the József Hild Medal. Source: Nagy László János Könyvtár és Szabadidő Központ, Ajka.

half of the 1950s this idea became the leading vision as new districts of industrial towns were constructed in Hungary. Then, after the post-Stalinist thaw and increasingly from the 1960s onward, Hungarian architects also adopted Western neighborhood models that they considered more up-to-date. In fact, the modernist model continued to exert an influence in Eastern Europe well into the 1980s, long after it had begun to be harshly criticized in Western Europe (P. Hall 1992, 43).

In Hungary, we find the system of housing blocks or units in each socialist city. The result was a large degree of homogeneity across the country. In an area of Tatabánya called Újváros (New City), six massive neighborhood units were established, and in Tiszaszederkény-Leninváros enormous units that housed 10,000 people were constructed. The completion of the necessary services (shops and public institutions) planned for each unit was almost always behind schedule, if completed at all. In the case of the last large-scale construction of a housing estate, in Dunaújváros in the early 1980s, the supply subcenter and a full infrastructure were in fact never realized.

Perhaps the most interesting aspect of the Hungarian story is that important elements of Western interwar urbanism, modern as they were at the time, arrived in Hungary through Soviet intervention. Or, more precisely, these modernist ideas actually *returned* to Hungary via the Soviet Union. A few city planning principles carved in stone at the beginning of the 1950s remained valid until the political changes of 1989. The creation of functional zones and the system of neighborhood units proved to be the most lasting concepts. We notice that in the development of socialist cities a sternly technocratic attitude prevailed—in the narrow and worst sense of the word. The focus of attention was never the city and its inhabitants, but always the industrial establishment. By continuously economizing, planners, especially in the 1950s and 1960s, ignored the facilities necessary to create a friendly atmosphere and a humane environment: such amenities as dividing belts, green space, social meeting places, and recreational institutions remained unrealized for many years in the majority of Hungarian socialist cities. Unfortunately, these desirable aspects of modern architecture and city planning did not find a ready home in Hungary.

Notes

1. See János Sebestyén, "Report on the Moscow Talks between 15 April and 15 May 1950," 18 May 1950, Magyar Országos Levéltár (Hungarian National Archive; henceforth, MOL) XIX-A-2-ii, 4. d.

2. Documentation of the Country Planning Department of TERINT, 21 December 1951, MOL XXVI-A-1, 17. d. Volume 11, p. 9.

3. Ibid., p. 12.

4. MOL XIX-A-16-b, 339. d.

5. Letter of the National Office of Construction to the regional planning department of the National Planning Office, 9 November 1953, MOL XIX-A-16-a, 490. d.

6. "Report on the Present State of Tatabánya and of the Plans for the New Tatabánya," d. n. MOL XIX-A-2-ii 38. d. 1423/1955, p. 3.

7. See the records of the committee's meeting of 21 October 1953, MOL M-KS 276. f. 53. cs. 142. ő. e. The proposal presented to the Committee, entitled "Plans for the Main Square of Sztálinváros," was reproduced in Prakfalvi (1992: 148).

12 Mediators of Modernity: Planning Experts and the Making of the "Car-Friendly" City in Europe

Per Lundin

On Saturday, October 16, 1954, readers of the Swedish morning daily *Svenska Dagbladet* learned how "the car society" was revolutionizing life and culture in the United States: when you returned books to the library in Miami, you did so from the sidewalk, and leaving shoes for repair in Detroit was equally convenient. Across the United States there were innumerable drive-in cinemas and drive-in restaurants; in Denver you could even attend Sunday service in a drive-in church. The paper enthused: "These are all ingredients of a society that has begun to adapt to the car" (*Svenska Dagbladet*, October 16, 1954).[1]

One might have assumed that the newspaper's foreign correspondent was reporting home on life in 1950s America. But this was not the case. *Svenska Dagbladet* was actually reporting on a lecture given by the Swedish traffic engineer and town planner Stig Nordqvist. On returning from a year's study of traffic engineering in the United States, Nordqvist discussed the social and cultural importance of the automobile. He had been awarded a "dollar scholarship" by the Swedish Road Association to obtain advanced training at Yale University's Bureau of Highway Traffic. The Swedish Road Association had arranged the scholarship through the International Road Federation (IRF), a lobby organization that since 1949 had been sending to the United States European engineers working in the field of road traffic. At Yale Nordqvist and his fellows studied traffic engineering, a newly founded scientific approach to the problem of traffic management. Nordqvist spent nine months at Yale and three months touring the United States.

Svenska Dagbladet repeated Nordqvist's observation that parking spaces and traffic facilities took up "tremendous areas" in the United States. He explained that a high-grade freeway required a width of 150 to 180 feet (50 to 60 meters), and that urban areas needed a lot of these wide strips. No less than 40 percent of Los Angeles was occupied by parking spaces, and the big new regional shopping centers might have parking areas five times larger than the shopping area itself. Nordqvist emphasized that the needs of traffic would dominate the land area of future cities, and that urban building could no longer continue at the accustomed densities, unless special pedestrian-only precincts were created, encircled by arteries for motor traffic. Nordqvist saw such shopping centers as "pedestrian islands in the car society."

Even though Nordqvist had been engaged in advanced technical studies, the news reporter's attention had been attracted chiefly by his social and cultural observations of the American automobile society. In the paper's view, Nordqvist painted an attractive picture. The car culture was a comfortable society, characterized by a "drive-in mentality"; extensive automobile use opened up residential areas and allowed more people to live in single-family homes. These new areas were indeed of such low density that it was not economically possible to serve them by bus or tram. The automobile also brought America's great national parks within the reach of ever more people, adding an entirely new dimension to recreation.

Svenska Dagbladet's article illustrates the way in which technical experts such as traffic engineers acted in the 1950s as purveyors of powerful visions. These dreams concerned the possibility of material prosperity, a higher standard of living, and a new life that combined the advantages of town and country. Across much of Europe, the public eagerly embraced this message and readily credited the experts with an almost prophetic role. There was an apparently insatiable demand for the experts' images of the United States and the knowledge and experience they had acquired there. The experts' pronouncements on social change were given credibility by their base of technical knowledge and their scientific approach. In fact, Nordqvist recalls that he was given the label of "traffic expert" only after his return home (Lundin 2004).

These "traffic experts" were a group who shared many values and had comparable educational backgrounds. Above all, they broadly agreed on what the problems of society were and how these could be addressed and solved. Their ideas circulated throughout Europe and the rest of the world; the IRF's scholarship program alone sponsored 216 engineers from 60 different countries in its first decade (Seely 2004). A reunion in the summer of 1955 in Wiesbaden, Germany, brought together eight European traffic engineers who had studied at Yale, among them Sweden's Nordqvist, and three highly placed officials from the IRF's European office (*Annual Report...* 1955).

This chapter examines how urban planning ideas and practices circulated among planning experts in postwar Sweden and between Sweden and other countries. I describe the circulation of knowledge from a Swedish perspective, and explore where the knowledge came from, how widely it was disseminated, and how and by whom it was spread. In particular, I analyze the content of the knowledge in terms of modernity, and show how knowledge about traffic engineering and urban planning was appropriated in the planning of Swedish towns and communities. While concentrating on Sweden, I also draw connections to other European countries. I argue that the public reception of the experts' messages can best be understood as a longing for modernity, and that the experts of the 1950s and 1960s succeeded in articulating that desire. They formulated it in terms of progress, science, efficiency, and rationality—expressions curiously devoid of all architectural, historical, political, and social references (Rabinow 1989). These planning experts can thus be seen as mediators of modernity.

Europe's Desire for Modernity

Postwar Europeans longed for modernity. The long years of war had left Europe in ruins. The civilian population suffered from the organized destruction, "total war," and systematic bombing raids on cities. The Dutch city of Rotterdam and British cities such as London, Sheffield, Coventry, and Plymouth were badly hit, and half of Germany's urban areas were destroyed—Dresden, Duisburg, Essen, and Bochum, to mention just a few. Warsaw and Budapest had systematically been laid waste by Nazi Germany (Hohenberg and Lees 1995).

During the first postwar years the war-ravaged countries of Europe concentrated on the reconstruction of cities and infrastructure. After the privations and barbarities of war, Europe largely turned its back on tradition and history and instead directed its gaze toward a brighter and more prosperous future. In this volume Hans Buiter describes how bomb-flattened Rotterdam became a tabula rasa for town planners, an opportunity to realize a modernist utopia. Ambitious modernist plans were also drawn up for the rebuilding of Berlin, Cologne, and London. Coventry, Sheffield, and Plymouth in Britain and Dortmund in Germany became virtual playgrounds for modernist planning. Yet most of these early grandiose projects got no further than the drawing board. For several years, a severe shortage of financial resources, building materials, and skilled labor prevented large-scale building. In some cities where the physical environment had a strong symbolic value, the demolished areas were reconstructed: Warsaw, Budapest, and Nuremberg were rebuilt as almost identical replicas of their former selves (Diefendorf 1990). While some cities tried to restore a lost past, others saw an opportunity to put the past behind. German cities especially found it desirable to espouse modernity; to embrace the future became a favored survival strategy.

The first decade or so after the war saw economic growth in Western Europe accelerate at an unprecedented rate, and the boom was to a great extent fueled by the reconstruction and reshaping of its destroyed cities. Government intervention and planning went hand in hand with the concentration, modernization, and automation of industry, and social safety nets were extended to create a welfare state. In Sweden, formally neutral during the war, economic development took the form of road-building and housing construction, the harnessing of hydroelectric power, and the rationalization of agriculture. The 1950s and 1960s came to be known in Sweden as *rekordåren*, or the boom years. Germans referred to the economic upturn as *das Wirtschaftswunder*, and in France there were *les trente glorieuses*, the thirty years of glorious growth. Eric

Hobsbawm (1994) describes the period from 1947 to 1973 as an exceptional golden age. Tony Judt (2005) places the economic takeoff somewhat later, in the 1950s, and emphasizes the variety of circumstances that existed within Europe, but largely agrees this was an unusual "age of affluence."

More than ever before, the United States represented the modern society to the rest of the world. Thomas Hughes (1989) has drawn attention to "the second discovery of America" in the 1920s, when it was seen as a full-fledged technological landscape and the world's most productive nation. But even if the United States was perceived as a technological nation made up of inventors, scientists, and industrialists, many Europeans still saw it as a cultural backwater; in conservative German circles in the 1920s, people spoke disparagingly of American materialism, rationalism, and mechanism as worrisome threats to German culture. Thirty years later, the United States had become a positive model to postwar Europe, both culturally and nationally. Not only was the United States the victorious nation that had survived the war largely unscathed; it was also the economic and cultural engine of the new world order. The organized economic support given by the Marshall Plan was a decisive factor in the economic recovery of Western Europe. In *Irresistible Empire*, Victoria De Grazia (2005) shows how the once-distinctive European middle-class lifestyle gradually gave way to American mass consumerism. For all these reasons, the United States came to be the embodiment of postwar modernity.

One tangible expression of this desirable modernity was the automobile, and the culture and values that accompanied it. The automobile embodied key defining attributes of modernity: freedom, mobility, democracy, material prosperity, comfort, masculinity. It is no coincidence that Roland Barthes (1957/2002) described the automobile as "the supreme creation of an era," modernity's equivalent of the cathedrals of the Middle Ages. The desire for the automobile may indeed be interpreted as a desire for modernity. Even before the First World War the Italian futurists hailed the automobile as a symbol of speed, beauty, and virility. With his machine-inspired modernistic urban visions, Le Corbusier became the most influential architect and town planner of the twentieth century (Wollen and Kerr 2002). In 1930 Le Corbusier devised a town planning scheme he called *la ville radieuse*, later published as *The Radiant City: Elements of a Doctrine of Urbanism to Be Used as the Basis of Our Machine-Age Civilization* (1933/1967). There he presented a city divided into distinct functional sectors: residential, civic, business, factories, and heavy industry. Of the remaining land, 85 percent was intended for use as parks, playing fields, and express motorways linking the city's various functional districts (see Misa, this volume). It was a city built for speed, with pedestrians and high-speed vehicles kept apart and slow traffic separated from rapid. But the time was not yet ripe for these ideas. Not until the 1950s and 1960s did it become financially feasible for workers in Europe to realize the dream of owning an automobile.

European Mass Motoring

Mass motoring in Europe arrived with the economic, material, and social development of the golden years. As automobile ownership percolated down and throughout the social structure, mass motoring became strongly associated with modernity. Georges Gallienne, chairman of the IRF's European office, asserted in 1950 that the automobile was "one of the best means of expressing individual liberty and one of the greatest factors in the development of the human personality." He depicted the automobile, in sharp contrast to the railway, as an inherently progressive and liberal mode of transport, "a powerful means of transport, set up as in a single unit, at the service of the mass of the people" (quoted in Blomkvist 2004, 283).

Automobile ownership by the masses also became a symbol of a free and united Europe. In 1950 the United Nations' Economic Commission for Europe (ECE) presented a plan to link the existing national road networks into a gigantic system of pan-European highways. The Marshall Plan worked along with the united European-American road lobby to achieve this vision. The road lobby believed that mass automobile ownership would break down ideological, cultural, and political boundaries. By 1990 this network of European highways covered more than 40,000 kilometers, and today it manifests the common European identity longed for by the postwar political visionaries (Mom 2005; Blomkvist 2006).

Even as some visionaries argued for a common European road network, the course of events across Europe varied according to local circumstances. Every country developed its own dynamics in adapting to mass motoring. Until at least 1930, the motorcar was mostly a middle-class phenomenon and often met with opposition from the labor movement. Automobiles were driven by wholesalers, engineers, doctors, and the urban middle class, and their values permeated motoring. After the Second World War the automobile became for many a symbol of a free, liberal, and economically flourishing society. One of the countries that wholeheartedly welcomed mass automobile ownership was the Federal Republic of Germany. By affirming mass motoring and the modernity it represented, West Germans tried to shake off their troubled past (Sachs 1992; Wolf 1996). But there were exceptions. In the United Kingdom the car persisted as a privilege of the few, and the unsympathetic attitude of the British labor movement reinforced this tendency (W. Plowden 1971). In Norway leading social democratic politicians expressed skepticism about the automobile throughout the 1950s, arguing that the automobile stood for the old, inequitable society. Norway actively regulated the import and sale of private vehicles, so that only in the early 1960s did the number of automobiles began to rise rapidly (Østby 2004).

In contrast to Norway and the United Kingdom, Sweden evinced a very broad-based support for the automobile. Between 1950 and 1960 the number of automobiles rose nearly fivefold, to more than a million, and Sweden achieved Europe's highest per capita automobile density. During the 1950s the automobile was desired by ordinary people, industrial and commercial leaders, and governing politicians. Together with atomic energy, automation, housing construction, and various welfare schemes, automobile ownership gave expression to a distinctive Swedish modernity (Johansson 2004).

Social democrats held power for the greater part of the postwar period, and their attitude toward the automobile conditioned how Sweden adapted to mass automobile ownership. During the 1940s and early 1950s the Swedish labor movement, like its British counterpart, was wary about automobile ownership; it too saw the car as an upper class luxury. But in the early 1950s the social democrats' view changed because the economy was growing and ever more families could afford to buy cars. At this point, the view of the automobile as an exclusive class symbol could no longer be sustained. By 1956 the social democratic minister of transport and communications, Sven Andersson (1956, 3), rhetorically asked whether mass motoring was "the most visible aspect of the democratization of our time." The social democrats even aligned the motorization of Sweden with the greater modernization project called *folkhemmet* (the people's home). Since both liberals and conservatives had already expressed their support for putting Swedes on wheels, mass car ownership did not generate any great political controversies (Lundin 2004). But even if there was a general political consensus in Sweden and in many other European countries, the rapid rise in the number of automobiles did not occur without friction. The most violent clash was between mass motoring and the European city.

Mass Motoring Comes to the City

Without doubt, the automobile is one of the most important forces that transformed the city in the twentieth century. Urban development in Europe over the last 150 years may conveniently be divided into two main phases. During the first period, from the mid-nineteenth to the mid-twentieth century, the structure and morphology of cities was changed by transport technologies (railway, tram, bus, and underground railway) and various municipal technologies (gas and electricity, water and sewage, and district heating), as Dieter Schott discusses in his chapter. In the second period, roughly from the 1950s onward, the automobile dramatically accelerated the scale and pace of changes initiated during the first period (H. Andersson 1981). In spatial terms, the growth of mass car ownership sharply reduced the density of urban building and sharply increased congestion in the city center.

The spatial and social consequences of mass motorization were first seen in the United States. Los Angeles was the archetype. As early as the 1920s Los Angeles experienced mass car ownership at levels not reached in other U.S. cities until the 1950s or in European cities until the 1980s. The implications of mass motoring found their clearest expressions in Los Angeles, with the result that the city became something of a laboratory for the rest of the world (P. Hall 1998). Architects, engineers, planners, and politicians from all over the globe visited this Californian cityscape—and were both enchanted and terrified.

One of these visitors was Arne S. Lundberg, undersecretary of state at the Swedish Ministry of Transport and Communications. He reported home to Sweden on how the U.S. city was changing: "The private car has made possible a previously unimaginable expansion of the urban area and it looks capable of completely changing the whole organization of urban development. I repeat that it is not the skyscraper but the far-reaching, widely spaced, low-level building that is the salient feature of American cities. Many find Los Angeles a deplorable form of city, but is it impossible that the car city of the future will find its shape and be better than the city of pedestrians and trams?" (A. Lundberg 1953, 42).

While the spatial dispersion of the motorized city was striking, congestion was the most obvious immediate consequence. The lack of space and the struggle for the street had long been a part of the urban scene, but with mass motoring the problem became acute. The automobile soon became its own worst enemy, with cars parked on crowded city streets obstructing the flow of moving vehicles. As U.S. city centers were jammed, many shop owners fled to new suburban shopping malls, with the result that vast tracts of inner cities turned into slums. In Chicago Lundberg witnessed the characteristic American solution of elevated expressways that sliced through the inner city to connect automobile drivers with the widely scattered suburban areas. Seeing new expressways being built at fantastic cost in all the major cities, he described them as breaking through the urban fabric "with the force of an atomic bomb" (Lundberg 1953, 42).

As car ownership and automobile use reached undreamed-of levels, these problems also became manifest in European cities. The meeting of the automobile and European cities, with their old centers and acute shortage of space, generated different conflicts than in the United States, where, after all, many cities had grown up with the car. City streets in Europe had to be widened to accommodate car traffic, and this process often involved the confiscation of land and the demolition of buildings. After attending the second European conference on traffic engineering in Bürgenstock, Switzerland, Swedish road engineer Nils von Matern (1954, 385) reported that the greatest problem for mass motoring in Europe was that "the old cultural milieu [of Europe] creates ... difficulties in implementing firm but rational interventions."

In the mid-1950s Sweden was clearly at a crossroads: Should motoring be restricted or embraced? Automobiles were causing congestion in old city areas, and doctors began to speak of fatal road accidents as the modern tuberculosis. Something had to be done. But even if the consequences of mass motoring were soon seen as serious social problems, there was no overarching or comprehensive political vision of the

Figure 12.1
"The car society is at the door" was the message to Swedish towns in 1958. Source: *Rädda Storgatan!* (Save Storgatan!), Handelskamrarnas nämnd (Stockholm, 1958).

shape a car-crowded future should take. Few people could formulate the problems and even fewer could solve them. This is where the experts stepped into the picture.

Architects of the Car Society

In the depoliticized Sweden of the 1950s, with its great faith in modern technology and science, technical experts enjoyed great freedom of maneuver. They had the ear of politicians and the public alike, and were given a mandate to formulate a vision of the future. By depicting "progress" as inevitable, these experts sidestepped several difficult choices. They asked only whether Sweden should adapt car use to the "old society" or should build a "car society" (Nordqvist 1955, 31). This new automobile society appeared inescapable, and restricting automobile use was never a realistic alternative.

This attitude was clearly demonstrated at the Swedish conference Bilstaden (The Car City). The meeting, held in Stockholm in September 1956, attracted enormous interest. Gathered in the crowded hall were architects, engineers, town planners, local government officers, and politicians from all over the country. Most leading Swedish technical experts were there, and the ideas presented in lectures, debates, and discussions circulated across the country. The conference was led by the architect Uno Åhrén, who had become Sweden's first professor of town planning in 1947. In the late 1920s and the 1930s Åhrén, a follower of Le Corbusier, became known as the prime theoretician of Swedish functionalism and one of its leading practitioners. He sounded the conference's keynote:

The enormous expansion in motoring has become the most acute problem of social planning. Forecasts of the future numbers of cars in our cities have had to be revised as time has gone by.... But the car must not be seen as something that causes trouble for social planning. We must focus on motoring as a positive factor of great significance in our way of life. We must therefore examine whether the car ought not to lead to a new kind of thinking in town planning. One good way of doing this is to study how to plan for a "car city," a new city that really is designed to take account of the car. (*Bilstaden* 1960, 3)

Åhrén's not-so-subtle use of "we" underlines a shared identity with a definite view of what the problems of society were and how they were to be solved. Curiously, the experts identified the city, and not the automobile, as the problem. From the viewpoint of the car society, the traditional city with its compact center was "an absurdity," and the experts defined their goal as remaking cities to fit the standards supposedly demanded by the automobile (Hasselquist 1960, 131). The car city was the radical and visionary solution expected to solve all problems in a single blow. The city was to make way for the car, not vice versa. Looking back, one is struck by the experts' self-assured tone; they assumed the role of authoritative and omnipotent problem solvers in society.

In their formulation the experts reduced town planning to a simple question. Town planning was equated with traffic planning, as Nordqvist (1965) made clear: "What is a city? A traffic system that has to be redesigned." Planning was thus a matter of satisfying the conditions imposed by the increase in automobile traffic. Of 100 Swedish planning and research projects in the transport field between 1960 and 1965, 93 were concerned with private motor traffic while only 7 dealt with public transport (*Samhällsplanering* 1969).

The Car City conference displayed ideas and approaches that were circulating across Europe. A series of similar expert-composed manifestos and visions of the motorized society circulated across Western Europe in the late 1950s and the early 1960s. Especially influential was the 1959 book by German town planner Hans Bernhard Reichow,

Die autogerechte Stadt: Ein Weg aus dem Verkehrs-Chaos (The car-friendly city: A way out of traffic chaos). It epitomized the approach of the traffic experts: congestion and road accidents were defined as the problems of the cities, and the solution was to build the "car-friendly" city. Similar conclusions were voiced by Danish traffic planner P. H. Bendtsen in *Town and Traffic in the Motor Age* (1961) and by British architect Geoffrey Jellicoe's *Motopia: A Study in the Evolution of Urban Landscape* (1961). Another influential publication was the British government report *Traffic in Towns* (1963), known as the Buchanan Report after its main author, traffic planner Colin Buchanan. Swedish traffic engineer Nils Rosén (1964, 1209) described Buchanan's thinking as "not new, but more clearly described, better argued and reported in greater detail than one is used to" (see also S. Plowden 1972). The Buchanan Report should be seen as a codification of knowledge and ideas that had already been circulating among European traffic experts, and as an endorsement of existing lines of thought.

How then did the traffic and urban planners mobilize the authority of science? Historian of science Lorraine Daston (1992) argues that an objective ideal of knowledge transfer emerged in the nineteenth century. As the scientific community grew in numbers and extent, so did the need to communicate methods and results. As scientists ventured beyond their own context, country, culture, and language, they needed to strip off references to specific cultural codes, traditions, and local conditions. Traffic and urban planners used many of these same strategies. Above all, planners justified their emerging discipline through an appeal to abstract, scientific knowledge, from which all subjective qualities had been erased in favor of apparently objective facts, generating universality as opposed to recognizing the uniqueness of the local. As representatives of the universal and the modern, these experts supposedly rose above the historical, cultural, political, and social specificity of the local setting. Science helped to reduce ideology, politics, and interests to quantitative and neutral elements that could easily be passed between individuals, institutions, regions, and countries. In this way the experts attained the role of conductors of "development," and in the 1950s and 1960s their formulation of problems came to prevail across most of Western Europe (Antila 2003; Blomkvist 2004; Cherry 1996; Flonneau 2002, 2006; Lundin 2004; Østby 2004; Schmucki 2001).

But did the experts deal in more than knowledge? Did they really bring modernity? The hypothesis of experts as agents of modernity leads us to the question of how they acquired, handled, and transmitted this modernity. The next step is therefore to examine how their knowledge spread.

The Circulation of Knowledge

One way of understanding how traffic planning ideas circulated in Sweden is to study the technical literature on road and street building and traffic planning. I carried out a bibliometric study covering the years 1946 to 1955 to discover when the literature arrived in Sweden, where it came from, and to which institutions it found its way.[2] When the traffic planning literature is classified by nation, a clear pattern emerges. Of a total of 1,063 books, conference papers, and journals obtained during the period of 1946 to 1955, the United States is the country of origin of 54 percent of the literature, followed by the United Kingdom with 19 percent and West Germany with 6 percent. This division remains relatively constant over this decade. Compared with the earlier period, the shift of emphasis away from Germany and to the United States is striking: in 1935, fully 39 percent of the imported literature came from Germany and only 17 percent from the United States. The shift from Germany to the Anglo-Saxon world can be understood by the crushing defeat of Nazi Germany in the Second World War; the knowledge base that existed in Germany was destroyed or scattered across the world. But the transition in Swedish cultural life from Germany to Britain, in particular, was discernible even around the time of the First World War, and the subsequent changes were part of a larger cultural movement toward the West (Sörlin 1994).

Considering individual institutions, I identified five major knowledge centers in highway engineering and traffic planning: the Bureau of Highway Traffic at Yale University in New Haven, Connecticut; the Eno Foundation in Saugatuck, Connecticut; the Highway Research Board at the Bureau of Public Roads in Washington, D.C.; the Institute of Traffic Engineers (ITE) in New Haven, Connecticut; and the Road Research Laboratory at the Department of Scientific and Industrial Research in London. Four of the five central institutions were thus American, and all five were Anglo-Saxon. This sample outlines the contours of an international network located in three geographical centers: Washington, D.C., the small state of Connecticut on the East Coast of the United States, and London.

Swedish engineering schools took part in this international network through their research libraries, which diligently kept up with the latest technical literature. The two main journals in traffic planning were the ITE journal *Traffic Engineering*, whose publication began in 1933, and the Eno Foundation's *Traffic Quarterly*, which first appeared in 1947. These journals found their way into Swedish research libraries within a year or two, together with the chief American books on traffic planning, such as the *Manual of Traffic Engineering Studies* (1945), the *Parking Manual* (1946), and the standard work of reference on the dimensioning of roads, the *Highway Capacity Manual* (1950).

But the fact that the knowledge was available does not necessarily mean that it was used. An unambiguous connection between the availability of literature and its actual influence is not easy to establish. Even if technical knowledge is freely available, it will be put to use only if there is someone who can find it, understand it, use it, and pass it on. And for this to happen, someone must have an interest in the knowledge concerned; and in our case, such a serious interest arose only in the 1950s. Of course the formal published literature was only one aspect of circulation; the traffic experts' study trips and conferences were another, more easily quantifiable type of circulation.

Before the Second World War, continental Europe was the principal destination for students of traffic and transport engineering and town planning. When Nazi Germany built a national network of motorways based on American models, even U.S. engineers went to Germany to study them (Seely 1987). Swedish links with Germany remained strong for years. For example, Torsten Åström, director of Stockholm Tramways and later holder of the Royal Institute of Technology chair in traffic and transport engineering, visited the United Kingdom, Belgium, the Netherlands, Germany, Switzerland, and northern Italy to study "transport and town planning questions." The influence of British town planning in Sweden was manifested by the exhibition "Replanning Britain," which took place in Stockholm in the spring of 1946. British "new towns" attracted particular notice. A study trip to the United Kingdom arranged by the Swedish Association of Municipal Technology in 1949 had 43 participants from 11 Swedish towns. The Swedish experts visited Ebenezer Howard's classic Welwyn Garden City and the planned new town of Stevenage near London (Eriksson 1949). Curiously, well into the 1950s some Swedish visitors commented on the "relative backwardness" of U.S. town planning (A. Lundberg 1953, 42). The shift in attention to the United States in the mid-1950s was due to one thing: the automobile.

As awareness heightened that the automobile was about to fundamentally change town and community planning, there was also a striking change in the pattern of experts' travel after the end of the Second World War. There are 45 study visits and conferences recorded in Swedish journals during the period 1946–1955.[3] More than half of these journeys had the United States as their chief destination, and across this period the annual journeys to this country doubled. Engineers, in particular, went to the United States to study road and street traffic. Toward the mid-1950s, architects and town planners also became interested in U.S. town planning.

Contemporary estimates suggest that there were about 50 traffic engineers in Sweden in the early 1950s. The First European Traffic Engineering Conference took place in The Hague in the summer of 1953, and had about 150 participants from 16 countries. There Sweden was represented by 21 participants, comprising perhaps 40 percent of the country's traffic engineers (Kullström 1953). The Second European Traffic Engineering Conference in Bürgenstock in 1954 attracted roughly 300 people from Western Europe, about 20 of them Swedes (Matern 1954). A third conference was held in Stresa, Italy, in the autumn of 1956 with nearly 450 participants from 33 countries (Nordqvist 1956a). Between study visits and conference participation, a high proportion of Swedish experts undertook educational trips.

These conferences helped create a Europe-wide community of traffic engineering experts that dealt with specific European questions about adapting cities to mass automobile use. There were, however, other organizations, beyond the expert groups themselves, active in the acquisition of traffic engineering knowledge. Earlier research has suggested the importance of national lobbying organizations and organized special interest groups on an international level in Sweden, Norway, and Finland (Antila 2003; Blomkvist 2004; Østby 2004). Indeed, the European traffic engineering conferences were arranged by three international special interest organizations: the World Touring and Automobile Organisation (OTA), the Permanent International Association of Road Congresses (PIARC), and the IRF. These organized special interests also helped establish a European expert community.

Bruce Seely (2004) points out that Americans made energetic contributions to the spread of highway and traffic engineering across the world, especially in postwar Europe. The three principal institutions were the Bureau of Public Roads (BPR), responsible for the U.S. federal highway system since 1916; the Bureau of Highway Traffic at Yale University; and the lobbying organization IRF. These institutions helped disseminate expert knowledge by financing, hosting, and training foreign highway and traffic engineers. I have already mentioned the IRF's active exchange program, and the BPR's operations were even more extensive. In 1953 alone the BPR hosted visits by 139 highway and traffic engineers from 35 countries; in 1954 there were an additional 170 visitors. A number of Swedish engineers took part in these exchange programs. The BPR and the IRF exchanges were financed largely under the American Marshall Plan, although Swedish engineers' participation in BPR's program was funded by Sweden. European institutions were active, too. In 1954, the Organization for European Economic Cooperation (OEEC) organized a study visit to the United States for 32 European traffic experts from 11 countries. The Swede Dag Blomberg went to the city of Milwaukee, Wisconsin, which had geographical conditions typical of those in Sweden (Blomberg 1954; *Traffic Engineering and Control in the USA* 1955).

The circulation of knowledge is difficult to grasp in its entirety, and with certain exceptions it did not show any definite form during the first postwar decade. It is not possible to identify one central actor. Even though trips were often singular events, many such visits occurred during these years. My literature survey indicates that traffic engineering seeped into Sweden via numerous different channels. Certainly, there was a very strong desire among Swedish engineers, companies, special interest organizations, and politicians to acquire highway and traffic engineering knowledge. In fact, the majority of Swedish study trips took place outside the formal international exchange programs. A large number of different government authorities, cooperative organizations, professional organizations, individual cities, private funds, and foundations helped fan this circulation of knowledge by financing and organizing various study trips. Just as there were many ways for traffic engineering knowledge to reach Sweden, there were many ways of getting Swedes to the United States.

The study trip may be regarded as a specific form of knowledge acquisition and transfer. The reports on the visits show how educational elements were interwoven with pure tourism; long holidays were frequently combined with the study visits. Another important element was the possibility of acquiring knowledge by doing practical work abroad. The trips also gave the travelers valuable social and cultural experience. What role did this experience play in forming the traffic experts' view of what was "modern"? By lifting the supposedly neutral scientific veil, we can obtain a picture of these ideological and cultural conceptions.

Encountering the United States

My analysis shows that the United States was the most frequent destination of Swedish architects, town planners, and civil engineers. These visits to the States aroused great interest at home. Indeed, I suggest that it was in the encounter with America that Swedish postwar planning ideology was formulated. Let us look more closely at how travelers regarded this interaction and their travels.

In twentieth-century Sweden, the United States came to represent the future—or at least a possible future. Discussions of America to a considerable extent shaped the way in which actors saw the Sweden of the future (Alm 2002). The media scholar Amanda Lagerkvist (2005) analyzes the conceptions of "Amerika" that prevailed in Swedes' accounts of their travels in the 1950s and stresses that an explicit purpose of these trips—visits supported by public authorities, organizations, companies, parishes, foundations, and the press—was to draw a blueprint for the future of Sweden. This directed learning is strikingly illustrated by traffic engineers' reports with such titles as "What Has the USA to Teach Us about Traffic Planning" (1954) and "Learning about Traffic in the USA" (S. Lundberg 1956). Traffic engineers sought to bring back home knowledge that could be applied to Swedish modernization; the travelogues of Swedish intellectual icons Alva and Gunnar Myrdal, particularly *Kontakt med Amerika* (Contact with America) (1941), set the tone. The Myrdals, both later Nobel Prize winners, saw a definite affinity between the two countries, in that both aimed for modernization and peace. But it should not be forgotten that the ties of kinship were not only ideological; Sweden also had strong blood ties with the United States after the great emigrations between the mid-nineteenth century and 1930, when more than a million Swedes crossed the Atlantic.

Running through Swedes' reports on their travels in the 1950s is their ambivalence with regard to the modern (Lagerkvist 2005). After all, this was a charged encounter between a middle-class European culture and the American way of life. Swedes' fascination with the United States did not go as far as unreserved approval, and the visitors to America brought back a complex picture. The returning architects, engineers, and town planners conveyed a view of the United States that oscillated between a naive optimism inspired by the country's economic and technical progress and a spirited rejection, particularly when faced with the purely capitalist competitive society. For instance, in *Amerika: Dröm och verklighet* (America: Dream and reality) (1960), the journalist Björn Ahlander painted a picture of technological modernity—motorways lined with motels, gas stations, and restaurants—but also of the impotence of the big city. He described the modern, abundant society with the same ambivalence that many Swedes felt when encountering the United States. And yet despite these conflicting emotions, Swedish visitors to America often reported feeling that its lifestyle would inevitably spread. After visiting the United States, the town planning architect Sven Tynelius declared: "The motor car is not a pastime of the better off, but *an inescapable factor in the progress of society*. Our whole way of life is sliding toward Americanization" (Tynelius 1956, 3).

These reports also offer the travelers' personal discoveries of the United States. Gunnar Kullström from the Stockholm Urban Planning Office took part in the 1954–1955 IRF-organized program in traffic engineering at Yale's Bureau of Highway Traffic. He was not impressed by the quality of the teaching. Several of the courses were "rather elementary in their design," especially for European traffic engineers who usually had many years of traffic engineering activity, and some of the teachers were not "entirely suited to university teaching" (Kullström 1955). Kullström believed the training involved "spending of a lot of time for very little result." But though he did not show any great enthusiasm for the formal training, his personal encounter with the American traffic culture was all the more notable. He vividly described the experience of "being able to live for so long among and personally experience American traffic and feel the sensations that arise from driving in U.S. traffic and oneself being able to see the most modern in the way of roads, road junctions and filling stations." The great value of study visits to the United States, Kullström emphasized, was in "gaining a personal perception of the most modern traffic constructions and also an understanding of the driving conditions that may be expected in Europe in the near future" (Kullström 1955, 161, 164, 165).

In a study of scientific travel, Sven Widmalm (2001) noted the travelers' feelings of immediacy. This characterized the experience of civil engineer Harry Bernhard during a seven-month-long study trip to North America during 1954–1955. On returning home, he observed: "I had previously discussed traffic problems in connection with clearance projects, etc., and, of course, seen statistics about traffic in the United States, heard about it, read about it, got a picture from films, etc., but I have to say that it was only the personal encounter on the spot that showed me how enormous American motor traffic really is" (Bernhard 1956, 10). Clearly, the experts' personal experiences of traffic, housing, and cities in the United States helped them form a clear impression of the consequences of full-scale mass motoring. They believed they had seen the future.

The Concept of Historical Repeatability

Interpreting the United States as a mirror of the future, the Western European traffic experts articulated their planning ideas against a background of American experiences, solutions, and mistakes. These experts implicitly believed that history follows a definite line of development: the United States had climbed a step higher on the staircase of progress, and now it was the turn of Sweden and other European countries to follow. The same conception typified Swedish encounters with the Third World. With their evolutionary perception of history as unfolding in stages, the travelers considered that the Third World was going through the same stages or history as had the West (Edman 2004). This conception of societal development goes back to Herbert Spencer's social Darwinist theories of the nineteenth century, which assumed that societies progressed from one "stage" to another. It was a hierarchic way of thinking that implied that developed cultures were superior to less developed ones.

In *Machines as the Measure of Men* (1989), Michael Adas shows how Europeans' perceptions of their own technical and scientific superiority contributed to the establishing of technology and science as yardsticks of progress in the eighteenth century. These conceptions were revived in the post-World War II campaigns for "modernization." For example, one of the leading postwar modernization theorists, economist Walt Whitman Rostow, distinguished five stages that led to economic breakthrough. His theory involved an evolutionary perception of societal development (Runeby 1998). The idea of evolutionary progress was also aptly expressed by the Swedish engineer Rolf Oom. After taking Åhrén's advanced course *Vad har USA att lära oss om trafikplaneringen?* [What can the USA teach us about traffic planning?], Oom wrote that "the advantageous situation of having up to now belonged to an 'underdeveloped' part of the world in the area of automobility" would make it possible for Europeans to learn from the solutions and mistakes of the United States (1954, 159). Just as the Third World was expected to learn successfully from the West, so Europe was expected to learn from the United States.

From this perspective, it is hardly surprising that the steady flow of reports on the United States by architects, town planners, and civil engineers during the first postwar decade attracted a lot of notice. Soon enough, local government officials and politicians also wanted to cross the Atlantic and see "the future" with their own eyes. Stockholm had just begun the reshaping of the central area of Nedre Norrmalm, soon known as "the transformation of the city center." One of Sweden's largest building and planning projects, it sprang from the urban visions energetically proposed by the Stockholm Master Planning Commission. From 1951 onward there was also discussion of the future of streetcars in Stockholm (Gullberg 2001). Developments in the United States aroused planners' curiosity about the nature of town planning and urban building there. In the States, public transportation, and streetcars in particular, had lost a lot of ground to the private automobile, with heavy loss of revenue as a result. Swedish city planners wondered how U.S. cities reacted to this development and what efforts, if any, they had made to reconcile public transportation and a modern, democratic society.

With the intention of studying these U.S. experiences, Stockholm city officials made one of the more publicized study visits to the United States in March 1956. The social democratic politician Helge Berglund (1956) observed that, since "in some respects development in America is several decades ahead of Sweden," it ought to be possible "to trace the course of development there, which could give an indication of what we have to look forward to here at home." The Stockholm delegation—including representatives of the City of Stockholm and Stockholm Tramways—started their study trip in New York, continued through Boston, Toronto, Chicago, Cleveland, Pittsburgh, Houston, New Orleans, Washington, D.C., and San Francisco, and finished in Los Angeles (Berglund 1956, 213). On the basis of its experiences in the United States, the Stockholm delegation thought it could draw certain conclusions on the future of Swedish urban traffic. Pointing to the "overwhelming importance of car use" in the United States, the group believed that the nation's community development was being driven by the motor car. But the growth of motoring had far-reaching implications for the city, both in the form of housing construction that created "incredibly" long distances between home and work, and also in a "heavy-handed" effect on the central areas of the city. The picture of the car-dominated U.S. cities was not entirely positive; cities with efficient subway systems were better organized and more "fortunately planned" than those that had left transport entirely to automobiles and buses, as the latter course of action had implied a "disorganization and splitting of the business of the city."

Despite these evident problems, the Stockholm delegation considered mass car use as an "irresistible development" to which Swedish cities had to adjust. In future city traffic plans, public transport would have to accommodate the automobile. Public transportation solutions would have to be arranged in harmony with the automobile, as a contribution to the solving of its problems, and in some cases even to overcome chaotic traffic situations. The politician Berglund went so far as to say that the task of public transportation was to carry "the carless proletariat" (1956, 213, 214, 219).

The concept of evolutionary progress led to a view of the autonomous development of motoring. Other kinds of traffic, such as public transportation, were regarded as reactive to the automobile. Even the supporters of public transport for the most part accepted the inevitability of car-driven "progress" and thus shared—with resignation—the view that there was nothing they could do but go along with it. The head of Stockholm Tramways, Hans von Heland, even posed the unsettling question of whether public transport in Sweden would entirely disappear in the near future (Heland 1956). The consensus on the automobile's effect on the city was so strong in the 1950s and early 1960s that it united both development optimists and development pessimists.

The Car City—Sprawl and Concentration

How, then, did these powerful currents affect town planning in postwar Europe? Modernist town planning visions were drawn up for reconstructing many of Europe's war-torn cities, but at first only a few were implemented. However, things changed quickly. The speed of economic growth in the 1950s came as a surprise, particularly to West Germans, who started talking about new construction (*Neubau*) in the mid-1950s instead of reconstruction (*Wiederaufbau*) (Diefendorf 1989). This economic growth, coupled with the explosive increase in the number of automobiles, created room for the experts to implement car-centered modernist town plans.

In Sweden, which had been spared the devastation of war, the economy soared, and by around 1960 the country had the highest car density in Europe. Here the "car city" was implemented to a greater extent than in many other European countries. When the Stockholm satellite town of Vällingby was inaugurated in 1954, contemporaries thought a social-democratic utopia had been realized, and planners from all over the world traveled there in search of inspiration (Cherry 1990; P. Hall 1998). Gradually, however, the social democratic planning visions were drained of their politico-ideological content, and gave way to the absolute "needs" of motoring and trade. The U.S. influence became increasingly prevalent, and the Stockholm satellites Farsta (1960–1966) and Täby (1964–1971) were geared even more toward motoring and commerce.

As a concept, the "car city" combined two striking and partly paradoxical patterns in modern urban building: a sprawl and dispersion of the urban built environment, but at the same time a certain concentration. The influential Swedish engineer and traffic planner Stig Nordqvist (1961, n.p.) described this new type of town with an apposite metaphor: "The natural organization of the car society is car-free islands with concentrated activity, surrounded by motorways and parking facilities. The car-free islands may be housing areas, shopping centers, office complexes, schools, etc." Note that it was *automobile traffic* that set the terms for the pedestrian traffic; the notion of "car-free islands" presumed, of course, a surrounding sea of automobiles.

City planning experts created car-free islands by transforming both city centers and residential areas into systems of pedestrian streets. At first these were restricted to individual streets such as the internationally noticed Sergelgatan in central Stockholm in the 1950s. These car-free islands were to be easily accessible to drivers, so "feeder roads" were planned around them, often in the form of a ring. At the destinations parking spaces were provided; often these were in multistory parking garages in larger cities. This car-centered model was the starting point for such projects as the 1967 plan for central Stockholm, *City 67*, in which main roads and car parks entirely surrounded

Figure 12.2
The district Sergelgatan in central Stockholm was made a pedestrian precinct in the 1950s, the prototypical "pedestrian island in the car society." Source: "Storstadens trafikproblem: Föredrag vid Transportforskningskommissionens årsmöte den 11 november 1954" (Traffic problems of the big city: Paper to the Annual Meeting of the Transport Research Commission, 11 November 1954), Royal Academy of Engineering Sciences Transport Research Commission, Meddelande 26 (Stockholm, 1955).

Huvudleder för motortrafik

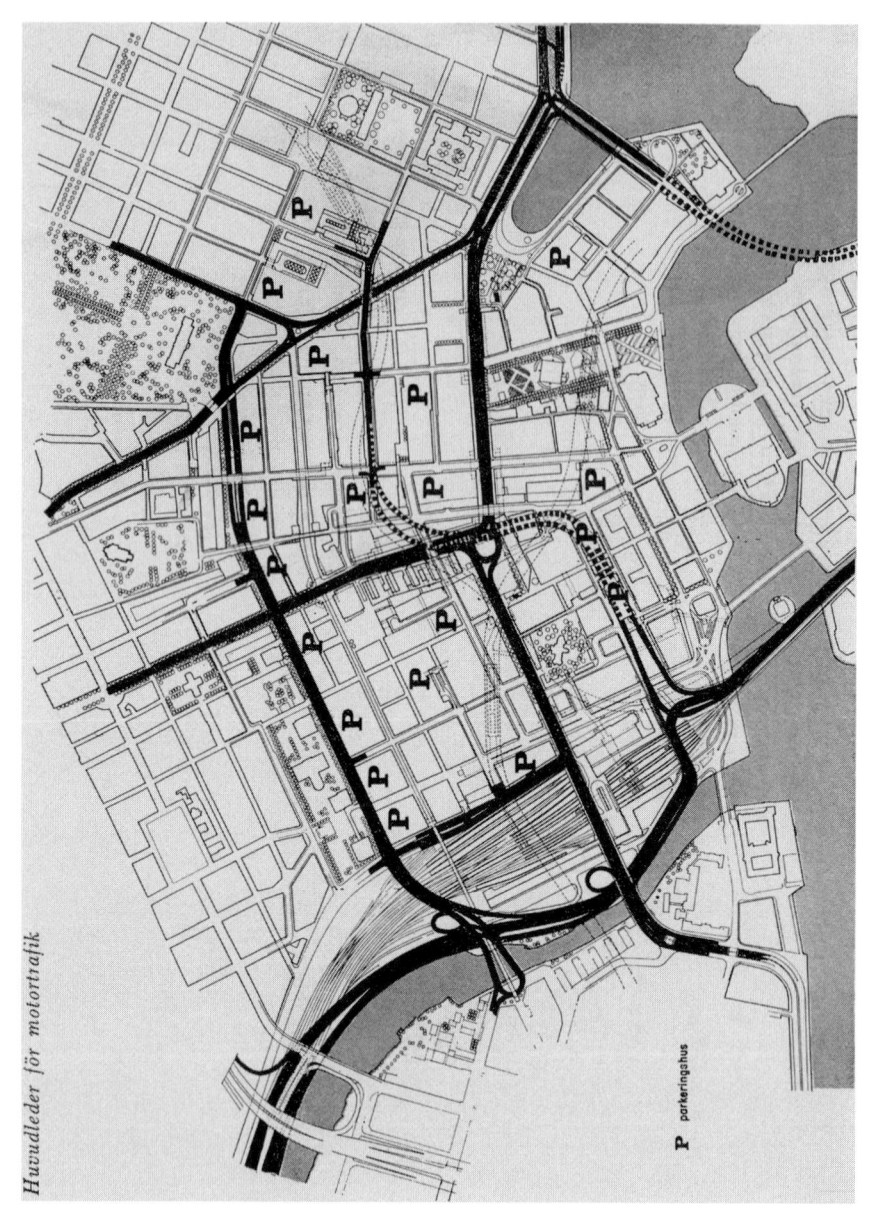

P *parkeringshus*

Figure 12.3
In "City 67" central Stockholm was designed as a pedestrian system surrounded by main roads and multistory car parks (P), combining "car-free islands" with a surrounding sea of automobiles. Source: *City 67: Stockholm: Principplan för den fortsatta citysaneringen framlagd i maj 1967* (Stockholm: Basic plan for the continued redevelopment of the city, presented in May 1967) (Stockholm, 1967).

the pedestrian areas. No point in the city center was to be farther than 280 yards (250 meters) from a multistory parking garage.

The size of the city was very relevant to the realization of a fully developed urban car society. Our knowledge about the adjustment of cities to the automobile concerns large cities, and in particular their central areas. The extensive postwar demolition of buildings in city centers to create car-friendly main roads and parking facilities has received enormous attention along with justified indignation. Priceless cultural and historical environments were lost. In Sweden approximately 100,000 housing units were demolished between 1963 and 1973 (Schönbeck 1994). Here the encounter between mass motoring and the city was most dramatic: the building of through roads, the widening of streets, and the provision of parking garages also left gaping wounds in the townscapes of such cities as Birmingham, Hamburg, Hannover, Kiel, and Paris (Cherry 1996; Diefendorf 1989; Flonneau 2002).

Small and medium-sized towns fared quite differently. For them, Nordqvist (1956b) saw "no other reasonable prospect than that the automobile will totally dominate transportation relatively soon." Nor was there any obstacle to "free motoring" on the fringes of the big cities. However, he believed that the central parts of the big cities, with their lack of space, would always depend on a certain amount of public transportation (Nordqvist 1956b, 1132). At the Car City conference in 1956, workshop discussions focused on towns with a population of around 100,000, with the belief that cities larger than this could not depend wholly on the automobile. A similar conclusion was reached by Colin Buchanan in *Traffic in Towns* (1963). And, in fact, modern Stockholm, for example, has an efficient public transportation system alongside its many large-scale traffic solutions and multistory car parks from the 1960s.

Therefore, the car city was fully implemented not in the central areas of the great cities of Europe, but rather in the urban landscape formed by their less densely built-up outer areas. The pattern one finds in Sweden is quite typical. During the period of 1956 to 1974, nearly half of Sweden's current stock of around four million homes was built. These housing areas were planned on the same car-friendly principles as the city centers, with continuous footpaths in the center of the residential enclaves (car-free islands) and, around them, roads from the surrounding sea of automobiles. Housing grew up in symbiosis with car-centered transportation systems, forming a sparse but interconnected urban landscape. It is here, in the anonymous urban landscapes consisting of planned housing schemes and industrial buildings, that we find Europe's car city (see fig. 12.4).

Conclusion

In his celebrated study of modernity, *All That Is Solid Melts into Air* (1982), Marshall Berman asks why it was so difficult to oppose New York's powerful planner Robert

Figure 12.4
In the early 1970s the suburb of Staffanstorp, near Lund in southern Sweden, came to stand as a symbol of the anonymous housing area with large-scale planning and total adaptation to the car. Photographer: Lars Mongs, May 15, 1973.

Moses. Berman's explanation is that Moses successfully appropriated the vision of the modern. Opposing his plans for bridges, tunnels, freeways, and demolition meant going against history, progress, and modernity itself.

I interpret the power of planning experts in postwar Western Europe in the same way. In this chapter I have shown that planning experts acted as agents of modernity in the 1950s and 1960s with their visions of a car society. Their view of the modern was shaped by their encounter with the American car society and spread across Europe, embodied in the abstract and apparently value-neutral language of science and technology. Because this language gave them a universality that transcended different local contexts, similar solutions—including ring roads, feeder roads, multistory car parks, and pedestrian precincts—arose all over Europe, albeit at different times and to differing extents. The experts expressed modernity in terms of progress, science, efficiency, and rationality and made themselves, for a time with the full blessing of society, the sole interpreters of the modern.

Yet, as we know, criticism of the car society arose toward the end of the 1960s. Organized protests arose against the "car-friendly" demolition of city centers. The undesirable effects of the motor car on the environment and on health had become apparent, congestion continued, and road accidents persisted. Motoring itself became politicized in the protests of the late 1960s. The criticism was accentuated and gained a firmer foothold in the 1970s when the public realized that the economic boom of the postwar decades was over.

A productive way of interpreting these confusing years is to see them as a growing criticism of modernity itself. Since the modernist experts were also subjected to criticism in the early 1970s, they lost their authoritative position of earlier decades. It certainly did not help the modernist cause that the problems accompanying mass motoring had not really been solved but rather had tended to increase. Criticism was particularly directed at the fact that the experts saw themselves as having an exclusive right to formulate the problems and to propose and implement solutions. The modernity that the experts claimed to represent was no longer unchallenged.

Notes

1. All translations from Swedish and German have been made by Bernard Vowles.

2. I have made use of the accession catalogues for foreign literature that the Royal Library in Stockholm began using in 1886. It is not feasible to continue this examination beyond 1955 because in that year the Royal Library began to organize its accession catalogues alphabetically instead of by subject.

3. I have gone through the main Swedish specialist journals for road and traffic engineers, town planners, and planning architects: *Byggmästaren, Kommunalteknisk tidskrift, Plan, Svenska vägföreningens tidskrift, SVR-tidningen/Väg- och vattenbyggaren*, and *Teknisk tidskrift*.

13 Greening the City: Urban Environmentalism from Mumford to Malmö

Andrew Jamison

In the summer of 2001, a housing exposition was held in the Western Harbor area of Malmö, a city of some 250,000 inhabitants located in southern Sweden, just across the Öresund straits from Copenhagen. Bo01, as the expo was called (the Swedish word *bo* has about the same meaning as "reside" in English), was meant to put a new image of Malmö on the map of Europe, both literally and figuratively. On the one hand, it was a commercial housing exhibition, with many intriguing examples of environmentally friendly architecture and interior design. On the other hand, and much more ambitiously, the exposition was meant to provide a symbolic vision of the "city of tomorrow."

In a part of the city that had lost its large shipbuilding operation and was in dire need of renewal, Bo01 became a site for the marketing of "sustainable urban development," a cluster of ideas and practices circulating around Europe since the late 1980s. With the help of a newfound environmental consciousness, Malmö would cast off its industrial past and become a postindustrial city of the future; the exposition was thus an attempt to fashion a green image for the city. The Bo01 displays were meant to inform and educate visitors about the various aspects of sustainable urban development, as the organizers—who represented both the municipal government and local housing, construction, and energy companies—understood it. Their idea was not merely to protect or save the natural environment, but to make money while doing so, to integrate environmental concern into economic development as a kind of "green business" (Jamison 2001). Sustainable urban development was defined as having four components: environmental, social, technical, and human. Bo01 thus included displays about homelessness and the plight of immigrants, as well as energy conservation and environmentally friendly construction. In addition, a few imaginative gardens designed by landscape architects showed what an ecological aesthetic might look like.

Within the exposition grounds was also a site for a building that had not yet begun to rise in this city of tomorrow. At the time, it was just a huge hole in the ground, but the pictures and information material on display in the nearby exhibition hall made it clear that this was to be no ordinary structure. The Malmö office of the national cooperative association HSB (Hyresgästernas Sparkasse- och Byggnadsförening) which had been established as a cooperative society in 1928 to provide affordable housing for its mostly working-class members, had given Spanish-born architect Santiago Calatrava the opportunity to turn a sculpture that he had made into a 55-story skyscraper. The building was to be called Turning Torso and would represent "a bold vision in modern advanced architecture creating new values," according to the exposition program; to live in it would be like living between "heaven and earth but closer to heaven" (Bo01 2001, 46, 108).

Like the *Twisting Torso* sculpture on which it was based, the skyscraper was to be a typically Calatravan combination of adventurous architecture and experimental engineering. Yet it was unclear in what way, if any, it could be considered ecologically sustainable. Although the Swedish state had given several million crowns to the project for various environmental measures regarding energy use and waste disposal, the Turning Torso was not particularly environment-friendly. Nor was the project described as being a part of the solution to any pressing social problems of the city. It was, rather, in its architectural vision—its ambition to mix technical innovation with artistic modernism—that one could perhaps read in a relation to sustainable urban development. By being a building that was meant to last, and, as such, to "sustain" a modernist architectural sensibility into the future, the Turning Torso could perhaps be seen as an example of what the exposition program termed "technical sustainability."

The building was expected to be ready for occupancy in the spring of 2003, but like the financial calculations of Bo01 as a whole, the time plan for the Turning Torso proved to be unrealistic. At the end of 2004, the building was still not completed, and very few of the extremely expensive apartments had been sold. Because of unusually large cost overruns, the director of HSB in Malmö, Jonny Öhrbäck, who had been the instigator of the project, had been forced to resign, and Calatrava's design had shown itself to be more complicated and more expensive to build than had been originally anticipated. When the grand official opening took place in 2005, however, Calatrava was hailed by city officials as a local hero, and the building was generally seen as the future symbol of Malmö that would help the city compete for the income to be derived from investors, tourists, and well-heeled new residents.

The tale of the Turning Torso provides a window into the world of sustainable urban development, and into the ways in which urban environmentalism has come to be appropriated into European cities. In the Western Harbor area of Malmö, Sweden, the city of tomorrow has met the architectural visions of the past, providing lessons for all who are concerned about the future of the city.

Setting the Stage

The history of urban environmentalism can conveniently be divided into three phases, or, in dramatic language, three acts. The first act recounts how a cluster of urban eco-

Figure 13.1
Calatrava's Turning Torso is now occupied, but still (in 2007) not all of the Bo01 exhibition area has been finished. Photographer: Mikael Hård.

logical ideas and practices emerged in an early wave of environmentalism in the late nineteenth and early twentieth centuries. Drawing on the romantic critiques of industrialization, as well as theories and concepts from geography and history, urban reformers and "progressive" writers fashioned a cluster of social, political, and scientific ideas and practices that received a kind of paradigmatic articulation in Lewis Mumford's book *The Culture of Cities*, published in 1938.

The second act begins in the 1960s, as part of the new environmental consciousness that developed in that era, when many of the earlier ideas were combined with new ones and circulated around the world. In many cities, experiments in such areas as environmentally friendly building construction, water management, park design, and energy use were carried out in housing blocks, neighborhoods, communes, and in many a previously derelict building or deserted field. In Berkeley, California, activists occupied an unused piece of land and declared it a "people's park," which brought on the police, as well as the wrath of the governor of California, the former film actor Ronald Reagan. In Copenhagen, a group of mostly young people moved into a deserted military base and soon declared it a free town named Christiania. During the ensuing decades, imaginative experiments in environmentally friendly living, in terms of design and architecture as well as lifestyle and behavior, would develop in this refuge.

Through their various information campaigns and protest activities, environmental organizations spread around Europe the ideas and practices of environmental protection and the vision of a green, or at least greener, city. In Berlin and Basel, in London and Amsterdam, as well as in many smaller cities of northwestern Europe, environmentally concerned architects and planners, as well as students and resident groups of various kinds, carried out experiments in "urban ecology," often in collaboration with squatters or other community activists. As an example of what was happening, we will later pay a brief visit to an exhibition that was staged during the summer of 1976 at the Modern Museum in Stockholm, where some of the experiments in energy conservation and appropriate technology that were taking place around the world were on display. Activists also spread the environmental message in Malmö in the 1970s, protesting against some of the negative effects of industrial production and, in particular, the nuclear energy plant at Barsebäck, just up the coast. At the Kockums shipbuilding company in the Western Harbor, which was one of the largest employers in the city, local groups organized discussions about "alternative production."

By the late 1980s, the plot had begun to thicken, or at least shift, as the greening of the city began to be reformulated into more instrumental terms, according to the doctrine of sustainable development. For cities, this doctrine was translated into a policy program of sustainable urban development. At the same time, new actors entered the scene, more or less upstaging the environmental movement "activists" who had played the leading roles in act two. The moral engagement of the activist was replaced by the more specialized competence of the expert, and urban environmentalism was transformed into a professional activity. The third act follows a story line of appropriation, and tells of the tensions and contentions that have been associated with the implementation of sustainable urban development in particular settings, often as part of commercially oriented urban redevelopment projects.

Much of the visionary idealism that was so evident in the 1970s gave way to a more institutionalized activity, and professional roles were constructed around the quest for sustainable development (see Jamison 2005). As the greening of the city became the task of the specialized expert rather than the "movement intellectual," the arenas or sites of activity became the municipal office and consulting firm rather than the open-ended public space of the environmental movement. Sustainable urban development was implemented primarily in and around those areas of municipal government and public administration, which were charged with effectuating Agenda 21, agreed upon at the World Conference on Environment and Development in Brazil in 1992.

As the ideas and practices of sustainable urban development have taken on a concrete form, and in many cities literally been brought to market in commercial redevelopment schemes of various kinds, the underlying contradictions have also become more visible. In Malmö, in the continuing unfolding of the tale of the Turning Torso, this is an act that has not yet ended; indeed, the entire "play" of urban environmentalism throughout Europe remains without an ending, since the tensions are yet to be resolved and the actual meaning of the efforts remains unclear and uncertain. Are cities, in the name of sustainable urban development, actually improving their environmental performance, or is "greening" in the process of becoming a marketing gimmick, helping cities compete for tourists and investments in an increasingly market-oriented and commercial age?

The Emergence of Urban Environmentalism

The roots of urban environmentalism—what later became known as sustainable urban development—can be traced back to the late nineteenth century, when reform-minded movements and individuals confronted the multifarious challenges of the industrializing city in both Europe and North America. While European critics often looked back to the cities of the past and sought to restore or preserve preindustrial elements within the contemporary city, no such options were available in most of the urban centers of North America. In England, for example, the romantic poet and craftsman William Morris entered politics in the 1860s as part of the "anti-Scrape" movement to save what could be preserved of London's preindustrial past. Whereas Morris and many other social critics in Europe objected to the ugliness of the industrial city and the loss of a sense of beauty in the industrial way of life, American reformers were, for the most part, more pragmatic in their critique of the industrializing city.

In North America, urban reform could not mean the glorification of the medieval past, for America had not had one. Rather, the rapid growth of the cities in the nineteenth century inspired what might be termed the "co-construction" of a new urban landscape. American cities were pioneers in their efforts to bring nature more intimately into the modern metropolis, and to keep these areas a bit more wild than those in the classic European cities. New York's Central Park became a model for other cities to follow, as did its mixture of ethnic neighborhoods, the myriad of cultural enclaves that were to give the modern or modernist city its quintessential ambiance, or what Raymond Williams (1977) has termed "structure of feeling." The environmental challenges faced by the industrializing American city would give rise to a new sort of urban philosophy that drew on ideas from Europe but related them to a new set of experiences (for the paradigmatic case of Chicago, see Cronon 1992).

As large numbers of immigrants fled to North America from oppression and poverty in Europe, such cities as Chicago and Detroit, where new industrial corporations were growing rapidly, became key sites for further reform efforts. Some of the older cities of the Northeast were also affected by the technical and organizational "innovations" that were serving to transform the American economy from a mainly rural and natural resource–based orientation to an urban and manufacturing orientation. The cities that developed were not simply in need of the greenery that parks and landscape architects could provide; they required, in addition, the expertise of new sorts of scientists and engineers who could deal with the pollution of the air and the water, and the variety of public health problems that were to be found both inside and outside the conglomerations of industrial production. As Jane Addams (1910/1961, 81–82) described Chicago, which she devoted her adult life to trying to reform:

> The streets are inexpressibly dirty, the number of schools inadequate, sanitary legislation unenforced, the street lighting bad, the paving miserable and altogether lacking in the alleys and smaller streets, and the stables foul beyond description. Hundreds of houses are unconnected with the street sewer.... Rear tenements flourish; many houses have no water supply save the faucets in the back yard, there are no fire escapes, the garbage and ashes are placed in wooden boxes which are fastened to the street pavements.

The efforts to deal with these social and environmental problems have been characterized by historians as an important part of the "progressive" movement. Opinions have differed as to what the movement was all about (Jamison 1998), but there can be no denying that, in the period between 1880 and 1940, there was a widespread interest in the United States, among geographers, planners, sociologists, social workers, as well as politicians and so-called muckraking journalists, in the environmental problems of the industrializing city (Gottlieb 1993).

In 1906 the writer Upton Sinclair published his extremely influential novel *The Jungle*, depicting the plight of immigrants who worked in the meat-packing industry in Chicago and providing a catalog of urban environmental problems. Meanwhile, at the University of Chicago, sociologist Robert Park developed an "ecological" approach to investigating the city that inspired a number of other reform-minded or action-oriented academics. One of those who was most strongly affected by these developments was the young writer Lewis Mumford, who came of age during the "progressive era" and served in the 1920s as secretary for the Regional Planning Association of America, one of the organizations that tried to act on the disclosures of the muckrakers and the theories of the human and urban ecologists.

Mumford played an important role in consolidating the ideas and practices of urban environmentalism. He combined an interest in planning and technology with a passion for cultural history, and throughout his long life, in his many writings about the city, he sought to apply an ecological perspective to urban issues. His own mentor, as he called him, was the Scottish geographer Patrick Geddes, but he was also influenced by the economist Thorstein Veblen, the philosopher John Dewey, and the regional planner Benton McKaye (Hughes and Hughes 1990). As a public intellectual, Mumford served to popularize a wide range of academic work, making the ideas of urban environmentalism accessible for a broader audience, as well for officials and authorities in cities around the world.

From his first book, *The Story of Utopias* (1922), to his magnum opus, *The City in History* (1961), Mumford provided a highly personal philosophy of urban environmentalism. In his critiques of modern architecture, highway construction, military technology, and functionalist urban planning, Mumford continually tried to argue for a better balance between the mechanical, instrumental approaches of the "modernists" and the concern for the organic, the natural, and the living that characterized romantic critics of contemporary society. As he put it in the introduction to *The Culture of Cities* (Mumford 1938, 9),

Today we begin to see that the improvement of cities is no matter for small one-sided reforms: the task of city design involves the vaster task of rebuilding our civilization. We must alter the parasitic and predatory modes of life that now play so large a part, and we must create region by region, continent by continent, an effective symbiosis, or co-operative living together. The problem is to co-ordinate, on the basis of more essential human values than the will-to-power and the will-to-profits, a host of social functions and processes that we have hitherto misused in the building of cities and politics, or of which we have never rationally taken advantage.

Mumford developed an approach to the city that attempted to place planning efforts in relation to the natural environment in which the city was located. The city was to be seen as a kind of living organism, and his lifelong effort was to bring the forces of life—as understood by biologists as well as artists—into the planning process. "Every living creature is part of the general web of life," he wrote, and that ecological understanding had to become a part of urban planning:

As our knowledge of the organism has grown, the importance of the environment as a cooperative factor in its development has become clearer; and its bearing upon the development of human societies has become plainer, too. If there are favorable habitats and favorable forms of association for animals and plants, as ecology demonstrates, why not for men? If each particular natural environment has its own balance, is there not perhaps an equivalent of this in culture? (Mumford 1938, 302)

Mumford's ideas about city planning were perhaps more influential in Europe than in his native United States. In the decades after the Second World War, American cities tended to metamorphose into urban sprawls, the greater metropolitan regions that included the surrounding newly built suburbs and the extensive highway networks that came to transform the environment in fundamental ways. With a column in *The New Yorker* and his many articles in architectural and planning journals, Mumford would be one of the more vocal public critics of the process. Biologist Rachel Carson, conservationist Fairfield Osborn, and Mumford can be considered some of the key "ecological intellectuals" who in their public educational activity would help plant the seeds for the eventual emergence of a new kind of environmental activism in the late 1960s (see Jamison and Eyerman 1994).

Urban Environmentalism in Circulation

The environmental movement was one of the outcomes of the tumultuous 1960s in both Europe and North America. As part of the "cognitive praxis" or knowledge production of the environmental movement, a wide range of urban environmentalist ideas and practices began to circulate, both across the Atlantic and across the national borders of Europe. Economic and social developments after the Second World War had brought with them a range of environmental problems—from the air pollution that contributed to the photochemical smog that Arie Haagen-Smit had identified in the skies of Los Angeles in the 1950s to the toxic waste sites that had spread like a plague across the urban landscape, as the middle classes had fled to the suburbs, to the poisoning of the rivers that had previously connected urban centers to one another. Many were the voices in the 1960s that sought to educate the public about this "environmental crisis," and by the end of the decade, a new kind of political agency had emerged in the form of activist organizations, concerned scientists, and engaged citizens (see Jamison 2001).

This process was very open-ended, both in terms of participation and in terms of the ideas and practices that were considered important to circulate. Through most of the 1970s, urban environmentalism was thus a multifaceted "social movement" that combined a variety of disparate activities and political interests in one all-encompassing and temporary public space (Eyerman and Jamison 1991).

In the summer of 1972, the United Nations held a conference in Stockholm on the human environment that is generally considered to be one of the formative moments of the international environmental "movement" (McCormick 1989). At the time, many national governments had begun to set up state agencies to deal with the new public issues of environmental protection, which were then characterized primarily as air and water pollution and the management of waste. Major research programs in environmental science and technology were also inaugurated at academic institutions and within intergovernmental and nongovernmental organizations. In North America and in many European countries, new laws were passed for regulating the environmental behavior of private and public citizens, and a number of environmental action groups were created around the world. The Stockholm conference served to legitimate and introduce to a much wider public these various ideas and practices that had begun to be carried out in the name of environmentalism, and helped stimulate the circulation of urban environmentalism around Europe.

On the eve of the conference, a development economist, Barbara Ward and an experimental biologist, René Dubos, published a book that helped set the agenda for the coming deliberations: *Only One Earth: The Care and Maintenance of a Small Planet* (1972). Also important in this period was the international bestseller *The Limits to Growth*, written by a multidisciplinary team of experts commissioned by the Club of Rome to forecast the future development of the world's resource and energy use, using some of the new-fangled computer modeling techniques that had emerged in the postwar period.

In many European cities, groups of activists with an interest in architecture and construction carried out experiments in what started to be called "urban ecology." There were also a number of sites for collective living in which recycling of waste products and workshops for alternative, environmentally friendly technology were established. In London, a group of such activists put out the journal *Undercurrents* for many years, and in the mid-1970s produced the anthology *Radical Technology* in which many of these alternative ideas and practices were presented (Boyle and Harper 1976). Both the journal and the book were central agents for circulating what David Dickson (1974) at the time termed "utopian technology." In relation to transportation, building construction, energy production and use, agriculture, and even communication, loosely organized activist networks tried to put into practice the vision of an ecological or "green" city as an intrinsic part of the larger movement protesting environmental degradation and opposing nuclear energy.

In Copenhagen, the activists who had settled in Christiania began to make ecological products and to undertake projects of ecological building construction and design, creating an alternative community in the middle of a modern city. Similar but smaller attempts to "liberate" parts of major cities from their environmentally deleterious surroundings took place in West Berlin, Amsterdam, and Vienna, as well as in many smaller cities. Environmentally concerned architects made major efforts at what was called ecological urban restructuring in West Berlin, where several neighborhoods and housing developments were subjected to ecological city planning (see Hahn 1987). Shut off from the countryside around them, biologists in West Berlin also pioneered the study of urban ecology; in their scientific activity, they combined an environmental consciousness with an interest in urban planning.

A brief look at the ARARAT exhibition (Alternative Research in Architecture, Resources, Art, and Technology) held in Stockholm in the summer of 1976 might help the reader remember the kinds of experiments and practices that were circulating around Europe at the time. On Skeppsholmen, the museum island just across the water from the royal castle and the national parliament, the alternative-technology movement put its practices on display for the entire summer (ARARAT 1976). There were geodesic domes and low-energy houses with construction and design techniques that were purported to produce more energy than they utilized; there were solar heating panels and wind power plants in various shapes and sizes; there were prototypes of community gardens and urban agriculture; and, not least, there were seminars and lectures on ecological design and architecture, focusing on the application of the concepts and principles of ecosystem ecology to the design and construction of buildings and other artifacts.

Many of the architects who planned the ARARAT exhibition had been associated with the Center for Interdisciplinary Studies at Göteborg University. One of the early institutional forums for the study and teaching of human ecology in Sweden, the center (now a department) still serves as a focal point within the university for cross-disciplinary fertilization. The ARARAT exhibition was a major event, as were many of the other displays of urban ecology in the second half of the 1970s. In the name of small-scale, appropriate technology, many development assistance projects helped to circulate environmental ideas in general and urban environmental ideas in particular from countries like Sweden to other parts of the world. Urban environmentalism had perhaps its main impact in the 1970s in this developmental context.

In Europe, much of this experimentation with urban ecology and utopian technology fell on hard times at the end of the 1970s, when conservative governments came into office and an ideological "backlash" was felt throughout Europe and North America. An economic downturn during the same period also made ecological experimentation less interesting for many European cities and governments, and so, in general terms, one can refer to the first half of the 1980s as a time of short-term decline for urban environmentalism. Instead of experimenting with urban ecology, the technically minded became fascinated by the potential of personal computing and what were seen as the enormous possibilities, both economic and social, of using the new information technologies. A kind of high-tech fever infected European society, especially in the cities (cf. Heßler, this volume), and it became difficult for urban environmentalism to gain new adherents.

In many of the northern European countries, the energy debates of the late 1970s had served to bring environmental politics into the established parliamentary and governmental arenas. Even though nuclear energy was more or less removed from the political agenda at the end of the 1970s either by parliamentary decisions or public referenda, the energy debates led to a transformation of the environmental movement into a much broader and more differentiated network of political actors. In the course of the 1980s, environmentalists thus came to take on more specialized and professionalized social roles and identities. New types of activism emerged, as well as more delimited kinds of professional expertise. Some 15 years after Stockholm, when the World Commission on Environment and Development, headed by the Norwegian medical doctor turned politician Gro Harlem Brundtland, published its report, *Our Common Future* (1987), the concept of "sustainable development" was launched as a new approach to environmental policy. By linking environmental concern with socioeconomic development, what had primarily been until then an idealistic political program of "saving the earth" was transformed into more specialized and instrumental projects of sustainable development. At the World Conference on Environment and Development, held in Rio de Janeiro in 1992, the concept of sustainable development was made the cornerstone of the document that most of the participating governments agreed upon (the president of the United States refused to sign the agreement, with the justification that "our life-style is not for negotiation"). It was with the signing of this document, the so-called Agenda 21, that the curtain falls on act two.

The Appropriation of Sustainable Urban Development

Only One Earth and *Our Common Future* were hybrid documents, in the sense that they combined the expertise and experience of different kinds of scientists and other experts into a new "synthesis." In *Only One Earth*, a biologist and an economist collaborated in trying to forge a new idea about environmental politics, linking the expertise of the social and natural sciences into a program of caring for the environment; and *Our Common Future*, the politicians, business people, scientists, and environmental activists on the Brundtland commission worked out another kind of political program, linking the language of economic development with the "environmentalisms" that had been articulated in the 1970s. The conferences at which the ideas were presented—Stockholm in 1972 and Rio in 1992—can similarly be considered hybrid events, in the sense that they brought people with different interests together for deliberative purposes. They provided contexts of mediation, or sites of hybridization, that offered new opportunities for environmentalists with more specialized interests and competence. Somewhat different professional roles were constructed in relation to the different facets of sustainable urban development. Following the distinctions made in the Bo01 program, we can characterize the forms of professional expertise as ecological, social/human, and technical.

Ecological sustainability calls for a natural science expertise, usually derived from systems ecology; the aim is to find particular ecosystems and ecological niches in the urban landscape where various practical measures can be taken to make the city "greener." Social sustainability involves combining environmental ideas with concepts from the social sciences and humanities. Especially in Europe, sociologists and political scientists have articulated a discourse of ecological modernization in order to provide an overarching theory to the various projects. Technical sustainability, on the other hand, has been articulated in new approaches to architecture, construction, technology, and design, and in new criteria of technological innovation, which have been characterized by such concepts as industrial ecology, cleaner production, environmental management, renewable energy, and eco-efficiency.

At an institutional or organizational level, agents of sustainable urban development have carved out spaces within the public and private spheres to carry out their projects. In Malmö and many other cities in Europe, these sites of mediation were originally associated with Local Agenda 21 activities that were established, following the Earth Summit in 1992, in order to promote various kinds of environment-friendly behavior at the local level (for an overview, see Lafferty 2001). In traffic planning, energy utilization, waste management, water use, and not least access to "green areas," cities across Europe have instituted a large number of projects of urban ecology, human/social ecology, and industrial ecology. Sustainable urban development has become a new organizational field, where specialized experts work at the interface between the public and private sectors to link environmental concern with socioeconomic development.

The people who are employed to work on these projects have been expected to apply their specialized training, usually in environmental science or engineering. Ecological sustainability projects have often involved the application of biological, ecosystems ecology to the activity area of one or another municipal authority (for traffic, construction, energy, parks, etc.). The practitioners have thus developed, in one way or another, a kind of professional identity as an environmental city planner, in an emerging field of expertise that can be characterized as urban ecology. In Malmö and other Swedish cities, the term "city ecologist" (*stadsekolog*, or *kommunekolog*) is sometimes used to refer to this professional role.

The pursuit of social or human sustainability has involved primarily communicative activities and what might be called a social ecological expertise. This kind of competence is fostered, among other places, in urban planning departments and "human ecology" programs at universities, as well as in the environmental subfields of the various social sciences (environmental sociology, environmental economics, etc.). In

Denmark, a special "green guide" program was established in the 1990s to place environmental change agents within public sector organizations and schools, and in Malmö, and other parts of Sweden, environmental communicators have emerged in the form of expert advisers to municipal authorities and private companies. The environmental consultants for Bo01 were recruited both from the landscape architecture and planning divisions of the nearby Agricultural University in Alnarp, as well as from the International Institute for Industrial Environmental Economics in Lund.

Technical sustainability has generally been pursued by instituting procedures and practices of environmental management in business firms. On the institutional level, this has taken the form of what are generally referred to as environmental management systems, or departments of environmental management. There have been a number of recent developments in such areas as cleaner technology, pollution prevention, waste minimization, ecological design, and architecture, as well as more specific forms of "green engineering" in some fields of science and technology (green chemistry, green biotechnology, renewable energy, energy planning, etc.). In Malmö, a green roofs initiative has been established to carry out a number of projects in ecological construction. A number of technical experiments with energy-saving, "sustainable" traffic planning and water-use management have also been undertaken in different parts of the city, especially in the area of Augustenborg.

In many European cities, some projects in sustainable urban development have occurred explicitly in the private sector, creating networks that link "environmental managers" from different companies with one another. These kinds of urban partnerships between the public authorities and private companies have envisioned a potential for growth and urban renewal in the quest for sustainable development. But in many such efforts the environmental aspects have tended to be downplayed while the commercial aspects have taken on the primary importance. One example of a private-sector sustainability endeavor is the Sustainable Business Hub established in Malmö during the 1990s, which arranges conferences, produces information material for local business people, and provides contacts with national and international networks, such as the World Business Council for Sustainable Development and the Greening of Industry Network. Within companies, the variety of technical activities has given rise to a range of "ecodesigners" or industrial ecologists at various points in the production process. But in many of these projects, as the events in Malmö well illustrate, it has proved difficult to achieve a balance between the economic and the environmental (see, e.g., Sandström 2002 on organizational approaches to greening).

Sweden set up a special fund in 1995 to support "local investment projects in sustainable development" (the so-called LIP money), and the city of Malmö was successful in obtaining many hundreds of millions of Swedish crowns to carry out projects in sustainable urban development. The Agenda 21 office was gradually transformed into a LIP office, and in the late 1990s, Malmö carried out a large number of different projects in sustainable urban development with the support of LIP money. There were also activities—both projects and networks—that were funded by the European Commission, in the form of partnerships with other cities in Europe—in particular, Newcastle, Venice, Tampere, The Hague, Leipzig, and Vienna. One such network, with the acronym PRESUD (Peer Review in European Sustainable Urban Development), has brought partners from other European cities to Malmö to evaluate its efforts in sustainable urban development. As could be expected, the visitors found many positive features but also a good many problems. Sustainable urban developers from Malmö have also participated in a number of looser networks, both within Sweden and internationally. In 1994, the so-called Aalborg Charter was formulated at a Sustainable Cities and Towns in Europe conference, and has provided guidelines for carrying out projects in sustainable urban development. By the late 1990s, when the upcoming Bo01 housing exposition began to be planned, the organizers had to find ways to bring these various activities in sustainable urban development into the planning process. It was time, we might say, for sustainable urban development to go to market.

The Tale of the Turning Torso

The Bo01 exposition was to be held when the new Öresund bridge connecting Malmö and Copenhagen was finished, and it seemed only natural that the exposition should be not merely a traditional housing exhibition but something bigger and more ambitious. The mayor of the city, the local leader of the Social Democratic party, Ilmar Reepalu, had been educated as an architect, and had already shown his political acumen by successfully convincing the national government to support the founding of a university college in Malmö.

Reepalu was not alone in seeing the upcoming exposition—and the reconstruction of the Western Harbor area, where the exposition was to be held—as an enormous opportunity for Malmö to improve its reputation. It had long been viewed as a drab, working-class city in the shadow of cosmopolitan Copenhagen. Because money from the LIP program could be obtained in greening the city, it was decided that the exposition would focus on sustainable urban development, and thereby encourage the exhibiting companies to think about projecting a more ecological image. A group of housing officials, municipal civil servants, and representatives of both companies and environmental organizations was set up to plan the exposition, which gradually grew into a major manifestation. Bo01 received some 250 million crowns (and Turning Torso about 5 million more) from LIP, and funds were collected from many other sources for what became an effort to market Malmö as the sustainable city of tomorrow. The budget could not be kept, however, which led to the bankruptcy of the planning committee soon after the exposition was closed down.

While other cities in the vicinity had recently held housing exhibitions—the suburban town of Staffanstorp (see figure 12.4) had held one in 1997 and Helsingborg, a larger city up the coast, had held one in 1999—Bo01 in Malmö was thus going to be something different. It was not only, or even primarily, to be about housing; it was going to be about sustainable urban development. Landscape architects were invited to create imaginative gardens, to bring out some of the more complicated natural processes involved in sustainable urban living (what happens to waste and how water is supplied and managed, for example). Unfortunately, the gardens were destroyed after the exposition, as were most of the ecological restaurants and cafés that dotted the grounds. There was an exhibit about social problems, with films and lectures portraying the dark sides of the contemporary city. Efforts were made to involve immigrant groups in the exposition, but it proved difficult to compete for their time due to their involvement in the annual Malmö festival, a free cultural manifestation and urban fair that is held for a week every August. Sydkraft, the local energy company, devoted an entire building to its work with renewable energy, which included demonstrations to

Figure 13.2
Some of the ecologically inspired housing that still remains in the Western Harbor area in Malmö, Sweden. Photographer: Mikael Hård.

explain how the city of tomorrow could be self-sufficient in terms of energy use. "Taking the year as a whole," Sydkraft wrote in the exposition program, "the number of kWh consumed equals the number generated" (Bo01, 48–49).

According to the advertising and marketing, the prime purpose of the Bo01 housing expo was public education (what in Swedish is called *folkbildning*) in the broadest sense. According to the program, Bo01's ambition was "to start a debate on how we wish to live in the city of tomorrow," and a wide range of educational institutions and cultural debaters had been invited to take part in discussing what the program called "active citizenship" (Bo01, 25). For anyone who visited the exposition, it was clear that the organizers wanted sustainable urban development to be a multifaceted cultural process that required the emergence of new ideas and practices, but, perhaps most importantly of all, new values and identities. A few years later, however, it is rather difficult to find those values embodied in the expensive and rather impressive housing that was built in the area. Instead of fostering new ecological values, the Western Harbor area has become most renowned for being the site of the Turning Torso.

In 1998, Santiago Calatrava, who had unsuccessfully bid for the design of the Öresund Bridge, attended a meeting at which plans for the bridge and the exposition were presented. Jonny Öhrbäck, the director of the Malmö office of the HSB housing company, which was one of the active partners in the Bo01 consortium, talked with Calatrava about the possibility of his making a contribution. A few months later, Öhrbäck and a group of city officials visited Calatrava in Switzerland in order to discuss further what his contribution might be. During that visit Öhrbäck saw a model of the *Twisting Torso* sculpture and had the idea that Calatrava should design a building based on it for the exposition site. Calatrava had not designed apartment buildings before but found the idea intriguing, and not long thereafter, the contract for the Turning Torso was signed.

Santiago Calatrava is one of the more outspoken modernist architects of our time, but he is not noted, in any of the voluminous literature about him, for an environmental sensibility or an interest in sustainability. Educated in architecture and engineering, he made his reputation by designing bridges and railway stations that combine artistic creativity and a fascination with processes of physical motion and what he has termed the "foldability" of metal materials (Jodidio 2003). Calatrava's constructions display an aesthetic interest in abstraction and a susceptibility to what Mikael Hård and I have elsewhere called "technological hubris": doing the impossible and transcending previously recognized limits (Hård and Jamison 2005). There is an experimental playfulness in his work that is reminiscent of Antoni Gaudí and Joan Miró, two of his main sources of inspiration. And there is certainly an impressive symbolic ambition in everything he does. As he has said: "I believe that one of the most important tasks is to reconsider the periphery of cities. Most often public works in such areas are purely functional, and yet even near railroad tracks, or spanning polluted rivers, bridges can have a remarkably positive effect. By creating an appropriate environment they can have a symbolic impact whose ramifications go far beyond their immediate location" (Jodidio 2003, 12). This interest in what Calatrava calls "transgression" has characterized modern architecture since it first emerged in the nineteenth century. In the words of Lewis Mumford (1963, 166), "the first erections of modern architecture, beginning with the Crystal Palace in 1851, rested on a firm foundation: the perception that the technology of the nineteenth century had immensely enriched the vocabulary of modern form and facilitated modes of construction that could hardly have been dreamed of in more ponderous materials." The natural environment was not so much to be respected and lived *in*; rather, the aesthetic ambition of modern architecture—especially as it grew into the modernism of Le Corbusier and others—became the transcendence of the natural surroundings and their replacement by the artificial and abstract "nature" visualized and given material form by the designs of the technologically infatuated architect. To live in the Turning Torso is to live in a work of art; it is certainly not to adapt one's way of living to the surrounding environmental conditions. It is perhaps not too surprising that there were not very many people in Malmö who were initially interested—or rich enough—to acquire one of the apartments; in a city afflicted with more than its share of homelessness and suffering from a high degree of cultural segregation in its housing "market," the building can be seen to represent a rather problematic interpretation of sustainable urban development.

Conclusions

The tale of the Turning Torso offers an illustrative example of the cultural tensions of sustainable development in general and sustainable urban development in particular. The quest for sustainable development has met a number of constraints since it was articulated as the cornerstone of a new international "agenda" in Brazil in 1992. The ambitions of sustainability have been countered by the commercial ambitions of the global marketplace, as well as by the aesthetic assumptions of modern architecture.

From such a perspective, the Turning Torso represents, in a particularly unholy alliance, the twin pressures of the market and the past. Together, commercialism and modernism have reduced sustainable urban development to the traditional urban planning project of constructing impressive buildings. Of course, the construction plans have paid lip service, conveniently enough with the support of LIP money, to environmental concerns—renewable energy and building materials were to be used, wherever feasible, and ecological design was to be encouraged in the individual apartments—but the discourse in which the building is presented is undeniably modernist. And as it towers over the rest of the city and indeed the rest of southern Sweden, the Turning Torso provides a symbol for the contemporary city that seems remarkably like the symbols of the past. Instead of the huge Kockums crane—Malmö's industrial symbol par excellence, which has been shipped off to South Korea along with the shipbuilding business—an artistic skyscraper reaches for the sky, telling an all too traditional story of technological hubris. The city of tomorrow has thus been captured by the modernist architecture of yesterday, and while the quest for sustainable urban development continues to take place in Malmö, the future seems to have been postponed.

References

Abû-Lughod, Janet. 1980. *Rabat: Urban Apartheid in Morocco*. Princeton: Princeton University Press.

Adas, Michael. 1989. *Machines as the Measure of Men: Science, Technology, and Ideologies of Western Dominance*. Ithaca: Cornell University Press.

Addams, Jane. 1910/1961. *Twenty Years at Hull-House*. New York: New American Library.

Agursky, Mikhail. 1987. *The Third Rome: National Bolshevism in the USSR*. Boulder, CO: Westview Press.

Ahlander, Björn. 1960. *Amerika: Dröm och verklighet* [America: Dream and reality]. Stockholm: Geber.

Aibar, Eduardo, and Wiebe Bijker. 1997. "Constructing a City: The Cerda Plan for Barcelona." *Science, Technology and Human Values* 22: 3–30.

Albers, Gerd. 1997. *Zur Entwicklung der Stadtplanung in Europa: Begegnungen, Einflüsse, Verflechtungen*. Braunschweig and Wiesbaden: Friedr. Vieweg & Sohn.

Aleksandrowicz, Dariusz. 1999. "The Socialist City and Its Transformation." Discussion Papers no. 10/99. Frankfurt an der Oder: Frankfurt Institute for Transformation Studies. Europa-Universität Viadrina.

Alexander, D. 1970. *Retailing in England during the Industrial Revolution*. London: Athlone.

Alm, Martin. 2002. *Americanitis: Amerika som sjukdom eller läkemedel: Svenska berättelser om USA under åren 1900–1939* [Americanitis: America as disease or cure: Swedish reports on the USA in the years 1900–1939]. Lund: Nordic Academic Press.

Amatori, F. 1989. *Proprietà e direzione: La Rinascente 1917–1969*. Milan: Angeli.

Andersson, Henrik O. 1981. "Tätbygdsutveckling och stadsförnyelse i Sverige" [Urban development and town renewal in Sweden]. In *Stadsförnyelse—kontinuitet, gemenskap, inflytande: Underlag för diskussion och fortsatt utredningsarbete: Publicerat av stadsförnyelsekommittén* [Town renewal: A basis for discussion and continued investigation]. Swedish Government Official Reports (SOU), 69–105. Stockholm: Liber.

Andersson, Sven. 1956. *Har vi råd med bilen?* [Can we afford the car?] Stockholm: Sveriges Socialdemokratiska Arbetareparti och Tidens Förlag.

Andor, Mihály, and Péter Hidy. 1986. *Város-szövevény—Kazincbarcika felfejtése* [City-jumble—untangling Kazincbarcika]. Budapest: Művelődéskutató Intézet.

Annual Report and Bulletin on Graduate Training 1954–1955. 1955. New Haven, CT: Yale University Bureau of Highway Traffic.

Antila, Kimmo. 2003. "Forgetting the Scale: International Ideas and Actors in Finnish Highway Building in the 1960's." Paper at Society for the History of Technology annual meeting, Atlanta, USA, 19 October 2003.

Apolinarski, Ingrid. 2005. "Die gesamtstädtischen Planungen für Eisenhüttenstadt in den Jahren 1950 bis 1989." In *Schönheit und Typenprojektierung: Der DDR-Städtebau im internationalen Kontext*, ed. Christoph Bernhardt and Thomas Wolfes, 321–339. Erkner: Institut für Regionalenwicklung und Strukturplanung.

ARARAT. 1976. *ARARAT: Alternative Research in Architecture, Resources, Art and Technology* (exhibition catalog). Stockholm: Modern Museum.

Arbeitsgruppe Stadtgeschichte, eds. 1999. *Erste sozialistische Stadt Deutschlands*. Berlin: Brandenburg Verlag.

Armstrong, John. 2000. "From Shillibeer to Buchanan: Transport and the Urban Environment." In *Cambridge Urban History of Britain*, Vol. III: *1840–1950*, ed. M. Daunton, 229–257. Cambridge: Cambridge University Press.

Arnold, David. 2005. "Europe, Technology, and Colonialism in the twentieth Century." *History and Technology* 21: 85–106.

Arvidsson, Claes, and Lars Erik Blomqvist, eds. 1987. *Symbols of Power: The Esthetics of Political Legitimation in the Soviet Union and Eastern Europe*. Stockholm: Almqvist & Wiksell International.

Åström, Torsten R. 1939. "Reseberättelse över studieresa avseende kommunikations- och stadsbyggnadstekniska spörsmål företagen under tiden den 20 juni–18 augusti 1939 genom England, Belgien, Holland, Tyskland, Schweiz och norra Italien" [Account of a study trip in connection with transport and planning matters made during the period 20 June–18 August 1939 to England, Belgium, Holland, Germany, Switzerland, and northern Italy]. Travel Writings 1939–40, Kungliga Tekniska Högskolan library archive.

Augé, Marc. 1992. *Non-lieux*. Paris: Seuil.

Auric, André Joseph. 1911. "Rapport général du service technique de la préfecture." *Génie Civil Ottoman* 1, no. 1: 1–5.

Austin, John Langshaw. 1962/1976. *How to Do Things with Words*. 2nd ed. Oxford: Oxford University Press.

Ayçoberry, Pierre, and Marc Ferro. 1981. "Chronologie." In *Une histoire du Rhin*, ed. P. Ayçoberry and M. Ferro, 11–13. Paris: Éditions Ramsay.

Bacia, Horst. 1994. "Ganz ohne Mythen geht es auch heute noch nicht: Die sozialistische 'Musterstadt' Magnitogorsk: Ein Abbild der Krise Russlands." *Frankfurter Allgemeine Zeitung*, no. 75 (30 March).

Bacon, Francis. 1624/2001. *Neu-Atlantis*. Stuttgart: Reclam.

Bacon, Mardges. 2001. *Le Corbusier in America: Travels in the Land of the Timid*. Cambridge: MIT Press.

Baer, T. 1928. "Baugeschichte der staatlichen Hafenanlagen in Mannheim." In *Die Mannheimer Hafen-Anlagen*, ed. Badischen Hafenverwaltung, 4–15. Koblenz: Rekord-Druckerei und Verlagsanstalt.

Bahrdt, Hans Paul. 1961. *Die moderne Großstadt*. Reinbek: Rowohlt.

Baldwin, Peter C. 1999. *Domesticating the Street: The Reform of Public Space in Hartford, 1850–1930*. Columbus: Ohio State University Press.

Banham, Reyner. 1960. *Theory and Design in the First Machine Age*. Cambridge: MIT Press, 1960.

———. 1986. *A Concrete Atlantis: U.S. Industrial Building and European Modern Architecture, 1900–1925*. Cambridge: MIT Press.

Baranowski, S. 2004. *Strength through Joy: Consumerism and Mass Tourism in the Third Reich*. Cambridge: Cambridge University Press.

Barke, M., and L. A. France. 1996. "The Costa del Sol." In *Tourism in Spain: Critical Issues*, ed. M. Barke, J. Towner, and M. T. Newton, 265–308. Wellingford: Cab International.

Barthes, Roland. 1957/2002. "The New Citroën." Reprinted in *Autopia: Cars and Culture*, ed. Peter Wollen and Joe Kerr, 340–341. London: Reaktion.

Basalla, George. 1984. "Science and the City before the Nineteenth Century." In *Transformation and Tradition in the Sciences: Essays in Honor of I. Bernard Cohen*, ed. Everett Mendelsohn, 513–529. Cambridge: Cambridge University Press.

Batur, Afife, ed. 1996. *Dünya Kenti Istanbul* [World City Istanbul]. Istanbul: Tarih Vakfı.

Batur, Enis. 2000. "Paris: Bir Fetiş-Mekan için Topografi Denemesi" [Paris: Topographical experiments on a fetish space]. In *Türk Edebiyatında Paris*, ed. Halil Gökhan and Timour Muhidine, 284–289. Istanbul: Yapı Kredi Yayınları.

Becher, Johannes R. 1959. *Vom Bau des Sozialismus*. Berlin: Verlag des Ministeriums für Nationale Verteidigung.

Becker, Sabina. 1993. *Urbanität und Moderne: Studien zur Großstadtwahrnehmung in der deutschen Literatur 1900–1930*. St. Ingbert: Röhrig.

Beckstein, Hermann. 1991. *Städtische Interessenpolitik: Organisation und Politik der Städtetage in Bayern, Preussen und im Deutschen Reich 1896–1923*. Düsseldorf: Droste Verlag.

Bektaş, Yakup. 2000. "The Sultan's Messenger: Cultural Constructions of Ottoman Telegraphy, 1847–1880." *Technology and Culture* 41: 669–696.

———. 2003. "Crossing Communal Boundaries: Technology and Cultural Diversity in the nineteenth-Century Ottoman Empire." In *Multicultural Science in the Ottoman Empire*, ed. Ekmeleddin Ihsanoğlu, Kostas Chatzis, and Efthymios Nicolaidis, 139–149. Turnhout: Brepols.

Bell, Daniel. 1973. *The Coming of Post-Industrial Society: A Venture in Social Forecasting*. New York: Basic Books.

———. 1976. *Die nachindustrielle Gesellschaft*. Frankfurt: Campus.

Bendtsen, P. H. 1961. *Town and Traffic in the Motor Age*. Copenhagen: Danish Technical Press.

Benjamin, Walter. 1982. *Das Passagen-Werk*. Frankfurt: Suhrkamp.

Ben-Joseph, Eran. 2005. *The Code of the City: Standards and the Hidden Language of Place*. Cambridge: MIT Press.

Benson, Timothy O., ed. 2002. *Central European Avant-Gardes: Exchange and Transformation, 1910–1930*. Cambridge: MIT Press.

Berglund, Helge. 1956. "Vilka praktiska lärdomar kan vi i Sverige draga ur stads- och trafikplaneringen i U.S.A.?" [What practical lessons can we in Sweden draw from town and traffic planning in the USA?]. *Svensk Lokaltrafik* 13, no. 4: 213–219.

Berkes, Niyazi. 1964/1998. *The Development of Secularism in Turkey*. London: Routledge.

Berman, Marshall. 1982. *All That Is Solid Melts into Air: The Experience of Modernity*. New York: Simon and Schuster.

Bernhard, Harry. 1956. "Stadsplane- och trafikfrågor i USA och Kanada I" [Town planning and traffic questions in the USA and Canada I]. *Plan* 1: 10–19.

Bernhardt, Christoph, ed. 2001. *Environmental Problems in European Cities of the 19th and 20th Century*. New York, Munich, and Berlin: Waxmann.

Bernstorff, Johann Heinrich von. 1911. *The Government of German Cities*. New York: City Club.

Bijker, Wiebe, and Karin Bijsterfeld. 2000. "Women Walking through Plans: Technology, Democracy and Gender Identity." *Technology and Culture* 41: 485–515.

"'Bilsamhället' revolutionerar bebyggelse och liv i USA" [The 'car society' is revolutionizing settlement and life in the USA]. 1954. *Svenska Dagbladet* (16 October).

Bilstaden [The car city]. 1960. Stockholm: Royal Institute of Technology.

Binder, Beate. 1999. *Elektrifizierung als Vision: Zur Symbolgeschichte einer Technik im Alltag*. Tübingen: Tübinger Vereinigung für Volkskunde.

Bloch, Ernst. 1985. *Christliche Philosophie des Mittelalters, Philosophie der Renaissance*. Leipziger Vorlesungen zur Geschichte der Philosophie, vol. 2. Frankfurt: Suhrkamp.

Blockmans, Wim. 2003. "Reshaping Cities: The Staging of Political Transformation." *Journal of Urban History* 30, no. 1: 7–20.

Blok, Johan. 1993. *Haagse trams: Het Haagse trammaterieel van 1864 tot heden.* Rijswijk: Elmar.

Blomberg, Dag. 1954. "Den europeiska delegationens trafikstudier i USA våren 1954" [Traffic studies of the European delegation, Spring 1954]. In *Vad har USA att lära oss om trafikplaneringen?* [What has the USA to teach us about traffic planning?]. Advanced Course in Town Planning at Kungliga Tekniska Högskolan, 13–15 September 1954.

Blomkvist, Pär. 2001. *Den goda vägens vänner: Väg- och billobbyn och framväxten av det svenska bilsamhället 1914–1959* [Friends of the good road: The road and car lobby and the growth of the Swedish car society]. Stockholm: Symposion.

———. 2004. "Transferring Technology—Shaping Ideology: American Traffic Engineering and Commercial Interests in the Establishment of a Swedish Car Society, 1945–1965." *Comparative Technology Transfer and Society* 2, no. 3: 273–302.

———. 2006. "Roads for Flow—Roads for Peace: Lobbying for a European Highway System." In *Networking Europe: Transnational Infrastructures and the Shaping of Europe, 1850–2000*, ed. Erik van der Vleuten and Arne Kaijser. Sagamore Beach, MA: Science History Publications.

Blotevogel, Hans Heinrich, ed. 1990. *Kommunale Leistungsverwaltung und Stadtentwicklung vom Vormärz bis zur Weimarer Republik.* Cologne and Vienna: Böhlau.

Blume, Mary. 1992. *Côte d'Azur: Inventing the French Riviera.* London: Thames & Hudson.

Bo01: Framtidstaden Mässkatalog. 2001. Exposition catalog. Malmö: City of Malmö.

Bodenschatz, Harald, and Christiane Post, eds. 2003. *Städtebau im Schatten Stalins: Die internationale Suche nach der sozialistischen Stadt in der Sowjetunion, 1929–1935.* Berlin: Braun Verlag.

Bodenschatz, Harald, Christiane Post, and Uwe Altrock. 2005. "Schmelztiegel internationaler Leitbilder des Städtebaus oder 'stalinistische Reaktion'? Die Sowjetunion auf der Suche nach der sozialistischen Stadt zwischen 1929 und 1936." In *Schönheit und Typenprojektierung: Der DDR-Städtebau im internationalen Kontext*, ed. Christoph Bernhardt and Thomas Wolfes, 17–59. Erkner: Institut für Regionalenwicklung und Strukturplanung.

Boenke, Susan. 1991. *Entstehung und Entwicklung des Max-Planck-Instituts für Plasmaphysik, 1955–1971.* Frankfurt and New York: Campus.

Bogdanowski, Janusz. 1996. "Urbanizacja krakowska w dobie PRL" [Kraków's urbanization process in the time of the People's Republic of Poland]. In *Kraków w Polsce Ludowej: Materiały sesji naukowej odbytej 27 maja 1995 roku* [Kraków in the People's Republic of Poland: Materials of a conference of May 27, 1995], ed. Towarzystwo Miłośników Historii i Zabytków Krakowa [Association of Friends of Kraków's History and Heritage]. Kraków: Wydawnictwo i Drukarnia DRUKROL.

Böhme, Gernot. 1993. *Am Ende des Baconschen Zeitalters: Studien zur Wissenschaftsentwicklung.* Frankfurt: Suhrkamp.

Böhme, Helmut. 2000. "Konstituiert Kommunikation Stadt?" In *Stadt und Kommunikation im digitalen Zeitalter*, ed. H. Bott, C. Hubig, F. Pesch, and G. Schröder, 13–39. Frankfurt and New York: Campus.

Borkowski, Tadeusz. 1976. "Stan badań nad mieszkańcami dzielnicy Nowa Huta oraz nad załoga Huty im. Lenina" [State of research on the citizens of Nowa Huta and the staff of the Lenin steelworks]. In *Huta im. Lenina i jej załoga* [The Lenin steelworks and its staff], ed. Antoni Stojak. Kraków: Zeszyty Naukowe Uniwersytetu Jagiellońskiego CCCCXLVII, Prace sojologiczne, Zeszyt 3 [Research papers of the Jagiellonien University CCCCXLVII, sociology papers, no. 3].

Bos, A. W. 1915. "Sewerage for Low Countries, with Special Regard to the Town of Amsterdam." *Transactions of the International Engineering Congress, 1915: Municipal Engineering*, 298–313. San Francisco: The International Engineering Congress.

Bossaglia, R. 1986. *Tra liberty e déco: Salsomaggiore*. Parma: Artegrafica Silva.

Bouneau, C. 1987. "La formation d'un marché régional: La consommation dans le grand Sud-Ouest de 1914 à 1946." In *L'Électricité et ses consommateurs*, ed. F. Cardot, 17–37. Paris: Press Universitaires de France.

Bourdieu, Pierre. 1979/1984. *Distinction: A Social Critique of the Judgement of Taste*. Cambridge: Harvard University Press.

Boyle, Godfrey, and Peter Harper, eds. 1976. *Radical Technology*. London: Wildwood House.

Braudel, Fernand. 1972. *The Mediterranean and the Mediterranean World in the Age of Philip II*. New York: Harper and Row.

Braun, Hans-Joachim. 1980. "Gas oder Elektrizität? Zur Konkurrenz zweier Beleuchtungssysteme, 1880–1914." *Technikgeschichte* 47: 1–19.

Breit, Gotthard. 1980. *Krakau und Nowa Huta: Alte und neue Städte in Polen*. Paderborn: Schöningh.

Brenner, Neil. 2004. *New State Spaces: Urban Governance and the Rescaling of Statehood*. Oxford: Oxford University Press.

Brooks, Michael W. 1997. *Subway City: Riding the Trains, Reading New York*. New Brunswick: Rutgers University Press.

Bruijn, Theo de, and Vicki Norberg-Bohm. 2005. *Industrial Transformation: Environmental Policy Innovation in the United States and Europe*. Cambridge: MIT Press.

Brummett, Palmira. 2000. *Image and Imperialism in the Ottoman Revolutionary Press*. New York: SUNY Press.

Buchanan, Colin. 1963. *Traffic in Towns: A Study of the Long Term Problems of Traffic in Urban Areas*. London: Minister of Transport.

Buchhaupt, Siegfried. 1994. "Umlandversorgung und Energiepolitik im Raum Mainz-Wiesbaden: Das Ausgreifen des städtischen Elektrizitätswerkes Mainz in die Region." In *Wege regionaler Elektrifizierung in der Rhein-Main-Neckar-Region*, ed. H. Böhme and D. Schott, 41–56. Darmstadt: THD Schriftenreihe Technik und Wissenschaft.

References

Buckow, Anjana. 2003. *Zwischen Propaganda und Realpolitik: Die USA und der sowjetisch besetzte Teil Deutschlands, 1945–1955*. Stuttgart: Franz Steiner Verlag.

Buiter, Hans. 1992. "De stad met de mooiste maquettes: Plannen voor Utrechts centrum en binnenstad." *Jaarboek Oud-Utrecht*, 5–43.

———. 2003. "Werken aan sanitaire en bereikbare steden." In *Techniek in Nederland in de twintigste eeuw*, vol. 6, ed. Johan W. Schot, 25–49. Zutphen: Walburg Pers.

———. 2005a. *Riool, rails en asfalt: 80 jaar straatrumoer in vier Nederlandse steden*. Zutphen: Walburg Pers.

———. 2005b. "Industrialisatie, internationalisatie en diversificatie: Van bouwbedrijf naar conglomeraat, 1947–1986." In *Bredero's bouwbedrijf: Familiebedrijf—mondiaal bouwconcern—ontvlechting*, ed. W. M. J. Bekkers, 59–107. Amsterdam: Dutch University Press.

Buiter, Hans, and Peter E. Staal. 2006. "City Lights: Regulated Streets and the Evolution of Traffic Lights in The Netherlands, 1920–1940." *Journal of Transport History* 27: 1–20.

Burke, Peter. 2002. *Papier und Marktgeschrei: Die Geburt der Wissensgesellschaft*. Berlin: Wagenbach.

Burnett, J. 1999. *Liquid Pleasures: A Social History of Drinks in Modern Britain*. London and New York: Routledge.

Bush, Vannevar. 1945. *Science—The Endless Frontier: Report to the President on a Program for Postwar Scientific Research*. Washington, DC: United States Government Printing Office.

Capuzzo, Paolo. 1998. *Vienna da città a metropoli*. Milan: Angeli.

———. 2003. "Gli spazi della nuova generazione." In *Genere, generazione e consume: L'Italia degli anni Sessanta*, ed. Paolo Capuzzo, 217–247. Rome: Carocci.

———. 2004. "Dalla città all'automobile e ritorno: Un percorso del Novecento." *Parolechiave* 32: 79–92.

Caron, François, ed. 1997. *Histoire générale de l'électricité en France. Vol. 1: Espoirs et conquêtes 1881–1918*. Paris: Fayard.

Castells, Manuel, and Peter Hall. 1994. *Technopoles of the World: The Making of Twenty-First-Century Industrial Complexes*. London and New York: Routledge.

Çelik, Zeynep. 1993. *The Remaking of Istanbul: Portrait of an Ottoman City in the Nineteenth Century*. Berkeley: University of California Press.

Cepl-Kaufmann, Gertrud, and Antje Johanning. 2003. *Mythos Rhein: Zur Kulturgeschichte eines Stromes*. Darmstadt: Wissenschaftliche Buchgesellschaft.

Chamberlain, Joseph Perkins. 1923. *The Regime of the International Rivers: Danube and Rhine*. New York: Columbia University.

Charney, Leo, and Vanessa R. Schwartz, eds. 1995. *Cinema and the Invention of Modern Life*. Berkeley: University of California Press.

Cherry, Gordon E. 1990. "Reconstruction: Its Place in Planning History." In *Rebuilding Europe's Bombed Cities*, ed. Jeffry M. Diefendorf, 209–220. London: Macmillan.

———. 1996. *Town Planning in Britain since 1900: The Rise and Fall of the Planning Ideal*. Oxford: Blackwell.

Choma, Maciej, and Mieczysław Gil. 1999. *50 lat. Huta im. Taduesza Sendzimira w Krakowie 1949–1999* [50 years of Sendzimir Steelworks Ltd. in Kraków, 1949–1999]. Kraków: Wydawnictwo Trans-Krak.

Chomątowski, Stanisław. 1999. "Życie gospodarcze Krakowa w Polsce Ludowej" [The economic life of Kraków in the People's Republic of Poland]. In *Kraków w Polsce Ludowej: Materiały sesji naukowej odbytej 27 maja 1995 roku* [Kraków in the People's Republic of Poland: Materials of a conference of May 27, 1995], ed. Towarzystwo Miłośników Historii i Zabytków Krakowa [Association of Friends of Kraków's History and Heritage], 45–64. Kraków: Wydawnictwo i Drukarnia DRUKROL.

Chwalba, Andrzej. 2004. *Dzieje Krakowa. Tom 6: Kraków w latach 1945–1989* [History of Kraków. Volume 6: Kraków in the years 1945–1989]. Kraków: Wydawnictwo Literackie.

Cioc, Mark. 2002. *The Rhine: An Eco-Biography, 1815–2000*. Seattle: University of Washington Press.

City 67: Stockholm: Principplan för den fortsatta citysaneringen framlagd i maj 1967 [Stockholm: Basic plan for the continued redevelopment of the city, presented in May 1967]. Stockholm: Esselte, 1967.

Cohen, Jean-Louis. 1995. *Scenes of the World to Come: European Architecture and the American Challenge, 1893–1960*. Paris: Flammarion; Montreal: Canadian Centre for Architecture.

Coles, T. 1999. "Department Stores as Retail Innovations in Germany: A Historical-Geographical Perspective on the Period 1870–1914." In Crossick and Jaumais 1999, 72–96.

Condit, Carl. 1977. *The Railroad and the City: A Technological and Urbanistic History of Cincinnati*. Columbus: Ohio State University Press.

———. 1980–1981. *The Port of New York*. 2 vols. Chicago: University of Chicago Press.

Connor, Walker. 1984. *The National Questions in Marxist-Leninist Theory and Strategy*. Princeton: Princeton University Press.

Coopey, Richard, ed. 2004. *Information Technology Policy: An International History*. Oxford: Oxford University Press.

Corbin, A. 1988. *Le territoire du vide: L'occident et le desir du rivage (1750–1840)*. Paris: Flammarion.

Cremers, E., F. Kaaij, and C. M. Steenbergen. 1981. *Bolwerken als stadsparken: Nederlandse stadswandelingen in de 19^e en 20^e eeuw*. Delft: Delftse Universitaire Pers.

Cronon, William. 1992. *Nature's Metropolis: Chicago and the Great West*. New York: W. W. Norton.

Cross, Gary. 1993. *Time and Money: The Making of Consumer Culture.* London and New York: Routledge.

Crossick, Geoffrey, and Serge Jumain, eds. 1999. *Cathedrals of Consumption: The European Department Store, 1850–1939.* Aldershot: Ashgate.

Curtin, P. D. 1984. *Cross-Cultural Trade in World History.* Cambridge: Cambridge University Press.

Czubiński, Antoni. 1992. *Dzieje najnowsze Polski 1944–1989* [The contemporary history of Poland 1944–1989]. Poznań: Wielkopolska Agencja Wydawnicza.

Dalen, R. van. 1979. "Ongewenste ontmoetingen in de Amsterdamse tram, klachten aan de Gemeente Tram Amsterdam in het begin van de twintigste eeuw." *Amsterdams Sociologisch Tijdschrift* 6: 449–472.

Daston, Lorraine. 1992. "Objectivity and the Escape from Perspective." *Social Studies of Science* 22, no. 4: 597–618.

Davies, Sarah. 1997. "'Us against Them': Social Identity in Soviet Russia, 1934–41." *Russian Review* 56, no. 1 (January): 70–89.

Davison, Roderic H. 1973. *Reform in the Ottoman Empire 1956–1976.* New York: Gordian Press.

De Bruijn, Theo, and Vicki Norberg-Bohm. 2005. *Industrial Transformation: Environmental Policy Innovation in the United States and Europe.* Cambridge: MIT Press.

De Certeau, M. 1980. *L'invention du quotidien: 1 Arts de faire.* Paris: Gallimard.

De Grazia, Victoria. 1981. *Consenso e cultura di massa nell'Italia fascista: L'organizzazione del dopolavoro.* Rome and Bari: Laterza.

———. 2005. *Irresistible Empire: America's Advance through Twentieth-Century Europe.* Cambridge, MA: Belknap Press.

Debord, G. 1992. *La société du spectacle.* Paris: Gallimard.

Delorme, Andrzej. 1995. *Antyekologiczna spuścizna totalitaryzmu: Polityka-Gospodarka-Środowisko naturalne* [Anti-ecologic heritage of totalitarianism: Politics-economy-environment]. Kraków: Wydawnictwo i drukarnia "Secesja."

Demangeon, Albert, and Lucien Febvre. 1935. *Le Rhin: Problèmes d'histoire et d'économie.* Paris: Librairie Armand Colin.

Derwig, Jan, and Erik Mattie. 1995. *Functionalism in the Netherlands.* Amsterdam: Architectura & Natura.

Deutinger, Stephan. 1999. "'Garching: Deutschlands modernstes Dorf': Die Modernisierung Bayerns seit 1945 unter dem Mikroskop." In *Neue Ansätze zur Erforschung der neueren bayerischen Geschichte,* ed. K. Weigand and G. Treffler, 223–247. Neuried: Ars Una.

Dickson, David. 1974. *Alternative Technology and the Politics of Technical Change.* Glasgow: Fontana.

Diefendorf, Jeffry M. 1989. "Artery: Urban Reconstruction and Traffic Planning in Postwar Germany." *Journal of Urban History* 15: 131–158.

———, ed. 1990. *Rebuilding Europe's Bombed Cities*. London: Macmillan.

Dinçkal, Noyan. 2004. *Istanbul und das Wasser: Zur Geschichte der Wasserversorgung und Abwasserentsorgung von der Mitte des 19. Jahrhunderts bis 1966*. Munich: Oldenbourg.

Dinçkal, Noyan, and Shahrooz Mohajeri. 2002. "Zentrale Wasserversorgung in Berlin und Istanbul: Einrichtung, Diffusions- und Akzeptanzprozesse im Vergleich." *Technikgeschichte* 69: 113–149.

Dluhosch, Eric, and Rostislav Švácha, eds. 1999. *Karel Teige, 1900–1951: L'Enfant Terrible of the Czech Modernist Avant-Garde*. Cambridge: MIT Press.

Doblas, Escorza. 2001. *El turismo de golf en la Costa del Sol: Análisis geográfico*. Málaga: Diputación Provincial de Málaga.

Driver, Felix, and David Gilbert, eds. 1999. *Imperial Cities: Landscape, Display and Identity*. Manchester: Manchester University Press.

Duben, Alan, and Cem Behar. 1991. *Istanbul Households: Marriage, Family and Fertility 1880–1940*. Cambridge: Cambridge University Press.

Dumont, Paul, and François Georgeon. 1985. "Un bourgeois d'Istanbul au début du XXe siècle." *Turcica: Revue d'Études Turques* 17: 127–188.

Durth, Werner, Jörn Düvel, and Niels Gutschow. 1998. *Architektur and Städtebau der DDR. Band 1. Ostkreutz: Personen, Pläne, Perspektiven*. Frankfurt and New York: Campus.

Eckert, Michael, and Maria Osietzki. 1989. *Wissenschaft für Macht und Markt*. Munich: Beck.

Eckstein, Barbara, and James A. Throgmorton. 2003. *Story and Sustainability: Planning, Practice, and Possibility for American Cities*. Cambridge: MIT Press.

Edman, Agneta. 2004. "'Herrgårdslandet'" [Manorial estate]. In *Hotad idyll: Berättelser om svenskt folkhem och kallt krig* [Threatened idyll: Stories of the Swedish *Folkhem* and the Cold War], ed. Kim Salomon, Lisbeth Larsson, and Håkan Arvidsson. Lund: Nordic Academic Press.

Edwards, Paul. 2003. "Infrastructure and Modernity: Force, Time, and Social Organization in the History of Sociotechnical Systems." In Misa et al. 2003, 185–225.

EKO Stahl GmbH., ed. 2000. *Einblicke: 50 Jahre EKO Stahl*. Eisenhüttenstadt: EKO Stahl GmbH.

Eldem, Edhem. 1992. "A Vision beyond Nostalgia: The Ethnic Structure of Galata." *Biannual Istanbul* 92: 28–34.

———, ed. 1999. *The Ottoman City between East and West: Aleppo, Izmir, and Istanbul*. New York: Cambridge University Press.

Emrence, Cem. 2001. "Istanbul Tramvayında Sınıf ve Kimlik (1871–1922)." [Class and identity in Istanbul's streetcars, 1871–1922]. *Toplumsal Tarih* 93: 6–14.

Engels, Friedrich. 1845/1987. *The Condition of the Working Class in England*. London: Penguin Books.

Enyedi, György. 1989. "Településpolitikák Kelet-Közép-Európában" [City policies in Eastern and Central Europe]. *Társadalmi Szemle* 10: 20–31.

Ergin, Osman Nuri. 1914–1922. *Mecelle-i Umur-u Belediye* [Book of Municipal Affairs]. 5 vols. Istanbul: Matbaa-yı Osmaniye.

Eriksson, Åke. 1949. "Engelsk fastighets- och bostadspolitik: Några intryck från kommunaltekniska föreningens resa i England sommaren 1949" [English property and housing policy: Some impressions from the Association of Municipal Technology's visit to England in the Summer of 1949]. *Kommunalteknisk Tidskrift* 15: 113–117.

Evans, Richard J. 1987. *Death in Hamburg: Society and Politics in the Cholera Years, 1830–1910*. Oxford: Clarendon Press.

Evin, Ahmet Ö. 1983. *Origins and Development of the Turkish Novel*. Minneapolis, MN: Bibliotheca Islamica.

Exertzoglou, Haris. 2003. "The Cultural Uses of Consumption: Negotiating Class, Gender, and Nation in the Ottoman Urban Centres during the Nineteenth Century." *International Journal of Middle East Studies* 35: 77–101.

Eyerman, Ron, and Andrew Jamison. 1991. *Social Movements: A Cognitive Approach*. Cambridge: Polity; University Park, PA: Pennsylvania State University Press.

Eyice, Semavi. 1973. "Istanbul'un Ortadan Kalkan Bazı Tarihi Eserleri" [The demise of some historical remnants in Istanbul]. *Tarih Dergisi* 27: 1–20.

Eysinga, W. J. M. van. 1935. *La Commission Centrale pour la Navigation du Rhin*. Leiden: A. W. Sijthoff.

Falk, P., and C. Campbell, eds. 1997. *The Shopping Experience*. London: Sage.

Faludi, Ádám. 2002. "Település a város helyén. Tatabánya látképe egy magán-történelemkönyvből" [A settlement in the place of a town: View of Tatabánya from a private history schoolbook]. *Magyar Napló* 14, no. 2: 51–65.

Faroqhi, Suraiya. 2000. *Subjects of the Sultan: Culture and Daily Life in the Ottoman Empire*. London: I. B. Tauris.

Faulhaber, Gerald R., and Gualtiero Tamburini, eds. 1991. *European Economic Integration: The Role of Technology*. Boston: Kluwer Academic.

Felt, Ulrike, Helga Nowotny, and Klaus Taschwer. 1995. *Wissenschaftsforschung: Eine Einführung*. Frankfurt and New York: Campus.

Fernandez, Alexandre. 1997. "La gestion des réseaux électriques par les grandes villes françaises, vers 1880–vers 1930." In *Energie und Stadt in Europa: Von der vorindustriellen Holznot bis zur Ölkrise der 1970er Jahre*, ed. D. Schott, 113–127. Stuttgart: Steiner.

Fischer, Wolfram, ed. 1992. *Die Geschichte der Stromversorgung*. Frankfurt: VDE.

Flonneau, Mathieu. 2002. *L'automobile à la conquête de Paris, 1910–1977: Formes urbaines, champ politique et représentations*. 3 vol. Paris: Université Panthéon-Sorbonne.

———. 2006. "City Infrastructures and City Dwellers: Accommodating the Automobile in Twentieth-Century Paris." *Journal of Transport History* 27, no. 1: 93–114.

Foucault, Michel. 1975/1977. *Discipline and Punish: The Birth of the Prison*. London: Allen Lane.

Fowler, D. 1995. *The First Teenagers: The Lifestyle of Young Wage-Earners in the Interwar Period*. London: Woburn Press.

French, Anthony R., and F. E. Ian Hamilton. 1979. "Is There a Socialist City?" In *The Socialist City: Spatial Structure and Urban Policy*, ed. Anthony R. French and F. E. Ian Hamilton, 1–21. New York: John Wiley & Sons.

Friedman, J. 1991. "Consuming Desires: Strategies of Selfhood and Appropriation." *Cultural Anthropology* 6: 154–164.

Frisnyák, Sándor, ed. 1979. *Kazincbarcika földrajza* [The geography of Kazincbarcika]. Kazincbarcika: Kazincbarcika Város Tanácsa.

Funck, Marcus, and Heinz Reif. 2003. Introduction to a special theme issue on European cities. *Journal of Urban History* 30, no. 1: 3–6.

Gaillard, J. 1975/1997. *Paris, la ville (1852–1870)*. Paris: L'Harmattan.

Garreau, Joel. 1991. *Edge City: Life on the New Frontier*. New York: Doubleday.

Gayko, Axel. 1999. *Investitions- und Standortpolitik der DDR an der Oder-Neiße-Grenze 1950–1970*. Frankfurt and Berlin: Peter Lang.

Gayler, H. J. 1984. *Retail Innovation in Britain: The Problems of Out-of-Town Shopping Centre Development*. Norwich: Geo Books.

Geist, J. F. 1969. *Passagen, ein Bautyp des 19. Jahrhunderts*. Munich: Prestel.

Gemeinde Garching bei München, ed. 1964. *Garching: Vom Heidedorf zum Atomzentrum*. Aßling, Pörsdorf: Verlag für Behörden und Wirtschaft.

Germuska, Pál. 2004. *Indusztria bűvöletében: Fejlesztéspolitika és a szocialista városok* [Under the spell of industry]. Budapest: 1956-os Intézet.

Giay, Frigyes. 1990. *Ajka*. Budapest: Építésügyi Tájékoztatási Központ.

Gibbons, Michael, et al. 1994. *The New Production of Knowledge: The Dynamics of Science and Research in Contemporary Societies*. London: Sage.

Gilfoyle, Timothy J. 1998. "White Cities, Linguistic Turns, and Disneylands: The New Paradigms of Urban History." *Reviews in American History* 26, no. 1: 175–204. The updated website version is a virtual bibliography of the subject: www.luc.edu/depts/history/gilfoyle/WHITECIT.HTM (accessed Sept. 2007).

Gilles, Pierre. 1988. *The Antiquities of Constantinople*. New York: Italica Press.

Gilson, Norbert. 1998. "Rationale Kalkulation oder prophetische Vision? Klingenbergs Pläne für die Elektrizitätsversorgung der 20er Jahre." In *Elektrizität in der Geistesgeschichte*, ed. K. Plitzner, 123–141. Bassum: Verl. für Geschichte der Naturwiss. und der Technik.

———. 1999. "Der Irrtum als Basis des Erfolgs: Das RWE und die Durchsetzung des ökonomischen Kalküls der Verbundwirtschaft bis in die 1930er Jahre." In *Elektrizitätswirtschaft zwischen Umwelt, Technik und Politik: Aspekte aus 100 Jahren RWE-Geschichte 1898–1998*, ed. H. Maier, 51–88. Freiberg: Technische Universität Bergakademie Freiberg.

Giuntini, Andrea, Peter Hertner, and Gregorio Núñez. 2004. *Urban Growth on Two Continents in the Nineteenth and Twentieth Centuries: Technology, Networks, Finance and Public Regulation*. Granada: Editorial Comares.

Giustino, Cathleen M. 2003. *Tearing Down Prague's Jewish Town: Ghetto Clearance and the Legacy of Middle-Class Ethnic Politics around 1900*. New York: Columbia University Press.

Göçek, Fatma Müge. 1996. *Rise of the Bourgeoisie, Demise of Empire: Ottoman Westernization and Social Change*. New York: Oxford University Press.

Göhner, Charles, and Emil Brumder. 1935. *Geschichte der räumlichen Entwicklung der Stadt Straßburg* Strasbourg: Heitz and Co.

Goodall, Francis. 1999. *Burning to Serve: Selling Gas in Competitive Markets*. Ashbourne: Landmark.

Goodman, David. 1999. *European Cities and Technology Reader: Industrial to Postindustrial Cities*. London: Routledge.

Goodman, David, and Colin Chant, eds. 1999. *European Cities and Technology: Industrial to Post-Industrial City*. London and New York: Routledge.

Gottlieb, Robert. 1993. *Forcing the Spring: The Transformation of the American Environmental Movement*. Washington, DC: Island Press.

Grahn, Ernst. 1904. "Die städtischen Wasserwerke." In Wuttke 1904, I: 301–344.

Gransche, Elisabeth, and Erich Wiegand. 1981. *Zur Wohnsituation von Arbeiterhaushalten zu Beginn des 20. Jahrhunderts*. Frankfurt: J. W. Goethe Universität.

Green, Nancy L. 1997. *Ready-to-Wear and Ready-to-Work: A Century of Industry and Immigrants in Paris and New York*. Durham: Duke University Press.

Greenhalgh, Paul. 1988. *Ephemeral Vistas: The Expositions Universelles, Great Exhibitions and World's Fairs, 1851–1939*. Manchester: Manchester University Press.

Gugerli, David. 1996. *Redeströme: Zur Elektrifizierung der Schweiz 1880–1914*. Zurich: Chronos.

Gül, Murat, and Richard Lamb. 2004. "Mapping, Regularizing and Modernizing Ottoman Istanbul: Aspects of the Genesis of the 1839 Development Policy." *Urban History* 31: 420–436.

Gullberg, Anders. 2001. C*ity—drömmen om ett nytt hjärta: Moderniseringen av det centrala Stockholm 1951–1979* [The city center—The dream of a new heart: The modernization of central Stockholm]. 2 vols. Stockholm: Stockholmia.

Habermas, Jürgen. 1962/1989. *Strukturwandel der Öffentlichkeit: Untersuchungen zu einer Kategorie der bürgerlichen Gesellschaft*. Frankfurt: Suhrkamp.

Hahn, Ekhart. 1987. *Ökologische Stadtplanung: Konzeption und Modelle*. Frankfurt: Haag + Herchen.

Hall, Peter. 1992. *Urban and Regional Planning*. 3rd ed. London and New York: Routledge.

———. 1996. *Cities of Tomorrow*. Rev. ed. Oxford: Blackwell.

———. 1998. *Cities in Civilization*. New York: Pantheon.

Hall, Thomas. 1997. *Planning Europe's Capital Cities: Aspects of Nineteenth-Century Urban Development*. London: Spon Press.

Hamlin, Christopher. 1990. *A Science of Impurity: Water Analysis in Nineteenth Century Britain*. Berkeley: University of California Press.

———. 1998. *Public Health and Social Justice in the Age of Chadwick: Britain, 1800–1854*. Cambridge: Cambridge University Press.

Haneda, Masashi, and Toru Miura, eds. 1994. *Islamic Urban Studies: Historical Review and Perspectives*. London and New York: Kegan Paul International.

Hannah, Leslie. 1979. *Electricity before Nationalisation: A Study of the Development of the Electricity Supply Industry in Britain to 1948*. London: Macmillan.

Hård, Mikael, ed. 2001. *History of Urban Technology in the 20th Century: Annotated Bibliography of Recent Literature*. ⟨www.histech.nl/tensphase2/Publications/Working/Hard%20Biblio.pdf⟩ (accessed September 2009).

———. 2003. "Zur Kulturgeschichte der Naturwissenschaft, Technik und Medizin: Eine internationale Literaturübersicht." *Technikgeschichte* 70: 23–45.

Hård, Mikael, and Andrew Jamison, eds. 1998. *The Intellectual Appropriation of Technology: Discourses on Modernity, 1900–1939*. Cambridge: MIT Press.

———. 2005. *Hubris and Hybrids: A Cultural History of Technology and Science*. New York: Routledge.

Hård, Mikael, and Thomas J. Misa, eds. 2003. *The Urban Machine: Recent Literature on European Cities in the Twentieth Century*. University of Minnesota, www.umn.edu/~tmisa/urban/biblios/urban-machine-complete.pdf (accessed September 2009).

Hardwick, M. Jeffrey. 2004. *Mall Maker: Victor Gruen, Architect of an American Dream*. Philadelphia: University of Pennsylvania Press.

Hardy, Anne I. 2005. *Ärzte, Ingenieure und städtische Gesundheit: Medizinische Theorien in der Hygienebewegung des 19. Jahrhunderts*. Frankfurt: Campus.

Harvey, D. 1994. *The Urban Experience*. Cambridge, MA, and Oxford: Blackwell.

Hasselquist, Erik. 1960. "Från bilprognos till trafikprognos" [From car forecast to traffic forecast]. In *Bilstaden*, 129–140. Stockholm: Royal Institute of Technology.

Hassenpflug, Dieter, ed. 2000. *Die Europäische Stadt: Mythos und Wirklichkeit*. Münster: Lit Verlag.

Haug, W. F. 1971. *Kritik der Warenästhetik*. Frankfurt: Suhrkamp.

Headrick, Daniel R. 1988. *The Tentacles of Progress: Technology Transfer in the Age of Imperialism, 1850–1940*. New York: Oxford University Press.

Heinelt, Hubert, and Daniel Kübler, eds. 2005. *Metropolitan Governance in the Twenty-first Century*. New York and London: Routledge.

Heland, Hans von. 1956. "Den kollektiva trafikens utveckling, dess ekonomiska betingelser och framtida perspektiv" [The development of public transport, its economic preconditions and future perspectives]. *Svensk Lokaltrafik* 13, no. 4: 220–227.

Hellige, Hans Dieter. 1986. "Entstehungsbedingungen und energietechnische Langzeitwirkungen des Energiewirtschaftsgesetzes von 1935." *Technikgeschichte* 53: 123–155.

Hellmann, Ulrich. 1990. *Künstliche Kälte: Die Geschichte der Kühlung im Haushalt*. Gießen: Anabas Verlag.

Hembry, P. 1990. *The English Spa, 1560–1815: A Social History*. London: Athlone Press.

Herzig, Thomas. 1992a. "Wirtschaftsgeschichtliche Aspekte der deutschen Elektrizitätsversorgung 1880 bis 1990." In *Die Geschichte der Stromversorgung*, ed. W. Fischer. Frankfurt: VDE.

———. 1992b. "Von der Werkstattzentrale zur Verbundwirtschaft." In *Technik und Wirtschaft*, ed. U. Wengenroth. Düsseldorf: VDI Verlag.

Heßler, Martina. 2001. *"Mrs. Modern Woman": Zur Sozial- und Kulturgeschichte der Haushaltstechnisierung*. Frankfurt and New York: Campus.

———. 2003. "Technopoles and Metropolises: Science, Technology and the City: A Literature Overview." In Hård and Misa 2003.

Hewett, Chris. 2001. *Power to the People: Delivering a Twenty-first-Century Energy System*. London: Institute for Public Policy Research.

Highway Capacity Manual: Practical Applications of Research. 1950. Washington, DC: Department of Traffic and Operations.

Hobsbawm, Eric. 1994. *Age of Extremes: The Short Twentieth Century 1914–1991*. London: Joseph.

Hobsbawm, Eric J., and Terence Ranger, eds. 1983. *The Invention of Tradition*. Cambridge: Cambridge University Press.

Hochreiter, Walter. 1994. "Vom Kommunalkraftwerk zum Großkraftwerk: Die Mannheimer Entwicklung (1910–1935)." In *Wege regionaler Elektrifizierung in der Rhein-Main-Neckar-Region*, ed. H. Böhme and Dieter Schott, 57–69. Darmstadt: THD Schriftenreihe Technik und Wissenschaft.

Hohenberg, Paul M., and Lynn H. Lees. 1995. *The Making of Urban Europe, 1000–1994*. Cambridge: Harvard University Press.

Hommels, Anique. 2005a. "Studying Obduracy in the City: Toward a Productive Fusion between Technology Studies and Urban Studies." *Science, Technology & Human Values* 30: 323–351.

———. 2005b. *Unbuilding Cities: Obduracy in Urban Sociotechnical Change*. Cambridge: MIT Press.

Hook, Karl. 1962. *Metropole im Rhein-Neckar Raum*. Mannheim: Statistisches Amt.

Horrocks, Ivan, Jens Hoff, and P. W. Tops, eds. 2000. *Democratic Governance and New Technology: Technologically Mediated Innovations in Political Practice in Western Europe*. London: Routledge.

Horsfall, Thomas Coglan. 1904. *The Improvement of the Dwellings and Surroundings of the People: The Example of Germany*. Manchester: Manchester University Press.

Horváth, Géza. 1972. A városépítés története [History of city planning]. In *Tatabánya története: Helytörténeti tanulmányok*, ed. Gábor Gombkötő et al., II: 192–210. Tatabánya: Városi Tanács Végrehajtó Bizottsága.

Horváth, Sándor. 2000. *Mentális térképek Sztálinvárosban* [Mental maps in Sztálinváros]. In *A mesterség iskolája: Tanulmányok Bácskai Vera 70. születésnapjára*, ed. Bódy Zsombor, Mátay Mónika, and Tóth Árpád, 450–474. Budapest: Osiris.

———. 2004. *A kapu és a határ: mindennapi Sztálinváros* [The gate and the border: The everyday Stalintown]. Budapest: MTA Történettudományi Intézete.

Howard, Ebenezer. 1898. *Tomorrow: A Peaceful Path to Real Reform*. London: Swan Sonnenschein.

Howe, Frederic Clemson. 1911. *How the German City Cares for Its People*. New York: City Club.

———. 1913. *European Cities at Work*. London: T. Fisher Unwin.

Hoy, Suellen. 1995. *Chasing Dirt: The American Pursuit of Cleanliness*. New York: Oxford University Press.

Hughes, Thomas P. 1983. *Networks of Power: Electrification in Western Society, 1880–1930*. Baltimore: Johns Hopkins University Press.

———. 1989. *American Genesis: A Century of Invention and Technological Enthusiasm, 1870–1970*. New York: Penguin.

Hughes, Thomas P., and Agatha C. Hughes, eds. 1990. *Lewis Mumford: Public Intellectual*. New York: Oxford University Press.

References

125 jaar verkeerspolitie. 1951. Amsterdam: Verkeerspolitie.

Hurd, Madeleine. 2000. "Class, Masculinity, Manners, and Mores: Public Space and Public Sphere in Nineteenth-Century Europe." *Social Science History* 24: 75–110.

Ibelings, Hans. 1995. *20th Century Architecture in the Netherlands*. Rotterdam: NAI Publishers.

———. 1997. *Americanism: Nederlandse Architectuur en het Transatlantische Voorbeeld*. Rotterdam: NAI.

Ihsanoğlu, Ekmeleddin. 2004. "Ottoman Science: The Last Episode in Islamic Scientific Tradition and the Beginning of European Scientific Tradition." In *Science, Technology and Learning in the Ottoman Empire: Western Influence, Local Institutions, and the Transfer of Knowledge*, ed. E. Ihsanoğlu, 11–48. Burlington, VT: Ashgate/Variorum.

Internationale Elektrotechnische Ausstellung. 1893. *Offizieller Bericht über die Internationale Elektrotechnische Ausstellung in Frankfurt am Main 1891*. Vol. 1, *Allgemeiner Bericht*. Frankfurt: Sauerländer.

Israel, Paul. 1992. *From Machine Shop to Industrial Laboratory*. Baltimore: Johns Hopkins University Press.

Jacobs, Allan B., Elizabeth Macdonald, and Yodan Rofé. 2002. *The Boulevard Book: History, Evolution, Design of Multiway Boulevards*. Cambridge: MIT Press.

Jacobs, Jane. 1961. *The Death and Life of Great American Cities*. New York: Random House.

Jager, Ida W. M. 2002. *Hoofdstad in gebreke: Manoeuvreren met publieke werken in Amsterdam 1851–1901*. Rotterdam: Uitgeverij.

Jajeśniak-Quast, Dagmara. 2001a. "Kommunalwirtschaftliche Kooperation geteilter Städte an Oder und Neiße." In *Grenzen im Ostblock und ihre Überwindung*, ed. Helga Schultz, 275–296. Berlin: Verlag Arno Spitz GmbH.

———. 2001b. "Soziale und politische Konflikte der Stahlarbeiter von Nowa Huta während der Sozialistischen Transformation." *Bohemia* 42, no. 2 (Veröffentlichungen des Collegium Carolinum, München).

———. 2002. "The 'European Coal and Steel Community' of the East: The Comecon and the Failure of Socialist Integration." In *National Borders and Economic Disintegration in Modern East Central Europe*, ed. Uwe Müller and Helga Schultz, 223–244. Berlin: Verlag Arno Spitz GmbH.

Jajeśniak-Quast, Dagmara, and Krystyna Lenczowska. 2000. "Zamknijmy przyszłość, myślmy o przyszłości, rozmowa z Michałem Sendzimirem Część I" [Let's close the past, let's think about the future: A conversation with Michael Sendzimir, part I]. In *Głos. Tygodnik Nowohucki*, no. 1 (456), 30 December 1999–1 January 2000; no. 2 (457), 7 January 2000.

Jajeśniak-Quast, Dagmara, and Katarzyna Stokłosa. 2000. *Geteilte Städte an Oder und Neiße: Frankfurt (Oder)-Słubice, Guben-Gubin und Görlitz-Zgorzelec 1945–1995*. Berlin: Berlin Verlag Arno Spitz GmbH.

James, K. 1999. "From Messel to Mendelsohn: German Department Store Architecture in Defence of Urban and Economic Change." In Crossick and Jumain 1999, 252–278.

Jamison, Andrew. 1998. "American Anxieties: Technology and the Reshaping of Republican Values." In Hård and Jamison 1998, 69–100.

———. 2001. *The Making of Green Knowledge: Environmental Politics and Cultural Transformation.* Cambridge: Cambridge University Press.

———. 2005. "Participation and Agency: Hybrid Identities in the European Quest for Sustainable Development." In *Managing Leviathan: Environmental Politics and the Administrative State*, ed. Robert Paehike and Douglas Torgerson. 2nd ed. Peterborough, Ontario: Broadview Press.

Jamison, Andrew, and Ron Eyerman. 1994. *Seeds of the Sixties.* Berkeley: University of California Press.

Janák, Pavel. 1911. "Hranol a pyramida" [The prism and the pyramid]. *Umělecký měsíčník* [Arts monthly] 1: 162–170.

Janik, Allan, and Stephen Toulmin. 1973. *Wittgenstein's Vienna.* New York: Simon and Schuster.

Jaumain, S. 1996. *Les petits commerçants belges face à la modernité (1880–1914).* Brussels: Editions de l'Université de Bruxelles.

Jazbinsek, Dietmar. 2003. "The Metropolis and the Mental Life of George Simmel." *Journal of Urban History* 30, no. 1: 102–125.

Jellicoe, Geoffrey Alan. 1961. *Motopia: A Study in the Evolution of Urban Landscape.* London: Studio.

Jenniskens, Antoon H. 1995. *Pak de bus: Openbaar vervoer in Maastricht 1884–1994.* Maastricht: Stichting Historische Reeks.

Jessen, Johann. 2001. "Suburbanisierung—Wohnen in verstädterter Landschaft." In *Villa und Eigenheim: Suburbaner Städtebau in Deutschland*, ed. T. Harlander, 316–329. Stuttgart and Munich: Deutsche Verlagsanstalt.

———. 2003. Editorial. *Die alte Stadt: Zeitschrift für Stadtgeschichte, Stadtsoziologie und Denkmalpflege* 30: 1–6.

Jiřík, Karel, et al. 1993. *Dějiny Ostravy* [History of Ostrava]. Ostrava: Nakladetelství Sfinga.

Jodidio, Philip. 2003. *Santiago Calatrava.* Cologne: Taschen.

Johansson, Alf W. 2004. "If You Seek His Monument, Look Around! Reflections on National Identity and Collective Memory in Sweden after the Second World War." In *The Swedish Success Story?* ed. Kurt Almqvist and Kay Glans. Stockholm: Axel and Margaret Johnson Foundation.

Johnson-McGrath, Julie. 1997. "Who Built the Built Environment? Artifacts, Politics, and Urban Technology." *Technology and Culture* 38: 690–696.

Josephson, Paul R. 1997. *New Atlantis Revisited: Akademgorodok, the Siberian City of Science.* Princeton: Princeton University Press.

Judt, Tony. 2005. *Postwar: A History of Postwar Europe since 1945*. New York: Penguin Press.

Jurdao Arrones, Francisco. 1990. *España en venta*. Madrid: Endymion.

Kaijser, Arne, and Marika Hedin, eds. 1995. *Nordic Energy Systems: Historical Perspectives and Current Issues*. Canton, MA: Watson.

Kaliński, Janusz, and Zbigniew Landau. 1998. *Gospodarka Polski w XX wieku* [The economy of Poland in the twentieth century]. Warsaw: Polskie Wydawnictwo Ekonomiczne.

Kalweit, H., ed. 1993. *Der Rhein unter der Einwirkung des Menschen—Ausbau, Schiffahrt, Wasserwirtschaft*. Lelystad: Internationale Kommission für die Hydrologie des Rheingebietes.

Kána, O., et al., eds. 1966. *Historie Výstavby a vzniku NHKG* [The history of the building and origin of the New Klement-Gottwald steelworks]. Prague: Práce.

Kargon, Robert H., Stuart W. Leslie, and Erica Schoenberger. 1992. "Far beyond Big Science: Science Regions and the Organization of Research and Development." In *Big Science: The Growth of Large-Scale Research*, ed. Peter Galison and Bruce Hevly, 334–354. Stanford: Stanford University Press.

Karnasiewicz, Jerzy Aleksander. 2003. *Nowa Huta: Okruchy życia i meandry historii* [Nowa Huta: The crumbs of life and the meandering of history]. Kraków: Wydawn. Tow. Słowaków w Polsce.

Karpat, Kemal H. 1974. "The Social and Economic Transformation of Istanbul in the Nineteenth Century." *Bulletin de l'Association Internationale d'Études du Sud-Est Européen* 12: 267–308.

———. 1985. *Ottoman Population 1830–1914: Demographic and Social Characteristics*. Madison: University of Wisconsin Press.

Kazgan, Haydar, and Sami Önal. 1999. *Istanbul'da Suyun Tarihi: Istanbul'un Su Sorununun Tarihsel Kökenleri ve Osmanlı'da Yabancı Su Şirketleri* [Water in Istanbul: The historical origins of Istanbul's water problem and the foreign water companies in Ottoman times]. Istanbul: Iletişim.

Kentgens-Craig, Margret. 1999. *The Bauhaus and America: First Contacts, 1919–1936*. Cambridge: MIT Press.

Keuning, H. J. 1948. "De economisch-geografische achtergrond van de Rijnvaart voor 1870." In Ligthart 1948.

Keyder, Çağlar, Eyüp Özveren, and Donald Quataert. 1993. "Port Cities in the Ottoman Empire: Some Theoretical and Historical Perspectives." *Review Fernand Braudel Center* 16, no. 4: 519–558.

Kieser, Clemens. 2003. "'Kein Strom oder Fluss hat mehrere Arme nöthig': Denkmale zum Gedenken an Johann Gottfried Tulla, den 'Bändiger des wilden Rheins.'" *Denkmalpflege in Baden-Württemberg: Das Nachrichtenblatt des Landesdenkmalamtes Baden-Württemberg* 32, no. 3: 231–234.

Kipping, Matthias, and Ove Bjarnar, eds. 1998. *The Americanisation of European Business: The Marshall Plan and the Transfer of U.S. Management Models*. London and New York: Routledge.

Kırsan, Ciler, and Gülen Çağdaş. 1998. "The Nineteenth-Century Row-houses in Istanbul: A Morphological Analysis." *Open House International* 23: 45–56.

Kirstein, Tatjana. 1984. *Die Bedeutung von Durchführungsentscheidungen in dem zentralistisch verfassten Entscheidungssystem der Sowjetunion: Eine Analyse des stalinistischen Entscheidungssystems am Beispiel des Aufbaus von Magnitogorsk (1929–1932)*. Wiesbaden: Harrassowitz.

Klette, Hermann. 1904. "Tiefbau." In Wuttke 1904, I: 370–419.

Knauer-Romani, Elisabeth. 2000. *Eisenhüttenstadt und die Idealstadt des 20. Jahrhunderts*. Weimar: VDG-Verlag.

König, Wolfgang. 1986. "Hochschullehrer und Elektrifizierungsberater Erasmus Kittler, das 'Darmstädter Modell,' und die frühe Elektrifizierung im Spiegel seiner Briefe aus den Jahren 1888/89." *Technikgeschichte* 54: 1–14.

Konvitz, Josef W., Mark H. Rose, and Joel A. Tarr. 1990. "Technology and the City." *Technology and Culture* 31: 284–294.

Körner, Klaus. 2002. *"Die rote Gefahr": Antikommunistische Propaganda in der Bundesrepublik 1950–2000*. Hamburg: Konkret-Literatur-Verlag.

Kóródi, József, and Géza Márton. 1968. *A magyar ipar területi kérdései* [The spatial questions of Hungarian industry]. Budapest: Kossuth Kiadó.

Kóródi, Jószef, and György Kőszegfalvi. 1971. *Városfejlesztés Magyarországon* [City development in Hungary]. Budapest: Kossuth.

Kotěra, Jan. 2001. *Jan Kotěra, 1871–1923: The Founder of Modern Czech Architecture*. Prague: Municipal House/Kant.

Kotkin, Stephen. 1992. *Steeltown, USSR*. Berkeley: University of California Press.

———. 1995. *Magnetic Mountain: Stalinism as a Civilization*. Berkeley: University of California Press.

Krabbe, Wolfgang R. 1979. "Munizipalsozialismus und Interventionsstaat: Die Ausbreitung der Städtischen Leistungsverwaltung im Kaiserreich." *Geschichte in Wissenschaft und Unterricht* 5: 265–283.

———. 1989. *Die deutsche Stadt im 19. und 20. Jahrhundert*. Göttingen: Vandenhoeck & Ruprecht.

———. 1990. "Städtische Wirtschaftsbetriebe im Zeichen des 'Munizipalsozialismus': Die Anfänge der Gas- und Elektrizitätswerke im 19. und frühen 20. Jahrhundert." In *Kommunale Leistungsverwaltung und Stadtentwicklung vom Vormärz bis zur Weimarer Republik*, ed. H. Blotevogel, 117–135. Cologne and Vienna: Böhlau.

Kreibich, Wolfgang. 1986. *Die Wissenschaftsgesellschaft: Von Galilei zur High-Tech-Revolution*. Frankfurt: Suhrkamp.

Krohn, Wolfgang. 1987. *Francis Bacon*. Munich: Beck.

Kronberg, Roger. 1924. "Le rôle economique du port de Strasbourg." Dissertation, Faculté du Droit, Université de Paris.

Kuban, Doğan. 1988. "Istanbul'un Batılılaşması" [The Westernization of Istanbul]. *Tarih ve Toplum* 59: 28–33.

———. 1996. *Istanbul, an Urban History: Byzantion, Constantinopolis, Istanbul.* Istanbul: Tarih Vakfı.

Kühl, Uwe. 1997. "Anfänge städtischer Elektrifizierung in Deutschland und Frankreich." In D. Schott 1997, 129–140.

———, ed. 2002. *Der Munizipalsozialismus in Europa / Le socialisme municipal en Europe.* Munich: Oldenbourg.

Kuiler, H. C. 1948. "De Rijn als hartader van Europa." In Ligthart 1948, 251–277.

Kullström, Gunnar. 1953. "Internationell kurs i trafikteknik i Haag" [International course in traffic engineering at The Hague]. *Svenska Vägföreningens Tidskrift* 40: 289–293.

———. 1955. "Kurs i traffic engineering vid Yale University" [Course in traffic engineering at Yale University]. *Kommunalteknisk Tidskrift* 21, no. 6: 161–166.

Kuroń, Jacek. 1989. *Wiara i wina: Do i od kumunizmu* [Faith and guilt: Once communist and back]. Warsaw: Niezależna Oficyna Wydawnicza.

Ladd, Brian. 1990. *Urban Planning and Civic Order in Germany, 1860–1914.* Cambridge: Harvard University Press.

———. 1997. *The Ghosts of Berlin.* Chicago: University of Chicago Press.

Lafferty, William, ed. 2001. *Sustainable Communities in Europe.* London: Earthscan.

Lagerkvist, Amanda. 2005. *Amerikafantasier: Kön, medier och visualitet i svenska reseskildringar från USA 1945–63* [Imaginary America: Gender, media and visuality in Swedish post-war travelogues]. Stockholm: Stockholm University.

Lancaster, B. 1995. *The Department Store: A Social History.* London: Leicester University Press.

Le Corbusier [Charles-Edouard Jeanneret]. 1933/1967. *The Radiant City: Elements of a Doctrine of Urbanism to Be Used as the Basis of Our Machine-Age Civilization.* London: Faber and Faber.

———. 1943/1973. *The Athens Charter.* New York: Grossman.

Leach, W. 1989. "Strategists of Display and the Production of Desire." In *Consuming Visions: Accumulation and Display of Goods in America, 1880–1920*, ed. S. Bronner, 99–132. New York: Norton.

Lebow, Katherine. 2001. "Public Works, Private Lives: Youth Brigades in Nowa Huta in the 1950s." *Contemporary European History* 10, no. 2: 199–219.

———. 2002. "Building 'People's Poland': Nowa Huta 1949–1956: Stalinism and the Transformation of Everyday Life in Poland's 'First Socialist City.'" Ph.D. diss., Columbia University.

———. 2005. "Socialist Leisure in Time and Space: Hooliganism and Bikiniarstwo in Nowa Huta, 1949–1956." In *Sozialgeschichtliche Kommunismusforschung: Tschechoslowakei, Poland, Ungarn und DDR 1948–1968*, ed. Christiane Brenner and Peter Heumos, 527–540. Munich: R. Oldenbourg Verlag.

Leerink, Johan A. 1938. *De verkeers-veiligheid op den weg: Een juridische, sociologische en verkeerstechnische studie*. Alphen aan den Rijn: Samsom.

Lees, Andrew. 1985. *Cities Perceived: Urban Society in European and American Thought, 1820–1940*. Manchester: Manchester University Press.

———. 1992. "Das Denken über die Großstadt um 1900: Deutsche Stellungnahmen zum urbanen Lebensraum im internationalen Vergleich." *Berichte zur Wissenschaftsgeschichte* 15: 139–150.

Lesnikowski, Wojciech, ed. 1996. *East European Modernism: Architecture in Czechoslovakia, Hungary, and Poland between the Wars, 1919–1939*. New York: Rizzoli.

Lewin, Moshe. 1985. *The Making of the Soviet System: Essays in the Social History of Interwar Russia*. New York: Pantheon Books.

Lewis, Bernard. 1968. *The Emergence of Modern Turkey*. Oxford: Oxford University Press.

Liernur, Charles Th. 1873. *Die pneumatische Canalisation in der Praxis: Deren Beschreibung, Entwickelungsgeschichte und gegenwärtige Ausdehnung, nebst Beleuchtung der Bedenken dagegen*. Frankfurt: Verlag der Ingenieur-Firma Liernur & De Bruyn-Kops.

Ligthart, T., ed. 1948. *Physisch- en economisch-geografische beschouwingen over de Rijn als Europese rivier*. Rotterdam: Van Kouteren's Uitgeversbedrijf.

Livet, Georges. 2003. *Histoire des routes et des transports en Europe: Des chemins de Saint-Jacques à l'âge d'or des diligences*. Strasbourg: Presses Universitaires de Strasbourg.

Lofland, Lyn H. 1973. *A World of Strangers: Order and Action in Urban Public Space*. New York: Basic Books.

Lombaerde, P. 1983. "La trasformazione di una città di mare: Ostenda 1865–1878." *Storia Urbana* 22–23.

Loos, Adolf. 1908/1998. *Ornament and Crime: Selected Essays*. Riverside, CA: Ariadne Press.

Looveren, W. van. 1948. "Een en ander over grenzen en waterscheidingen in en om het stroomgebied van de Rijn, die reeds overwonnen en nog te overwinnen zijn." In Ligthart 1948, 89–250.

Loreth, Zbigniew. 1999. "Szanse (Głos w dyskusji)." [The opportunities (Part of a discussion). In Towarzystwo Miłośników Historii i Zabytków Krakowa 1999, 171–177.

Löwer, Wolfgang. 1992. "Rechtshistorische Aspekte der deutschen Elektrizitätsversorgung von 1880 bis 1990." In Fischer 1992, 169–215.

Lozac'h, Valérie, ed. 1999. *Eisenhüttenstadt*. Leipzig: Leipziger Universität Verlag.

Luckin, Bill. 1986. *Pollution and Control: A Social History of the Thames in the Nineteenth Century*. Bristol: Adam Hilger.

———. 1990. *Questions of Power: Electricity and Environment in Interwar Britain*. Manchester: Manchester University Press.

Ludwig, Andreas. 2000. *Eisenhüttenstadt: Wandel einer industriellen Gründungsstadt in fünfzig Jahren*. Potsdam: Druckerei Andreas Arnold.

Lundberg, Arne. 1953. "Stadsplaneringen i USA" [Town planning in the USA]. *Plan* 2: 35–43.

Lundberg, Sven. 1956. "Att lära om trafik i USA" [Learning about traffic in the USA]. *Kommunalteknisk Tidskrift* 22, no. 5: 145–154.

Lundin, Per. 2004. "American Numbers Copied! Shaping the Swedish Postwar Car Society." *Comparative Technology Transfer and Society* 2, no. 3: 303–337.

———. 2006. "Controlling Space: Scientific Planning in Sweden between 1930 and 1970." In *Taking Place: The Spatial Contexts of Science, Technology, and Business Studies*, ed. Enrico Baraldi, Hjalmar Fors, and Anders Houltz, 107–150. Sagamore Beach, MA: Science History Publications.

Lyotard, Jean-François. 1994. *Das postmoderne Wissen*. Vienna: Passagen.

Magyary, Zoltán, and István Kiss. 1939. *A közigazgatás és az emberek* [Public administration and people]. Budapest: Egyetemi Könyvkiadó és Nyomda.

Marcuse, Herbert. 1964. *One-Dimensional Man*. Boston: Beacon Press.

Mardin, Şerif. 1974. "Super Westernization in Urban Life in the Ottoman Empire in the last Quarter of the Nineteenth Century." In *Turkey: Geographic and Social Perspectives*, ed. Peter Benedict, Erol Tümertekin, and Fatma Mansur, 403–446. Leiden: E. J. Brill.

Marek, Michaela. 2005. "Die Idealstadt im Realsozialismus: Versuch zu den Traditionen ihrer Notwendigkeit." In *Sozialgeschichtliche Kommunismusforschung: Tschechoslowakei, Poland, Ungarn und DDR 1948–1968*, ed. Christiane Brenner and Peter Heumos, 425–480. Munich: R. Oldenbourg Verlag.

Marroyo, F. Sánchez. 2003. *La España del siglo XX: Economia, demografia y sociedad*. Madrid: Istmo.

Matern, Nils von. 1954. "Kurs i trafikteknik i Bürgenstock" [Course in traffic engineering at Bürgenstock]. *Svenska vägföreningens tidskrift* 41: 385–390.

May, Ruth. 1999. *Planstadt Stalinstadt: Ein Grundriß der frühen DDR—aufgesucht in Eisenhüttenstadt*. Dortmund: Kolander & Poggel.

McCormick, John. 1989. *Reclaiming Paradise: The Global Environmental Movement*. Bloomington: Indiana University Press.

McKay, John P. 1976. *Streetcars and Trolleys: The Rise of Urban Mass Transport in Europe*. Princeton: Princeton University Press.

McShane, Clay. 1994. *Down the Asphalt Path: The Automobile and the American City*. New York: Columbia University Press.

Meggyesi, Tamás. 1985. *A városépítés útjai és tévútjai* [Ways and errors of city planning]. Budapest: Műszaki Könyvkiadó.

Meikle, Jeffrey L. 1995. "Domesticating Modernity: Ambivalence and Appropriation, 1920–40." In *Designing Modernity: The Arts of Reform and Persuasion 1885–1945*, ed. Wendy Kaplan, 143–167. New York: Thames and Hudson.

Meller, H. 1998. "Nizza e Blackpool: Due città balneari all'inizio del Novecento." *Contemporanea* 4: 651–680.

Melosi, Martin, ed. 1980. *Pollution and Reform in American Cities, 1870–1930*. Austin: University of Texas Press.

Melosi, Martin. 1981. *Garbage in the Cities: Refuse, Reform, and the Environment: 1880–1980*. College Station: Texas A & M University Press.

———. 1993. "The Place of the City in Environmental History." *Environmental History Review* 17: 1–23.

———. 2000. *The Sanitary City: Urban Infrastructure in America from Colonial Times to the Present*. Baltimore: Johns Hopkins University Press.

Merlin, Pierre. 2000. *New Towns and European Spatial Development*. Paris, ESF Report No. 413, 5 September 2000.

Metz, Friedrich. 1927. "Zur Kulturgeographie des nördlichen Schwarzwaldes." *Geographische Zeitschrift* 33, no. 4/5: 194.

Miller, D. 1987. *Material Culture and Mass Consumption*. Oxford: Blackwell.

Miller, M. B. 1981. *The Bon Marché: Bourgeois Culture and the Department Store, 1869–1920*. Princeton: Princeton University Press.

Millward, Robert. 2000. "The Political Economy of Public Utilities." In *Cambridge Urban History of Britain. Vol. III 1840–1950*, ed. M. Daunton, 315–349. Cambridge: Cambridge University Press.

Misa, Thomas J. 2004. *Leonardo to the Internet: Technology and Culture from the Renaissance to the Present*. Baltimore: Johns Hopkins University.

Misa, Thomas, Philip Brey, and Andrew Feenberg, eds. 2003. *Modernity and Technology*. Cambridge: MIT Press.

Misa, Thomas J., and Johan Schot. 2005. "Inventing Europe: Technology and the Hidden Integration of Europe." *History and Technology* 21: 1–19.

Mom, Gijs. 2005. "Roads without Rails: European Highway-Network Building and the Desire for Long-Range Motorized Mobility." *Technology and Culture* 46: 745–772.

Moravánszky, Ákos. 1998. *Competing Visions: Aesthetic Invention and Social Imagination in Central European Architecture, 1867–1918.* Cambridge: MIT Press.

Morris, J. 1993. *The Political Economy of Shopkeeping in Milan, 1886–1922.* Cambridge: Cambridge University Press.

Müller, Wolfgang, and Ruth Rohr-Zänker. 2001. "Amerikanisierung der 'Peripherie' in Deutschland?" In *Suburbanisierung in Deutschland,* ed. K. Brake, J. S. Dangschat, and G. Herfert, 27–39. Opladen: Westdeutscher Verlag.

Mullin, John Robert. 1976–1977. "American Perceptions of German City Planning at the Turn of the Century." *Urbanism Past and Present* 2: 5–15.

Mumford, Eric. 2000. *The CIAM Discourse on Urbanism, 1928–1960.* Cambridge: MIT Press.

Mumford, Lewis. 1938. *The Culture of Cities.* New York: Harcourt, Brace and Jovanovich.

———. 1961. *The City in History.* New York: Harcourt, Brace and World.

———. 1963. "The Case Against 'Modern Architecture.'" In Mumford, *The Highway and the City.* New York: Harcourt, Brace and World.

———. 1985. *A város a történelemben* [The city in history]. Létrejötte, változásai és jövőjének kilátásai. Budapest: Gondolat.

Muthesius, Stefan. 2000. *The Postwar University: Utopianist Campus and College.* New Haven: Yale University Press.

Myllyntaus, Timo. 1991. *Electrifying Finland: The Transfer of a New Technology to a Late Industrialising Economy.* London: Research Institute of the Finnish Economy.

Myrdal, Alva, and Gunnar Myrdal. 1941. *Kontakt med Amerika* [Contact with America]. Stockholm: Bonnier.

Nägelke, Hans-Dieter. 2003. "Einheitswunsch und Spezialisierungszwang: Stadt und Universität im 19. Jahrhundert." *Die alte Stadt: Zeitschrift für Stadtgeschichte, Stadtsoziologie und Denkmalpflege* 30: 7–19.

Nevzgodine, Ivan V. 2002. "Urban History of the Stalinist Company Towns in Ural and Siberia." Paper for the Sixth International Conference on Urban History: Power, Knowledge and Society in the City. Edinburgh, 5–7 September 2002.

Niethammer, Lutz, Alexander Plato, and Dorothee Wierling, eds. 1990. *Die volkseigenen Erfahrungen: Eine Archäologie des Lebens in der Industrieprovinz der DDR.* Berlin: Rowohlt.

Nieuwenhuis, John. 1955. *Mensen maken de stad, 1855–1955: Uit de geschiedenis van de dienst van gemeentewerken te Rotterdam.* Rotterdam: Gemeente-secretarie.

Niftrik, J. G. van. 1869. "Over het reinigen van steden." *Bouwkundige bijdragen* 19: 221–246, 379–408.

———. 1906. "Geschiedenis van de Publieke Werken van Amsterdam van 1864–1901." Unpublished manuscript, Municipal Archives, Amsterdam.

Noort, J. van den. 1999. "Een quaestie van geloof: waterproject en waterleiding." *Rotterdams Jaarboekje* 77: 215–247.

Nordqvist, Stig. 1955. "Bör vi anpassa bilismen till det gamla samhället eller bygga ett bilsamhälle?" [Should we adapt motoring to the old society or build a car society?]. *Industria* 51, no. 3: 31–35.

———. 1956a. "Trafikteknik" [Traffic engineering]. *Väg-och vattenbyggaren* 2: 113–115.

———. 1956b. "Stadstrafiken och dess konflikter" [Urban traffic and its conflicts]. *Teknisk Tidskrift* 86: 1129–1136.

———. 1961. "Tänk på år 2000" [Think of the year 2000]. Interview on Swedish Radio, April 1961, draft 6 March 1961, Stig Nordqvist Archive.

———. 1965. "Vad är en stad? Ett trafiksystem som måste omformas" [What is a city? A traffic system that has to be redesigned]. Talk on Swedish Radio, 8 March 1965, Stig Nordqvist Archive.

Norton, Peter D. 2007. "Street Rivals: Jaywalking and the Invention of the Motor Age Street." *Technology and Culture* 48, no. 2: 331–359.

Nowack. 1904. "Die öffentliche Gesundheitspflege." In Wuttke 1904, I: 446–460.

Nye, David. 1997. *Narratives and Spaces: Technology and the Construction of American Culture*. Exeter: University of Exeter Press.

Olsen, Donald J. 1986. *The City as a Work of Art: London, Paris, Vienna*. New Haven: Yale University Press.

Oom, Rolf. 1954. "Fortbildningskurs om trafikproblem" [Advanced course on traffic problems]. *Plan* 5: 159–160.

Oppenheim, Gary M. 2001. "The Sotsgorod Projekt: Western Utopias of the Eastern Bloc." Manuscript, Language and Cognitive Neuroscience Lab, University of Wisconsin-Madison.

O'Reilly, K. 2000. *The British on the Costa del Sol: Transnational Identities and Local Communities*. London: Routledge.

Orland, Barbara. 1990. *Haushalts(t)räume: Ein Jahrhundert Technisierung und Rationalisierung im Alltag*. Königstein: Langewiesche.

Ortaylı, Ilber, and Akıllıoğlu, Tekin. 1986. "Le Tanzimat et le modèle français: mimétisme ou adaptation?" In *L'Empire ottoman, la République de Turquie et la France*, ed. Hâmit Batu and Jean-Louis Bacqué-Grammont, 197–208. Istanbul/Paris: ISIS.

Ortaylı, Ilber. 1974. *Tanzimattan Sonra Mahalli Idareller, 1840–1878*. Ankara: T.O.D.A.I.E. Yayını.

Østby, Per. 2004. "Educating the Norwegian Nation: Traffic Engineering and Technological Diffusion." *Comparative Technology Transfer and Society* 2, no. 3: 247–272.

Osterhammel, Jürgen. 2002. "Gesellschaftsgeschichtliche Parameter chinesischer Modernität." *Geschichte und Gesellschaft* 28: 71–108.

Paletschek, Sylvia. 1997. *Die permanente Erfindung einer Tradition*. Stuttgart: Steiner.

Pamer, Nóra. 2001. *Magyar építészet a két világháború között* [Hungarian architecture between the world wars]. Budapest: Terc. Második, bővített kiadás.

Papayanis, Nicholas. 1996. *Horse-Drawn Cabs and Omnibuses in Paris: The Idea of Circulation and the Business of Public Transit*. Baton Rouge: Louisiana State University Press.

Paquier, Serge, and Jean-Pierre Williot, eds. 2005. *L'industrie du gaz en Europe aux XIXe et XXe siècles: L'innovation entre marchés et collectivités publiques*. Brussels, Bern, and Berlin: Peter Lang.

Pawlitta, Manfred. 1979. *Die "Sozialistische Industrialisierung" in Polen unter besonderer Berücksichtigung der Entwicklung in der eisenschaffenden Industrie*. Ph.D. diss., University of Oldenburg.

Pehnt, Wolfgang. 2005. *Deutsche Architektur seit 1900*. Munich: Deutsche Verlags-Anstalt.

Peichl, Gustav, and Vladimir Slapeta. 1987. *Czech Functionalism, 1918–1938*. London: Architectural Association.

Pemble, J. 1987. *The Mediterranean Passion: Victorians and Edwardians in the South*. Oxford: Clarendon.

Perényi, Imre. 1952. *A szocialista városépítés: Várostervezésünk néhány kérdése* [Socialist city planning]. Budapest: Népszava.

Plitzner, Klaus. ed. 1998. *Elektrizität in der Geistesgeschichte*. Bassum: Verlag für Geschichte der Naturwissenschaft und der Technik.

Plowden, Stephen. 1972. *Towns against Traffic*. London: Deutsch.

Plowden, William. 1971. *The Motor Car and Politics in Britain*. London: Bodley Head.

Pohl, Hans. 1992. *Vom Stadtwerk zum Elektrizitätsgroßunternehmen: Gründung, Aufbau und Ausbau der "Rheinisch-Westfälischen Elektrizitätswerk AG" (RWE) 1898–1918*. Stuttgart: Steiner.

Polivka, Heinz. 1996. "Basel—Cultural, Trade and Traffic Centre." In *The Rhine Valley: Urban, Harbour and Industrial Development and Environmental Problems*, ed. H. Heineberg et al. Leipzig: Institut für Länderkunde.

Porter, Dale H. 1998. *The Thames Embankment: Environment, Technology and Society in Victorian London*. Akron: University of Akron Press.

Prakfalvi, Endre, ed. 1992. *Építészet és tervezés Magyarországon 1945–1956* [Architecture and planning in Hungary 1945–1956]. Budapest: Országos Műemléki Hivatal Magyar Építészeti Múzeum.

———. 2004. "A hatvanas évek építészetéről" [On the architecture of the 60s]. In *"Hatvanas évek" Magyarországon*, ed. Rainer M. János, 474–505. Budapest: 1956-os Intézet.

Prodi, P., and A. Wandruszka, eds. 1996. *Il luogo di cura nel tramonto della monarchia d'Asburgo.* Bologna: Il Mulino.

Provoost, Michelle. 1996. *Asfalt: Automobiliteit in de Rotterdamse stedebouw* [Asphalt: Automobility in Rotterdam's urban development]. Rotterdam: 010.

Put, M. J. M. 2001. *Energiek Maastricht: Anderhalve eeuw nutsvoorzieningen, 1850–2000.* Maastricht: Nutsbedrijven Maastricht.

Rabinow, Paul. 1989. *French Modern: Norms and Forms in the Social Environment.* Cambridge: MIT Press.

Rappaport, Erika D. 2000. *Shopping for Pleasure: Women in the Making of London's West End.* Princeton: Princeton University Press.

Rapport van de commissie inzake de rioleering van waterverversching van Amsterdam. Amsterdam: Municipality of Amsterdam, 1906.

Rath, Gernot. 1969. "Die Hygiene der Stadt im 19. Jahrhundert." In *Städte-, Wohnungs- und Kleidungshygiene des 19. Jahrhunderts in Deutschland,* ed. Walter Artelt et al., 70–83. Stuttgart: Ferdinand Enke.

Reece, Susanna, and Gerrylynn Roberts. 1998. "'This Electric Age Is Woman's Opportunity': The Electrical Association for Women and Its All-Electric House in Bristol, 1935." *Local Historian* 28, no. 2: 94–107.

Reesema, A. Siewertsz van. 1902. *Het Liernurstelsel in Nederland.* The Hague: De Economist.

Reichow, Hans Bernhard. 1959. *Die autogerechte Stadt: Ein Weg aus dem Verkehrs-Chaos.* Ravensburg: Maier.

Reid, Donald. 1991. *Paris Sewers and Sewermen: Realities and Representations.* Cambridge: Harvard University Press.

Reimer, Michael J. 1995. "Urban Regulation and Planning Agencies in Mid-Nineteenth Alexandria and Istanbul." *The Turkish Studies Association Bulletin* 19: 1–27.

Replanning Britain: England bygger upp. 1946. An exhibition of British social planning arranged by the British Council and the Royal Institute of British Architects in consultation with the National Association of Swedish Architects. Liljevalch's Art Hall, Stockholm, 21 May–10 June 1946.

Reulecke, Jürgen. 1982. "Zur städtischen Finanzlage in den Anfangsjahren der Weimarer Republik." *Archiv für Kommunalwissenschaften* 21: 199–219.

———. 1985. *Geschichte der Urbanisierung in Deutschland.* Frankfurt: Suhrkamp.

Ribhegge, Wilhelm. 2003. "City and Nation in Germany from the Middle Ages to the Present: Origins of the Modern Civil Society in the Urban Tradition." *Journal of Urban History* 30, no. 1: 21–36.

Richter, Jenny, Heike Förster, and Ulrich Lakemann. 1997. *Stalinstadt—Eisenhüttenstadt: Von der Utopie zur Gegenwart: Wandel industrieller, regionaler und sozialer Strukturen in Eisenhüttenstadt.* Marburg: Schüren.

Rigby, T. H., and F. Fehér. 1982. *Political Legitimation in Communist States*. New York: St. Martin's Press.

Ritter, G. A., M. Szöllösi-Janze, and H. Trischler, eds. 1991. *Antworten auf die amerikanische Herausforderung: Forschung in der Bundesrepublik und der DDR in den "langen" siebziger Jahren*. Frankfurt and New York: Campus.

Roberts, Gerrylynn K., ed. 1999. *The American Cities and Technology Reader*. London and New York: Routledge.

Roberts, Gerrylynn K., and Philip Steadman. 1999. *American Cities and Technology: Wilderness to Wired City*. London and New York: Routledge.

Roche, D. 1989. *La culture des apparences*. Paris: Fayard.

Rodgers, Daniel T. 1998. *Atlantic Crossings: Social Politics in a Progressive Age*. Cambridge, MA: Belknap Press.

Rohkrämer, Thomas. 1999. *Eine andere Moderne? Zivilisationskritik, Natur und Technik in Deutschland 1880–1933*. Paderborn: Ferdinand Schönigh.

Rose, Mark H. 2004. "Technology and Politics: The Scholarship of Two Generations of Urban-Environmental Historians." *Journal of Urban History* 30: 769–785.

Rosén, Nils. 1964. "Ett brittiskt trafiksaneringsprogram" [A British traffic redevelopment program]. *Teknisk Tidskrift* 94: 1209–1214.

Rosenthal, Steven. 1980a. "Foreigners and Municipal Reform in Istanbul: 1855–1865." *International Journal for Middle Eastern Studies* 11: 227–245.

———. 1980b. *The Politics of Dependency: Urban Reform in Istanbul*. Westport, CT: Greenwood Press.

Roura, J. R. Cuadrado, and E. Torres Bernier. 1978. "El sector turístico y su entorno socioeconómico: Una aproximación al caso de la Costa del Sol." *Información Comercial Española* 533: 82–105.

Rubey, Norbert, and Peter Schoenwald. 1996. *Venedig in Wien: Theater- und Vergnügungsstadt der Jahrhundertwende*. Vienna: Ueberreuter.

Ruisz, Rezső. 1954. "A gazdaságföldrajz szerepe a városrendezésben" [The role of economic geography in city development]. *Földrajzi Értesítő* 3, no. 1: 26–49.

Runeby, Nils, ed. 1998. *Framstegets arvtagare: Europas idéhistoria 1900-talet* [Heirs of progress: Europe's history of ideas in the twentieth century]. Stockholm: Natur och kultur.

Sachs, Wolfgang. 1992. *For Love of the Automobile: Looking Back into the History of Our Desires*. Berkeley: University of California Press; originally published in German, 1984.

Salwiński, Jacek. 1999. "Decyzje o lokalizacji Nowej Huty pod Krakowem: Stan wiedzy" [The decision about the location of Nowa Huta near Kraków: The research state]. In Towarzystwo Miłośników Historii i Zabytków Krakowa 1999, 77–94.

Samhällsplanering: Katalog över forsknings- och utvecklingsarbete i Sverige 1960–65 [Social Planning: Catalogue of Research and Development Work in Sweden 1960–65]. 1969. Stockholm: Statens planverk.

Sandström, Johan. 2002. *Organizational Approaches to Greening: Technocentrism and Beyond*. Umeå, Sweden: Umeå University.

Šarapov, N. P. 1976. "Deutsche Arbeiter in der Sowjetunion: Über die Teilnahme deutscher Arbeiter und Spezialisten am sozialistischen Aufbau in der UdSSR 1930–1933." *Zeitschrift für Geschichtswissenschaft* 24, no. 10: 1110–1130.

Sassatelli, R. 2004. *Consumo, cultura e società*. Bologna: Il Mulino.

Sassen, Saskia. 1991. *The Global City: New York, London, Tokyo*. Princeton: Princeton University Press.

Saxenian, AnnaLee. 1994. *Regional Advantage: Culture and Competition in Silicon Valley and Route 128*. Cambridge: Harvard University Press.

Sayer, Derek. 1998. *The Coasts of Bohemia: A Czech History*. Princeton: Princeton University Press.

Schattenberg, Susanne. 2002. *Stalins Ingenieure: Lebenswelten zwischen Technik und Terror in den 1930er Jahren*. Munich: R. Oldenbourg Verlag.

Schivelbusch, Wolfgang. 1983. *Lichtblicke: Zur Geschichte der künstlichen Helligkeit im 19. Jahrhundert*. Munich and Vienna: Hanser.

Schlegel-Matties, Kirsten. 1995. *"Im Haus und am Herd": Der Wandel des Hausfrauenbildes und der Hausarbeit 1880–1930*. Stuttgart: Steiner.

Schlögel, Karl. 1995. "Magnitogorsk, die Pyramiden des 20. Jahrhunderts." In *Go East oder die zweite Entdeckung des Osten*, ed. Karl Schlögel, 169–184. Berlin: Siedler Verlag.

Schmal, H. 1990. "De tram in Nederland." *Historisch-geografisch Tijdschrift* 7: 84.

Schmale, Wolfgang. 2000. *Geschichte Europas*. Vienna: Böhlau.

Schmucki, Barbara. 2001. *Der Traum vom Verkehrsfluss: Städtische Verkehrsplanung seit 1945 im deutsch-deutschen Vergleich*. Frankfurt: Campus.

Schönbeck, Boris. 1994. *Stad i förvandling: Uppbyggnadsepoker och rivningar i svenska städer från industrialismens början till idag* [City in transformation: Development epochs and demolition in Swedish cities from the start of industrialism to the present day]. Stockholm: Statens råd för byggnadsforskning.

Schott, Dieter, ed. 1997a. *Energie und Stadt in Europa*. Stuttgart: Steiner.

———. 1997b. "Stadtentwicklung—Energieversorgung—Nahverkehr: Investitionen in die technische Vernetzung der Städte am Beispiel von Mannheim mit Ausblicken auf Darmstadt und Mainz." In *Investitionen der Städte im 19. und 20. Jahrhundert*, ed. K. H. Kaufhold, 149–179. Cologne: Böhlau Verlag.

References

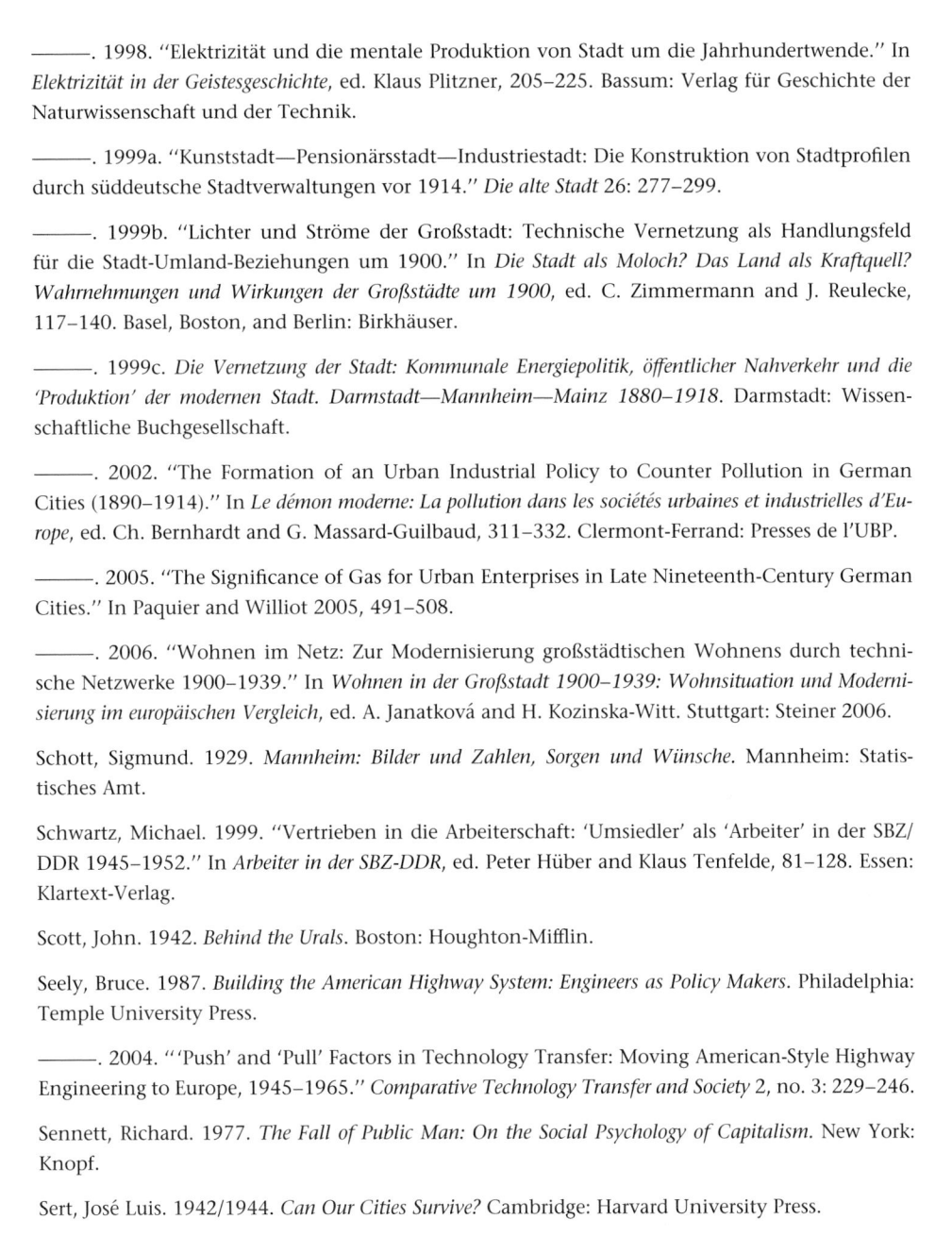

———. 1998. "Elektrizität und die mentale Produktion von Stadt um die Jahrhundertwende." In *Elektrizität in der Geistesgeschichte*, ed. Klaus Plitzner, 205–225. Bassum: Verlag für Geschichte der Naturwissenschaft und der Technik.

———. 1999a. "Kunststadt—Pensionärsstadt—Industriestadt: Die Konstruktion von Stadtprofilen durch süddeutsche Stadtverwaltungen vor 1914." *Die alte Stadt* 26: 277–299.

———. 1999b. "Lichter und Ströme der Großstadt: Technische Vernetzung als Handlungsfeld für die Stadt-Umland-Beziehungen um 1900." In *Die Stadt als Moloch? Das Land als Kraftquell? Wahrnehmungen und Wirkungen der Großstädte um 1900*, ed. C. Zimmermann and J. Reulecke, 117–140. Basel, Boston, and Berlin: Birkhäuser.

———. 1999c. *Die Vernetzung der Stadt: Kommunale Energiepolitik, öffentlicher Nahverkehr und die 'Produktion' der modernen Stadt. Darmstadt—Mannheim—Mainz 1880–1918*. Darmstadt: Wissenschaftliche Buchgesellschaft.

———. 2002. "The Formation of an Urban Industrial Policy to Counter Pollution in German Cities (1890–1914)." In *Le démon moderne: La pollution dans les sociétés urbaines et industrielles d'Europe*, ed. Ch. Bernhardt and G. Massard-Guilbaud, 311–332. Clermont-Ferrand: Presses de l'UBP.

———. 2005. "The Significance of Gas for Urban Enterprises in Late Nineteenth-Century German Cities." In Paquier and Williot 2005, 491–508.

———. 2006. "Wohnen im Netz: Zur Modernisierung großstädtischen Wohnens durch technische Netzwerke 1900–1939." In *Wohnen in der Großstadt 1900–1939: Wohnsituation und Modernisierung im europäischen Vergleich*, ed. A. Janatková and H. Kozinska-Witt. Stuttgart: Steiner 2006.

Schott, Sigmund. 1929. *Mannheim: Bilder und Zahlen, Sorgen und Wünsche*. Mannheim: Statistisches Amt.

Schwartz, Michael. 1999. "Vertrieben in die Arbeiterschaft: 'Umsiedler' als 'Arbeiter' in der SBZ/DDR 1945–1952." In *Arbeiter in der SBZ-DDR*, ed. Peter Hüber and Klaus Tenfelde, 81–128. Essen: Klartext-Verlag.

Scott, John. 1942. *Behind the Urals*. Boston: Houghton-Mifflin.

Seely, Bruce. 1987. *Building the American Highway System: Engineers as Policy Makers*. Philadelphia: Temple University Press.

———. 2004. "'Push' and 'Pull' Factors in Technology Transfer: Moving American-Style Highway Engineering to Europe, 1945–1965." *Comparative Technology Transfer and Society* 2, no. 3: 229–246.

Sennett, Richard. 1977. *The Fall of Public Man: On the Social Psychology of Capitalism*. New York: Knopf.

Sert, José Luis. 1942/1944. *Can Our Cities Survive?* Cambridge: Harvard University Press.

Seton-Watson, Hugh. 1964. *Nationalism and Communism: Essays 1946–1963*. New York: Praeger.

Sewing, Werner. 2002. "Suburbia Takes Command: Aspekte der Suburbanisierungsforschung in den USA." *Informationen zur modernen Stadtgeschichte* 2: 29–38.

Shaw, Stanford. 1979. "The Population of Istanbul in the Nineteenth Century." *International Journal of Middle Eastern Studies* 10: 265–277.

Shaw, Stanford, and Ezel Kural Shaw. 1977. *History of the Ottoman Empire and Modern Turkey.* 2 vols. Cambridge: Cambridge University Press.

Shore, Cris. 2000. *Building Europe: The Cultural Politics of European Integration.* London: Routledge.

Simmel, Georg. 1903. "The Metropolis and Mental Life." In Simmel, *Gesamtausgabe*, ed. Otthein Rammstedt. Frankfurt: Suhrkamp, 1989–2005.

Simon, Mariann. 2000. "...Megteremteni elméletünk és gyakorlatunk egységét..." [Creating the concord between theory and practice]. *Elmélet és gyakorlat 1952–1956.* Architectura Hungariae, vol. 2. ⟨arch.eptort.bme.hu/doc/simonm05.doc⟩ (accessed June 2007).

——— . 2001. "Minták és módszerek: a hetvenes évek hazai építészete és a karakter" [Models and methods: Hungarian architecture of the 70s and character]. *Építés-Építészettudomány* 29, no. 3–4: 347–360.

Simson, John von. 1983. *Kanalisation und Städtehygiene im 19. Jahrhundert.* Düsseldorf: VDI-Verlag.

Sinclair, Upton. 1906/1960. *The Jungle.* New York: New American Library.

Skarbowski, Jan. 1971. *Nowa Huta, pierwsze sozialistyczne miasto w Polsce* [Nowa Huta, the first socialist city in Poland]. Kraków: Wydawnictwo Literackie.

Smilor, Raymond W., George Kozmetsky, and David V. Gibson, eds. 1988. *Creating the Technopolis: Linking Technology Commercialization and Economic Development.* Cambridge, MA: Ballinger.

Smith, David M. 1996. "The Socialist City." In *Cities after Socialism: Urban and Regional Change and Conflict in Post-Socialist Societies*, ed. Gregory Andrusz, Michael Harloe, and Iván Szelényi, 70–99. Oxford: Blackwell.

Smith, Terry. 1993. *Making the Modern: Industry, Art, and Design in America.* Chicago: University of Chicago Press.

Sörlin, Sverker. 1994. *De lärdas republik: Om vetenskapens internationella tendenser* [The republic of the learned: On the international tendencies of science]. Malmö: Liber-Hermod.

Spechtenhauser, Klaus, and Daniel Weiss. 1999. "Karel Teige and the CIAM: The History of a Troubled Relationship." In Dluhosch and Švácha 1999, 216–255.

Spiewak, Martin. 2003. "Im Hirn der Republik." *Die Zeit: Wissen-Spezial* (17 July): 30.

Stalin, Iosif V. 1950. *A leninizmus kérdései* [Concerning questions of Leninism]. Budapest: Szikra.

Steen, Jürgen. 1991. *"Eine neue Zeit...!" Die Internationale Elektrotechnische Ausstellung 1891.* Frankfurt: Historisches Museum.

Stein, Erwin, et al., eds. 1913. *Monographien deutscher Städte: Darstellung deutscher Städte und ihrer Arbeit in Wirtschaft, Finanzwesen, Hygiene, Sozialpolitik und Technik. Band III. Darmstadt*. Oldenburg: Gerhard Stalling.

Sternberg, Rolf. 1998. *Technologiepolitik und High-Tech Regionen—ein internationaler Vergleich*. Münster: Lit Verlag.

Stier, Bernhard. 1999. *Staat und Strom: Die politische Steuerung des Elektrizitätssystems in Deutschland 1890–1950*. Weiher-Ubstadt: Verlag Regionalkultur.

Stine, Jeffrey K., and Joel A. Tarr. 1998. "At the Intersection of Histories: Technology and the Environment." *Technology and Culture* 39: 601–640.

Stippak, Marcus. 2007. *Wasserversorgung und Kanalisation in Darmstadt 1870–1914: Diskussion-Einführung-Entwicklung*. Darmstädter Schriften, vol. 90. Darmstadt: Justus-von-Liebig-Verlag.

"Stort stipendium för trafikstudier" [Large scholarship for traffic studies]. 1953. *Svenska Dagbladet* (25 April).

Stradling, David. 1999. *Smokestacks and Progressives: Environmentalists, Engineers, and Air Quality in America, 1881–1951*. Baltimore: Johns Hopkins University Press.

Strasser, Susan, Charles McGovern, and Matthias Judt, eds. 1998. *Getting and Spending: European and American Consumer Societies in the Twentieth Century*. Cambridge: Cambridge University Press.

Stremmel, Ralf. 1994. "Städtische Selbstdarstellung seit der Jahrhundertwende." *Archiv für Kommunalwissenschaften* 2: 234–264.

Stumm, Ingrid. 1999. "Kernfusionsforschung und politische Steuerung: Erste Großexperimente am Max-Planck-Institut für Plasmaphysik (IPP)." In Ritter et al. 1991, 217–237.

Sutton, Antony C. 1968. *Western Technology and Soviet Economic Development 1917 to 1930*. Stanford: Stanford University Press.

———. 1971. *Western Technology and Soviet Economic Development 1930 to 1945*. Stanford: Stanford University Press.

Švácha, Rostislav, ed. 1990. *Devětsil: The Czech Avant-Garde of the 1920s and 30s*. Oxford: Museum of Modern Art.

———. 1995. *The Architecture of New Prague 1895–1945*. Cambridge: MIT Press.

Szabó, Zoltán. 1938/1986. *Cifra nyomorúság* [Genteel poverty]. Budapest: Akadémiai Kiadó-Kossuth-Magvető.

Szczepański, Marek S. 1993. "A 'New Socialist City' in the Upper Silesian Industrial Region: A Sociological Case Study." In *Dilemmas of Regionalism and the Region of Dilemmas: The Case of Upper Silesia*, ed. Marek S. Szczepański, 142–161. Katowice: Universytet Slaski.

Szelényi, Iván. 1983. *Urban Inequalities under State Socialism*. New York: Oxford University Press.

Szíj, Rezső. 1967/1981. Építészeti problémák: Töredékes jegyzetek Dunaújvárosról [Problems in architecture: Notes on Dunaújváros]. In *Új építészet, új társadalom 1945–1978. Válogatás az elmúlt évtizedek építészeti vitáiból, dokumentumaiból*, ed. Major Máté and Judit Osskó, 277–292. Budapest: Corvina.

Szirmai, Viktória. 1988. *"Csinált" városok* ["Unnatural" cities]. Budapest: Magvető.

Szöllösi-Janze, Margit. 1999. "Einleitung." In Ritter et al. 1991, 43–49.

Tanpınar, Ahmet Hamdi. 1969. *Beş Şehir* [Five cities]. Istanbul: Dergâh Yayınları.

Tarr, Joel A. 1979. "The Separate vs. Combined Sewer Problem: A Case Study in Urban Technology Design Choice." *Journal of Urban History* 5: 308–339.

———. 1996. *The Search for the Ultimate Sink: Urban Pollution in Historical Perspective*. Akron, OH: Akron University Press.

———. 2001. "Urban History and Environmental History in the United States: Complementary and Overlapping Fields." In Bernhardt 2001, 25–39.

Tarr, Joel A., and Gabriel Dupuy, eds. 1988. *Technology and the Rise of the Networked City in Europe and America*. Philadelphia: Temple University Press.

Taverne, E. R. M. 1990. "The Lijnbaan (Rotterdam): A Prototype of a Postwar Urban Shopping Centre." In Diefendorf 1990, 146–154.

Taverne, Ed, Cor Wagenaar, and Martien de Vletter, eds. 2001. *J. J. P. Oud: Poëtisch functionalist, 1890–1963*. Rotterdam: NAi Uitgevers.

Teaford, Jon C. 1984. *The Unheralded Triumph: City Government in America, 1870–1900*. Baltimore: Johns Hopkins University Press.

Teige, Karel. 2000. *Modern Architecture in Czechoslovakia and Other Writings*. Los Angeles: Getty Research Institute.

Tekeli, Ilhan. 1994. *The Development of the Istanbul Metropolitan Area: Urban Administration and Planning*. Istanbul: Kent Basımevi.

Tekeli, Ilhan, and Selim Ilkin. 1996. "The Public Works Program and the Development of Technology in the Ottoman Empire in the Second Half of the Nineteenth Century." *Turcica: Revue d'études Turques* 28: 195–234.

Templ, Stephan. 1999. *Baba: The Werkbund Housing Estate Prague, 1932*. Basel, Boston, and Berlin: Birkhäuser Verlag.

Terlecki, Ryszard, Marek Lasota, and Jarosław Szarka, eds. 2002. *Nowa Huta—miasto walki i pracy* [Nowa Huta—the city of struggle and work]. Kraków: IPN.

Thompson, William. 2000. "'The Symbol of Paris': Writing the Eiffel Tower." *French Review* 73, no. 6 (May): 1130–1140.

Thue, Lars. 1995. "Electricity Rules: The Formation and Development of the Nordic Electricity Regimes." In Kaijser and Hedin 1995, 11–29.

Todd, Edmund. 1999. "Von Essen zur regionalen Stromversorgung, 1886 bis 1920: Das Rheinisch-Westfälische Elektrizitätswerk." In *Elektrizitätswirtschaft zwischen Umwelt, Technik und Politik: Aspekte aus 100 Jahren RWE-Geschichte 1898–1998*, ed. H. Maier, 17–49. Freiberg: Technische Universität Bergakademie Freiberg.

Tokarski, Peter. 1999. *Die Wahl wirtschaftspolitischer Strategien in Polen nach dem Zweiten Weltkrieg bis 1959*. Marburg: Herder-Institut.

Tolxdorff, Leo. 1961. *Der Aufstieg Mannheims im Bilde seiner Eingemeindungen (1895–1930)*. Stuttgart: Kohlhammer.

Towarzystwo Miłośników Historii i Zabytków Krakowa [Association of Friends of Kraków's History and Heritage]. 1999. *Narodziny Nowej Huty: Materiały sesji naukowej odbytej 25 kwietnia 1998 roku* [The birth of Nowa Huta: Materials of a conference of April 25, 1998]. Kraków: Drukarnia Uniwersytetu Jagiellońskiego.

Traband, André. 1966. *Villes du Rhin: Strasbourg et Mannheim-Ludwigshafen. Étude de géographie comparée*. Paris: Société d'édition Les Belles Lettres.

Traffic Engineering and Control in the USA. 1955. Organisation for European Economic Co-operation (OEEC), Project no. 149. Paris: OECD.

Triani, G. 1988. *Pelle di sole, pelle di luna: Nascita e storia della città balneare 1700–1946*. Padova: Marsilio.

Tynelius, Sven. 1956. "Garaget—ett ofrånkomligt komplement till morgondagens bostad" [The garage—an inescapable supplement to the home of tomorrow]. *Kommunalteknisk Tidskrift* 22: 1–4.

Ubicini, Jean Henri Abdolonyme. 1855. *La Turquie actuelle*. Paris: Hachette.

Vad har USA att lära oss om trafikplaneringen? [What has the USA to teach us about traffic planning?]. 1954. Advanced Course in Town Planning at Kungliga Tekniska Högskolan, 13–15 September.

Vleuten, Erik van der, and Arne Kaijser. 2005. "Networking Europe." *History and Technology* 21: 21–48.

Vogelbacher, Urs. 2004. "Die Bedeutung des Rheins für Basel und die Schweiz." *Revue d'Allemagne et des pays de langue allemande* 36: 113–122.

Vögele, Jörg. 1998. *Urban Mortality Change in England and Germany, 1870–1913*. Liverpool: Liverpool University Press.

Wagenaar, Michael F. 1998. *Stedebouw en burgerlijke vrijheid: De contrasterende carrières van zes Europese hoofdsteden*. Bussum: Thoth.

Wakeman, Rosemary. 2003. "Planning the New Atlantis: Science and the Planning of Technopolis, 1955–1985." *Osiris* 18: 255–270.

Walker, L. 1995. "Vistas of Pleasure: Women Consumers of Urban Space in the West End of London 1850–1900." In *Women in the Victorian Art World*, ed. Clarissa Campbell Orr, 70–85. Manchester: Manchester University Press.

Walsh, C. 1999. "The Newness of the Department Store: A View from the Eighteenth Century." In Crossick and Jumain 1999, 46–71.

Walton, J. K. 1996. "Leisure Towns in Wartime: The Impact of the First World War in Blackpool and San Sebastian." *Journal of Contemporary History* 31: 603–618.

Ward, Barbara, and René Dubos. 1972. *Only One Earth: The Care and Maintenance of a Small Planet*. London: Andrew Deutsch; New York: W. W. Norton.

Warncke, Carsten-Peter. 1998. *De Stijl 1917–31*. Cologne: Taschen.

Weber, R. 1928. "Die Rheinschiffahrt." In *Die Mannheimer Hafen-Anlagen*, ed. Badische Hafenverwaltung, 16–22. Koblenz: Rekord-Druckerei und Verlagsanstalt.

Wecławowicz, G. 1992. "The Socio-spatial Structure of the Socialist Cities in East-Central Europe: The Case of Poland, Czechoslovakia and Hungary." In *Urban and Rural Geography Papers from the Sixth Italian-Polish Geographical Seminar: Venice, September 15–23, 1990*, ed. F. Lando, 129–140. Venice: Cafoscarina.

Weiner, Tibor. 1951. *Sztálinváros, szocialista város: A városépítés módszere* [Sztálinváros, the socialist city: The method of city planning]. *Építés-Építészet* 11–12: 589–598.

Weiner, Tibor, Károly Valentiny, and Miklós Visontai. 1959. *Sztálinváros, Miskolc, Tatabánya*. Budapest: Műszaki Kiadó.

Weingart, Peter. 2001. *Die Stunde der Wahrheit? Zum Verhältnis der Wissenschaft zu Politik, Wirtschaft und Medien in der Wissensgesellschaft*. Weilerswist: Velbrück.

Wengenroth, Ulrich. 1989. "Motoren für den Kleinbetrieb: Soziale Utopien, technische Entwicklung und Absatzstrategien bei der Motorisierung des Kleingewerbes im Kaiserreich." In *Prekäre Selbständigkeit: Zur Standortbestimmung von Handwerk, Hausindustrie und Kleingewerbe im Industrialisierungsprozeß*, ed. Ulrich Wengenroth, 177–205. Stuttgart: Steiner.

———. 1993. "Die Technische Hochschule nach dem Zweiten Weltkrieg: Auf dem Weg zu High-Tech und Massenbetrieb." In *Die Technische Universität München: Annäherungen an ihre Geschichte*, ed. Ulrich Wengenroth, 261–298. Munich: Technische Universität München.

Widmalm, Sven. 2001. *Det öppna laboratoriet: Uppsalafysiken och dess nätverk 1853–1910* [The open laboratory: Uppsala physics and its networks 1853–1910]. Stockholm: Atlantis.

Wiedfeldt, Otto. 1904. "Städtische Betriebe." In Wuttke 1904, I: 181–197.

Williams, Raymond. 1977. *Marxism and Literature*. Oxford: Oxford University Press.

Williams, Rosalind H. 1982. *Dream Worlds: Mass Consumption in Nineteenth-Century France*. Berkeley: University of California Press.

Wilson, Elizabeth. 1991. *The Sphinx in the City: Urban Life, the Control of Disorder, and Women.* Berkeley: University of California Press.

Wit, Wim de, ed. 1983. *The Amsterdam School: Dutch Expressionist Architecture, 1915–1930.* Cambridge: MIT Press.

Wolf, Winfried. 1996. *Car Mania: A Critical History of Transport.* Trans. Gus Fagan. London and East Haven, CT: Pluto. Originally published in German, 1988.

Wollen, Peter, and Joe Kerr, eds. 2002. *Autopia: Cars and Culture.* London: Reaktion.

Womack, James P., Daniel T. Jones, and Daniel Roos. 1990. *The Machine that Changed the World.* New York: Rawson Associates.

Wuttke, Robert, ed. 1904. *Die deutschen Städte: Geschildert nach den Ergebnissen der ersten deutschen Städteausstellung zu Dresden 1903.* Vols. I–II. Leipzig: Brandstetter.

Yaffa, C. Draznin. 2001. *Victorian London's Middle-Class Housewife: What She Did All Day.* Westport, CT: Greenwood Press.

Yerasimos, Stéphane. 1992. "Occidentalisation de l'espace urbain: Istanbul 1839–1871. Les textes réglementaires comme sources d'histoire urbaine." In *Les villes de l'Empire Ottoman: activités et sociétés*, vol. 1, ed. Daniel Panzac, 97–120. Marseille: CNRS.

———. 1999. "Tanzimat'ın Kent Reformları Üzerine." In *Modernleşme Sürecinde Osmanlı Kentleri*, ed. Paul Dumont and Francois Georgeon, 1–19. Istanbul: Tarih Vakfı. First published in French in 1992.

Zängl, Wolfgang. 1989. *Deutschlands Strom: Die Politik der Elektrifizierung von 1866 bis heute.* Frankfurt: Campus.

Zarecor, Kimberly J. (Elman). 2002. "Becoming-Minor: Karel Teige's Architectural Narrative for the Nation." Manuscript, Columbia University, Graduate School of Architecture.

Zaremba, Marcin. 2001. *Komunizm, legitymacja, nacjonalizm: Nacjonalistyczna legitymacja władzy komunistycznej w Polsce* [Communisms, legitimations, nationalisms: Nationalistic legitimating of the communist power in Poland]. Warsaw: Wydawnictwo TRIO, Instytut Studiów Politycznych PAN.

Zblewski, Zdzisław. 2000. *Leksykon PRL-u* [Lexicon of the People's Republic of Poland]. Kraków: Wydawn Znak.

Zeitlin, Jonathan, and Gary Herrigel, eds. 2000. *Americanization and Its Limits: Reworking American Technology and Management in Post-War Europe and Japan.* New York and Oxford: Oxford University Press.

Żerebecki, Tadeusz. 1955. "Budownictwo wielkoblokowe na terenie Kombinatu" [The building of the large-scale blocs in the area of the combine]. *Biuletyn Ekonomiczny* [Economic bulletin], no. 2.

Zimmermann, Clemens, ed. 1997. *Europäische Wohnungspolitik in vergleichender Perspektive 1900–1939 / European Housing Policy in Comparative Perspective 1900–1939*. Stuttgart: IRB.

———. 2000. *Die Zeit der Metropolen: Urbanisierung und Großstadtentwicklung*. Frankfurt: Fischer.

Zon, Henk van. 1986. *Een zeer onfrisse geschiedenis: Studies over niet-industriële vervuiling in Nederland, 1850–1920*. The Hague: Ministerie van Volkshuisvesting, Ruimtelijke Ordening en Milieubeheer.

Zwick, Peter. 1983. *National Communism*. Boulder, CO: Westview Press.

Zysman, John, and Andrew Schwartz, eds. 1998. *Enlarging Europe: The Industrial Foundations of a New Political Reality*. University of California International and Area Studies Digital Collection, Research Series no. 99 ⟨repositories.cdlib.org/uciaspubs/research/99⟩ (accessed June 2007).

Contributors

Hans Buiter is a researcher at the Foundation for the History of Technology at the Eindhoven University of Technology in the Netherlands, where he also received his Ph.D. He was trained as a historian at the University of Utrecht. His main fields of interest are history of technology, urban history, and transportation history. His publications include "City Lights: Regulated Streets and the Evolution of Traffic Lights in The Netherlands, 1920–1940," written with P. E. Staal (*Journal of Transport History*, Fall 2006).

Paolo Capuzzo is a professor of contemporary history at the University of Bologna, Italy. He has been a research fellow at the universities of Vienna, Berlin, Brussels, and Leicester. His main fields of interest are urban history and the history of consumption, and his most recent publication is *Culture del consumo* (Bologna: Il Mulino, 2006).

Noyan Dinçkal is assistant professor of the history of technology at Darmstadt University of Technology, Germany. His research focuses on urban environmental history and the cultural history of technology. Dinçkal is the author of *Istanbul und das Wasser: Zur Geschichte der Wasserversorgung und Abwasserentsorgung von der Mitte des 19. Jahrhunderts bis 1966* (Munich: R. Oldenbourg, 2004). He is currently working on a book about the history of sports technology and science.

Cornelis Disco is assistant professor in the department of science, technology, and health policy studies at the University of Twente, the Netherlands. He holds a B.A. from Yale University and a Ph.D. from the University of Amsterdam, both in sociology. Broadly interested in the field of social and historical studies of science and technology, Disco has recently published on the history of urban technology and the history of Dutch and European hydraulic engineering. He is now writing a technological history of the Rhine as a European commons.

Pál Germuska is research fellow at The 1956 Institute in Budapest, Hungary, and holds a Ph.D. from Eötvös Loránd University, Budapest. His main fields of interest are urban history and the history of the military industry. His main publication is *Under the Spell of Industry: Development Policy and the Socialist Cities* (in Hungarian; Budapest: 1956 Institute, 2004).

Mikael Hård is professor of the history of technology at Darmstadt University of Technology, Germany, and holds a Ph.D. from Gothenburg University, Sweden. His main fields of interest are the cultural and intellectual history of science, technology, and medicine, as well as urban history. His publications include *Hubris and Hybrids: A Cultural History of Technology and Science* (New York: Routledge, 2005), written with Andrew Jamison.

Martina Heßler is professor of cultural history and the history of technology at Offenbach University of Art and Design, Germany. She holds a Ph.D. from Darmstadt University of Technology, and her main fields of interest are the cultural history of science and technology, as well as urban history and the history of consumption. Her publications include *Mrs. Modern Woman: Zur Sozial- und Kulturgeschichte der Haushaltstechnisierung* (Frankfurt: Campus, 2001).

Dagmara Jajeśniak-Quast is research fellow at the Centre for Research in Contemporary History, Potsdam, Germany. She holds a Ph.D. from the European University Viadrina in Frankfurt an der Oder, Germany. Her research interests are economic and social history, especially of Central and Eastern Europe, business history, and the history of technology. Her publications include *Willkommene Investoren oder nationaler Ausverkauf? Ausländische Direktinvestitionen in Ostmitteleuropa im 20. Jahrhundert* (Berlin: Berliner Wissenschaftsverlag, 2006), coedited with Jutta Günther.

Andrew Jamison holds a B.A. from Harvard University and a Ph.D. in the theory of science from Gothenburg University, Sweden. He is professor of technology and society at the department of development and planning, Aalborg University, Denmark. His books include *The Intellectual Appropriation of Technology: Discourses on Modernity, 1900–1939*, coedited with Mikael Hård (MIT Press, 1998) and *The Making of Green Knowledge: Environmental Politics and Cultural Transformation* (Cambridge University Press, 2001).

Per Lundin is post-doctoral fellow at the division of history of science and technology at the Royal Institute of Technology in Stockholm, Sweden, and holds a Ph.D. from this university. His published dissertation *Bilsamhället: Ideologi, expertis och regelskapande i efterkrigstidens Sverige* (Stockholm: Stockholmia förlag, 2008) was awarded the Johan Nordström and Sten Lindroth prize for outstanding scholarly work in history of ideas/history of science/history of technology.

Thomas J. Misa is ERA Land-Grant professor of the history of technology at the University of Minnesota, where he directs the Charles Babbage Institute. He teaches in the program for the history of science, technology, and medicine, and holds a faculty appointment in the department of electrical and computer engineering. Misa's recent books include *Modernity and Technology*, coedited with Philip Brey and Andrew Feenberg (MIT Press, 2003), and *Leonardo to the Internet: Technology and Culture from the Renaissance to the Present* (Baltimore: Johns Hopkins University Press, 2004).

Dieter Schott is professor of modern history at Darmstadt University of Technology, Germany, and holds a Ph.D. from the University of Konstanz, Germany. His main field of interest is the urban and environmental history of European cities. His publications include *Resources of the City: Contributions to an Environmental History of Modern Europe* (Burlington, VT: Ashgate 2005), coedited with Bill Luckin and Geneviève Massard-Guilbaud, and *Die europäische Stadt und ihre Umwelt* (Darmstadt: Wissenschaftliche Buchgesellschaft, 2008), coedited with Michael Toyka-Seid.

Marcus Stippak holds an M.A. and Ph.D. in history from Darmstadt University of Technology, Germany. His fields of research include the history of technology, urban history, environmental history, and contemporary church history. Currently he is working as a freelance historian at the Central Archive of the Protestant Church in Hesse and Nassau. His publications include *Wasserversorgung und Kanalisation in Darmstadt 1870–1914. Diskussion—Einführung—Entwicklung* (Darmstadt: Liebig, 2007).

Inside Technology
edited by Wiebe E. Bijker, W. Bernard Carlson, and Trevor Pinch

Janet Abbate, *Inventing the Internet*

Atsushi Akera, *Calculating a Natural World: Scientists, Engineers, and Computers during the Rise of US Cold War Research*

Charles Bazerman, *The Languages of Edison's Light*

Marc Berg, *Rationalizing Medical Work: Decision-Support Techniques and Medical Practices*

Wiebe E. Bijker, *Of Bicycles, Bakelites, and Bulbs: Toward a Theory of Sociotechnical Change*

Wiebe E. Bijker and John Law, editors, *Shaping Technology/Building Society: Studies in Sociotechnical Change*

Stuart S. Blume, *Insight and Industry: On the Dynamics of Technological Change in Medicine*

Pablo J. Boczkowski, *Digitizing the News: Innovation in Online Newspapers*

Geoffrey C. Bowker, *Memory Practices in the Sciences*

Geoffrey C. Bowker, *Science on the Run: Information Management and Industrial Geophysics at Schlumberger, 1920–1940*

Geoffrey C. Bowker and Susan Leigh Star, *Sorting Things Out: Classification and Its Consequences*

Louis L. Bucciarelli, *Designing Engineers*

H. M. Collins, *Artificial Experts: Social Knowledge and Intelligent Machines*

Paul N. Edwards, *The Closed World: Computers and the Politics of Discourse in Cold War America*

Herbert Gottweis, *Governing Molecules: The Discursive Politics of Genetic Engineering in Europe and the United States*

Joshua M. Greenberg, *From Betamax to Blockbuster: Video Stores and the Invention of Movies on Video*

Mikael Hård and Thomas J. Misa, editors, *Urban Machinery: Inside Modern European Cities*

Kristen Haring, *Ham Radio's Technical Culture*

Gabrielle Hecht, *The Radiance of France: Nuclear Power and National Identity after World War II*

Kathryn Henderson, *On Line and on Paper: Visual Representations, Visual Culture, and Computer Graphics in Design Engineering*

Anique Hommels, *Unbuilding Cities: Obduracy in Urban Sociotechnical Change*

David Kaiser, editor, *Pedagogy and the Practice of Science: Historical and Contemporary Perspectives*

Peter Keating and Alberto Cambrosio, *Biomedical Platforms: Reproducing the Normal and the Pathological in Late-Twentieth-Century Medicine*

Eda Kranakis, *Constructing a Bridge: An Exploration of Engineering Culture, Design, and Research in Nineteenth-Century France and America*

Christophe Lécuyer, *Making Silicon Valley: Innovation and the Growth of High Tech, 1930–1970*

Pamela E. Mack, *Viewing the Earth: The Social Construction of the Landsat Satellite System*

Donald MacKenzie, *An Engine, Not a Camera: How Financial Models Shape Markets*

Donald MacKenzie, *Inventing Accuracy: A Historical Sociology of Nuclear Missile Guidance*

Donald MacKenzie, *Knowing Machines: Essays on Technical Change*

Donald MacKenzie, *Mechanizing Proof: Computing, Risk, and Trust*

Maggie Mort, *Building the Trident Network: A Study of the Enrollment of People, Knowledge, and Machines*

Nelly Oudshoorn and Trevor Pinch, editors, *How Users Matter: The Co-Construction of Users and Technology*

Shobita Parthasarathy, *Building Genetic Medicine: Breast Cancer, Technology, and the Comparative Politics of Health Care*

Paul Rosen, *Framing Production: Technology, Culture, and Change in the British Bicycle Industry*

Susanne K. Schmidt and Raymund Werle, *Coordinating Technology: Studies in the International Standardization of Telecommunications*

Wesley Shrum, Joel Genuth, and Ivan Chompalov, *Structures of Scientific Collaboration*

Charis Thompson, *Making Parents: The Ontological Choreography of Reproductive Technology*

Dominique Vinck, editor, *Everyday Engineering: An Ethnography of Design and Innovation*

Index

Please note that some important phenomena recur so often in the book that it makes no sense to list them in the index, e.g., "architecture," "automobile," "city," "environment," "technology" (without a particular attribute), and "urban planning."

Aalborg Charter, 294
Aberdeen, 161
Addams, Jane, 126, 286
Agenda 21, 285, 291–293
Åhrén, Uno, 265
Ajka, 233, 244, 246, 248, 250, 252, 253 (fig. 11.4)
Akademgorodok, 212, 214, 220, 221, 224, 225
Alkmaar, 158
American influence, 159, 190
Americanization, 11, 17, 83, 99, 159, 216, 270, 271
American model, 18, 150, 159, 213, 216, 217, 227, 228, 268
Amersfoort, 158
Andersson, Sven, 262
Antwerp, 26, 72, 72 (fig. 4.1), 106
Appropriation (of knowledge, expertise, or technology), 12, 15, 17, 52, 59, 64, 66, 67, 72, 94, 101, 109, 124, 135, 162, 176, 184, 188, 197, 217, 252, 285
Arcade, 6, 15, 64, 102, 103, 103 (fig. 5.1), 108, 151, 158
Aristocracy, 111–113, 118
Arnhem, 145
Athens, 76, 77, 90

Athens Charter, 74, 76, 77, 92, 218, 235. *See also* CIAM
Austin, John L., 136
Avant-garde, 76, 77, 87, 89, 121, 237

Baba housing project, 90–92
Bacon, Francis, 211, 213, 220–224, 227, 228
Bacteriology, 171
Bangalore, 212
Barcelona, 6, 74, 77, 78
Barthes, Roland, 260
Basel, 18, 24–28, 31, 32, 35, 36, 38, 42–46, 65, 284
Bath, UK, 103, 112, 113, 115
Bath, public, 126, 139, 146
Bauhaus, 76, 91, 235, 240
Beirut, 64
Belgrade, 236
Beneš, Edvard, 194
Benjamin, Walter, 100–102
Berkeley, California, 227, 284
Berlage, H. P., 79, 81, 82, 85, 93
Berlin, 3, 6, 7, 10, 14, 16, 38, 76, 106–108, 111, 124–126, 146, 149, 157, 184, 203, 210 (n. 34), 213, 216, 235, 242, 259, 284, 290
Biarritz, 115
Bicycle, 8, 141, 155–158

Birmingham, 126, 167, 277
Bochum, 259
Boston, 6, 127, 136, 273
Boulevard, 1, 10, 58, 61, 83, 112–114, 136, 141, 143, 145, 151–153, 155, 162
Bourdieu, Pierre, 19
Brasília, 17, 74, 78
Bremen, 149
Brighton, 19, 113
Brno, 88, 90, 95 (n. 2)
Brundtland, Gro Harlem, 291
Brussels, 3, 103, 106, 151, 240
Buchanan, Colin, 266, 277
Budapest, 3, 6, 77, 87, 141, 149, 151, 235, 236, 240, 248, 259
Building code, 56
Bülow, Bernhard Fürst von, 130, 131
Bus, 10, 71, 99, 104, 105, 115, 141, 157, 258, 262, 273
Bush, Vannevar, 223, 224

Calatrava, Santiago, 18, 282, 283 (fig. 13.1), 296, 297
Cambridge, Massachusetts, 78, 212
Cambridge, UK, 212
Campanella, Tommaso, 211
Canal du Rhône au Rhin, 35, 45, 46
Cannes, 19, 113, 213
Car-friendly city, 17, 18, 266
Carson, Rachel, 288
Centralization, 51
Cesspool, 122, 144–147
Chadwick, Edwin, 125, 146, 147
Chester, 112
Chicago, 6, 14, 72, 78, 88, 108, 126, 127, 155, 172, 192, 237, 263, 273, 286, 287
Churchill, Winston, 204
CIAM (Congrès Internationaux d'Architecture Moderne), 73–79, 82, 83, 86, 88–90, 92–94, 95 (n. 2), 218, 235, 236
Circulation (of knowledge, expertise, or technology), 5, 7, 10, 11, 12, 15, 19, 64, 65, 67, 72, 90, 94, 123, 126, 127, 141, 165,
170, 187, 205–207, 228, 233, 234, 258, 268–269
Civilization, 32, 49, 52, 59, 222, 260, 287
Cleveland, 202, 237, 273
Co-construction of technology and society, 141–142
Coevolution of technologies, 142–144
Cogeneration, 179, 186
Cold War, 3, 7, 190–192, 202, 204–206
Cologne, 26–30, 134, 156, 177, 259
COMECON (Council for Mutual Economic Assistance), 202, 203, 205, 206
Commercial area, 104, 108
Commercialism, 16, 100–102, 105, 222, 297
Company town, 188
Concession, municipal, 167
Conservatism, 183, 260, 290
Construction materials, 73, 82–83, 102, 178
Consumption, 4, 5, 7, 12, 19, 62, 67, 99–107, 109–111, 115, 120, 165, 169–172, 177, 178, 183, 184, 220, 251
Copenhagen, 18, 156, 281, 284, 289, 294
Court life, 29–30
Coventry, 147, 259
Cracow. *See* Kraków
Crystal Palace, 104, 297
Cultural history, 15, 105, 287
Cultural studies, 15

Darmstadt, 29, 46, 121, 122, 124, 139, 140, 171
Deindustrialization, 201
Delft, 82
Department store, 3, 6, 15, 80, 84, 87, 88, 105–108, 110, 111, 155, 167, 189, 243, 251, 252
Depression, Great, 179, 202, 237
Dessau, 76, 77, 151
Detroit, 74, 77, 78, 159, 237, 257
Development, urban, 14, 23, 46, 51, 56, 62, 111, 116, 126–128, 140, 185, 187, 206, 214, 217, 262, 263, 281–285, 291–298
Dewey, John, 287
District heating, 179, 262
Donawitz, 205

Dordrecht, 28, 148, 149
Dortmund, 259
Dresden, 126, 131, 138, 141, 156, 166, 259
Duisburg, 43, 206, 259
Dunapentele. *See* Sztálinváros
Dunaújváros. *See* Sztálinváros
Durkheim, Émile, 3
Düsseldorf, 137, 138, 240

Ecological footprint, 120
Ecology
 human, 290, 292
 urban, 284, 289, 290, 292
Economic development, 23, 42, 111, 208 (n. 10), 228, 259, 281, 291, 292
Education, engineering, 66
Eiffel Tower, 16, 71, 104
Eisenhüttenstadt, 187–192, 194–198, 201, 203, 205–207
Electrification, 10, 14, 19, 40, 106, 126, 142, 150, 153, 161, 165, 168–171, 173, 174, 176, 178, 180 (fig. 8.3), 182–185
Elevator, 40, 75, 104, 108
Engels, Friedrich, 3, 125, 236
Engineering
 civil, 32, 131, 133, 146, 270–272
 social, 16
 traffic, 5, 10, 99, 143, 159, 257, 258, 263, 266–271, 279 (n. 3)
 urban, 16, 54, 141, 144, 147, 161
Entertainment, 111, 118, 121
Environmentalism, 282, 284, 285, 287–291
Environmental problems, 14, 133, 139, 168, 169, 179, 248, 286, 287, 288
Essen, 120, 131, 134, 259
European Commission, 294
European integration, 3, 4, 7, 94
Europeanism, 92
European Union, 1, 7
Exhibition, 8, 15, 19, 49, 64, 88, 103, 104, 127, 128, 130, 131, 137, 139, 165, 166, 169, 170, 205, 268, 281, 284, 290, 294, 295
Expansion, urban, 86–87

Expertise, 132, 135, 170, 258
 engineering, 65–66
Expressionism, 79, 93

Fair. *See* Exhibition
Febvre, Lucien, 25, 26
Feces, 144–150
Feuchtinger, M. E., 159
Film, image of the city in, 71
Fire protection, 51, 56
First World War, 39, 40, 45, 79, 87, 88, 93, 106, 123, 155, 167, 177, 178, 181, 182, 260, 267
Foucault, Michel, 8
Fountain, public, 49, 67
Frankfurt am Main, 4, 19, 28, 31, 71, 75, 75 (fig. 4.2), 77, 78, 84, 90, 94, 126, 138, 149, 165, 166, 168–170, 184, 185, 202, 235, 236
Frankfurt an der Oder, 192, 203
Functionalism, 16, 74, 77, 78, 82, 84, 85, 88, 89, 91, 92, 92 (fig. 4.5), 93, 94, 106, 195, 213, 218, 225, 235–239, 243, 265, 287
Functions
 integration of, 225–226
 separation of, 74, 77, 213, 219–220, 238, 260
Fürstenberg an der Oder, 191, 201
Futurism, 7, 89, 260

Garching, 18, 213–227
Garden city, 80, 81 (fig. 4.4), 88, 124, 235, 268
Garnier, Tony, 235
Gary, Indiana, 17, 137, 237
Gas network, 166–168, 170
Gasworks, 6, 14, 15, 43, 121, 131, 134, 166, 167, 171, 172, 175, 177, 178
Gaudí, Antoni, 296
Geddes, Patrick, 123, 133, 287
Gender, 13, 66–68, 101, 109, 198–199
Genoa, 77
Gibraltar, 115, 119
Giedion, Sigfried, 74, 75, 79, 89, 90
Glasgow, 65, 126
Globalization, 1, 3, 4, 24, 99, 100
Granada, 4

Granpré Molière, M. J., 80–82
Gropius, Walter, 73, 74, 78, 89

Haarlem, Netherlands, 145, 147
Habermas, Jürgen, 142
Hague, The, 80, 142, 150, 151 (fig. 7.2), 151–153, 162, 268, 294
Hamburg, 13, 145, 146, 148, 156, 235, 277
Hannover, 161, 169 (fig. 8.1), 277
Harbor, 9, 14, 23–28, 30–32, 33 (fig. 2.1), 35–45, 47, 53, 77, 111, 117, 134, 144, 147, 153, 171, 177, 180 (fig. 8.3), 191, 281, 282, 284, 294, 296
Hartford, Connecticut, 143
Haussmann, Georges Eugène, 10, 16, 52, 55, 56, 108
Heisenberg, Werner, 219, 221
Helsingborg, 295
Helsinki, 178
Heritage, 57, 94
History
  cultural, 15, 105, 124, 287
  intellectual, 13
  social, 12–13, 115, 145, 187
  urban, 6, 14–15, 124, 143, 190
History of consumption, 99, 100, 105
History of science, 13, 266
History writing, 5, 6
Hoffmann, Josef, 8
Homogenization, 1, 4, 10–12, 19, 20, 110, 112
Honecker, Erich, 203
Horsfall, Thomas C., 133–135, 140
Housing, public, 16, 17, 74, 75, 78, 80, 183, 235
Houston, Texas, 273
Howard, Ebenezer, 81, 235, 268
Howe, Frederic C., 136, 137, 138, 140
Humboldt, Wilhelm von, 216, 223
Hydropower, 45, 46, 176–179
Hygiene, 62, 64, 67, 86, 121, 133, 134, 139, 149, 171, 182

Identity, 6, 49, 67, 118, 128, 130, 219, 222, 224, 239, 261, 265, 291, 292, 296

Image, 7, 10, 12, 66, 71, 78, 93, 102, 106, 108, 109, 113–115, 118–124, 127, 128, 130, 136, 138, 140, 146, 213, 226, 258, 281, 294
Industrial city, 39, 77, 120, 121, 125, 134, 137, 187, 197, 206, 233–235, 245, 252, 254, 281, 285, 286
Industrialization, 1, 3, 26, 31, 39, 42, 112, 125, 131, 139, 185, 188, 201, 202, 207, 237, 240, 284–286
  socialist, 187, 189, 191, 193, 196–198, 202, 233
Infrastructure, 3, 12, 14, 19, 35, 40, 42, 46, 50, 54, 55, 62, 64–66, 101, 102, 104, 105, 109–112, 116, 117, 121, 141, 166, 189, 217, 254, 259
International agreements, 3, 24, 29, 32, 35, 38, 204, 291
Internationalism, 74
International Style, 17, 74
Interurban connections, 23–24
Iron Curtain, 187, 190, 202, 207
Istanbul, 7, 49–69, 99
Izmir, 65

Jacobs, Jane, 224
Jugendstil, 85, 106, 121

Kahn, Albert, 237
Karlsruhe, 32, 38, 39, 45
Kazincbarcika, 233, 240, 242, 244, 246, 247, 250, 251
Kehl, 34, 36, 39, 41
Khrushchev, Nikita, 205
Kiel, 277
Kittler, Erasmus, 170
Klerk, Michel de, 79, 94
Komló, 233, 240, 242, 244, 248, 250, 252
Kotěra, Jan, 85, 86, 95 (n. 3)
Kraków, 4, 111, 189, 193, 201, 203
Kunčice, 187, 189, 193, 194, 198, 201, 203, 204, 206, 207

Lang, Fritz, 3
Las Vegas, 104

Le Corbusier (Charles Edouard Jeanneret-Gris), 4, 17, 73–77, 87–92, 94, 235–237, 240, 260, 265, 297
Legitimation, 127, 139, 140, 187, 188, 206, 289
 historical, 130
Leiden, 148, 149
Leipzig, 156, 205, 294
Leisure, 41, 74, 77, 104, 105, 111, 112, 121, 173, 185, 206, 217, 220
Leninváros. *See* Tiszaújváros
Liberalism, 28, 30, 131, 134, 190, 204
Liernur, Charles, 147–150
Life, urban, 1, 6, 8, 10, 12, 13, 15, 60, 67, 68, 84, 99, 101, 103, 104, 114 (fig. 5.3), 128, 131, 161, 167, 198, 211, 218, 225, 228
Lighting
 electric, 104, 108, 155, 166, 172–174, 176
 gas, 60, 102, 108, 155, 166, 167, 171
Lindley, William, 146
Linear city, 235, 237
Linz, 205
Liverpool, 146
Load factor, 151, 170, 173
Lock-in, technological, 170
Loghem, Johannes B. van, 237
London, 3, 7, 14, 35, 53, 66, 68, 77, 100, 103, 104, 107–109, 112, 124, 127, 143, 145–149, 151, 157, 204, 211, 259, 267, 268, 284, 285, 289
Loos, Adolf, 85, 90
Los Alamos, New Mexico, 212, 214
Los Angeles, 74, 75, 77, 78, 257, 262, 263, 273, 288
Ludwigshafen, 29, 38
Lundberg, Arne S., 263, 268

Madrid, 6, 116
Magnitogorsk, 202, 203, 207, 233, 237, 238, 242
Maier-Leibnitz, Hans, 221
Mainz, 24, 26–30, 32, 34, 35, 37, 41, 46, 47
Málaga, 115, 116, 119
Malmö, 18, 281, 282, 284, 285, 292–298

Manchester, 131, 133, 134, 182, 183, 227
Mann, Thomas, 112
Mannheim, 7, 18, 24, 26, 27, 29–32, 33 (fig. 2.1), 34–42, 44–47, 170, 179, 180 (fig. 8.3), 186
Marseille, 7, 34, 76, 77
Marshall Plan, 11, 158, 192, 202, 260, 261, 269
Marx, Karl, 3, 236
May, Ernst, 73, 202, 235, 237, 238, 240
McKaye, Benton, 287
Medicine, 12, 113, 115
Megacity, 2, 49
Mendelsohn, Erich, 106, 107 (fig. 5.2)
Merin, Walt, 8
Metaphor, 67, 138, 221, 225, 228, 274
Metro. *See* Subway
Meyer, Hannes, 78, 240
Miami, 257
Miasma theory, 144, 146
Middle class, 16, 39, 53, 84, 87, 100, 105, 106, 108, 109, 111–113, 120, 121, 126, 128, 130, 131, 134, 142, 143, 153, 167, 171, 173, 175, 182, 185, 260, 261, 270, 288
Mies van der Rohe, Ludwig, 78, 88
Mietskaserne, 11 (fig. 1.4), 125
Migration, 198, 201, 207, 270
Milan, 102, 108, 112, 148
Miró, Joan, 296
Mobile, Alabama, 147
Mobility, 17, 100, 102, 110, 168, 188, 260
Modernism, 16, 17, 19, 71–78, 92–94, 240, 243, 282, 297
 Czech, 73, 75, 84–92
 Dutch, 73, 75, 79–84
Modernity, 1, 7, 15–17, 49, 50, 59, 62, 67, 69, 73, 87, 99, 104, 108, 122, 128, 135, 168, 258–262, 266, 271, 277–279
Modernization, 7, 50–52
Mondrian, Piet, 8
Monopoly, 34, 45, 53, 105, 167, 174–177
Monte Carlo, 113
Morris, William, 285
Moscow, 8, 17, 74, 76, 77, 78, 89, 192, 202, 205, 238–240

Moses, Robert, 277, 278
Motorization, 158, 262
Movement
 environmental, 284, 285, 288, 291
 labor, 261, 262
Mumford, Lewis, 3, 77, 78, 83, 158, 248, 284, 287, 288, 297
Munich, 18, 29, 146, 155, 172, 213, 214, 216–218, 221, 226, 228
Municipalization, 127, 170, 175
Muthesius, Hermann, 85
Myrdal, Alva, 270
Myrdal, Gunnar, 270

Nagy, Imre, 242
Naples, 65, 106
Napoleon, 30, 34, 131, 145
Napoleon III, 10
Narrative, 6, 47, 122, 124, 128, 130–132, 135, 138–140
Nationalism, 17, 74, 86–88, 93, 94, 130, 131, 187
Nationalization, 174, 181, 182, 185
Nazism, 42, 76, 181, 259, 267–268
Neighborhood unit, 238, 239, 243, 251, 252, 254
Network, technical, 41, 133, 143
Newcastle upon Tyne, 294
New Haven, Connecticut, 267
New Orleans, 273
New town, 233, 234, 236, 268
New York, 17, 53, 74, 78, 108, 124, 127, 155, 172, 205, 239, 273, 277, 286
Nice, 4, 19, 112–115, 213
Niemeyer, Oscar, 78
Nordqvist, Stig, 257, 258, 265, 274, 277
Novels, images of the city in, 106
Nowa Huta, 187, 188, 189, 190, 192, 194, 196–199, 200 (tab. 9.1), 201–207
Nuremberg, 107 (fig. 5.2), 129 (fig. 6.1), 259

OEEC (Organization for European Economic Cooperation), 269

Omnibus. *See* Bus
Order, spatial, 8–9, 141, 184, 234, 251
Organism, city as, 133, 138, 139, 287
Oroszlány, 233, 243–245, 247, 250
Osborn, Fairfield, 288
Oslo, 178
Ostend, 113
Ostrava, 189, 193, 194, 197, 201, 203
Oud, J. J. P., 81
Ownership
 private, 137, 167, 174, 176
 public, 9, 122, 134, 136, 138, 174, 175, 181, 185
Oxford, 212
Ózd, 233, 244, 246, 247, 248, 249 (fig. 11.3), 250, 252

Palma de Mallorca, 116
Paris, 1, 3, 7, 8, 10, 16, 17, 34, 35, 41, 51, 52, 55, 56, 59, 61, 65, 66, 71, 74, 77, 85, 86, 89, 100–102, 104–108, 112, 115, 124, 127, 131, 143, 145, 150, 151, 153, 171, 176, 211, 240, 277
Park, Robert, 287
Park
 amusement, 117
 business, 3
 industrial, 227
 public, 87, 112, 126, 134, 142, 143, 247, 253 (fig. 11.4), 258, 284, 286, 292
Parking, 3, 17, 78, 141, 155, 159, 161, 251, 257, 267, 274, 276 (fig. 12.3), 277, 278
Parma, 112
Pedestrian area, 4, 53, 161, 277
Perényi, Imre, 252
Perry, Clarence, 239
Pettenkofer, Max von, 146
Philadelphia, 78, 104, 172, 212
Pittsburgh, 201, 273
Planning
 central, 195
 ideals, socialist, 239–244
 regional, 78
 traffic, 265, 266, 267, 272, 292, 293

Plymouth, UK, 259
Policy, research, 18, 218, 222–224
Polis, 213, 225, 228
Politics, 1, 18, 24, 42, 45, 77, 78, 82, 87–89, 93, 94, 128, 138, 219, 221, 223, 240, 245, 266, 285, 287, 291
Population growth, 39, 50, 51, 128
Porto, 65
Postmodernism, 94, 117
Power station, 15, 40, 46, 121, 131, 151, 152, 170, 171, 173–179, 181, 185, 198, 246–248, 250, 290
Prague, 3, 84–91, 94, 95 (n. 2), 106, 111, 131, 148
Prefabrication, 91, 192, 196, 197, 197 (fig. 9.3), 198, 243, 244
Private capital, 54, 64–65, 102, 105
Professionalization, 125–127, 134–135, 140
Progressivism, 11, 16, 40, 52, 68, 121, 123, 124, 126, 127, 130, 131, 135–137, 139, 140, 183, 185, 188, 212, 240, 261, 284, 286, 287, 291, 292
Property, private, 60, 175, 176
Protest, 4, 149, 159, 183, 199, 204, 279, 284, 289
Public health, 7, 11, 12, 130, 132, 141, 144, 145, 147, 149, 171, 175, 182
Public-private partnership, 177
Public space, 55, 141–144, 153, 167, 252, 285, 288
Public work, 54, 55, 65, 66, 67, 82, 139, 141, 145, 146, 149, 151, 153, 155, 159, 296

Rákosi, Mátyás, 142, 243, 250
Rayon, 238, 239, 252
Recreation, 104, 121, 142, 143, 173, 235, 236, 239, 254, 258
Reform, social, 11, 52, 90, 93, 122, 168, 172
Regulation
 absence of, 117
 building, 106, 125
 street, 50, 55, 56, 57 (fig. 3.2), 58, 126, 142–144
 traffic, 153, 155, 156, 157

Regulation of nature, 36–38, 44, 109, 116
Reichow, Hans Bernhard, 265
Residential area, 3, 6, 8, 75, 121, 125, 189, 197, 218, 236, 241, 243, 245, 247, 248, 250, 251, 258
Resort, seaside, 19, 111, 112, 115, 116, 150
Resources, natural, 111, 112, 115, 118–120, 286
Rhine, 4, 18, 19, 23–32, 33 (fig. 2.1), 34–47, 137, 170, 176, 180 (fig. 8.3)
Rio de Janeiro, 291
Romanticism, 26, 284, 285
Rome, 3, 19, 26, 77, 85, 111, 289
Rostow, Walt Whitman, 272
Rotterdam, 7, 26, 27, 30, 35, 41, 42, 77, 80–84, 94, 142, 144–146, 148, 150, 153, 154 (fig. 7.3), 155–159, 161, 162, 259
Ruhr area, 26, 31, 32, 41, 42, 120, 125, 176
Ruisz, Rezső, 244, 245

St. Petersburg, 3, 131, 149
Salgótarján, 233, 244–248, 250, 252
Salonika, 65
Salsomaggiore, 112
San Francisco, 227, 273
Sanitation, 7, 12, 13 (fig. 1.5), 16, 51, 58, 62, 64, 65, 125, 128, 132, 139, 141, 142, 144, 146, 149, 150, 162, 286
Sanremo, 113
San Sebastián, 115, 116
Scholten, A. W., 145, 146
Schumacher, Fritz, 235
Schütte-Lihotzky, Grete, 184, 240
Science city, 121, 212–221, 224, 227
Second World War, 5, 11, 17, 109, 115, 118, 140, 142, 158, 171, 189, 190, 192, 194, 207, 212, 214, 222, 223, 227, 234, 236, 238, 239, 261, 267, 268, 272, 288
Segregation, 9, 53, 59, 62, 64, 143, 158, 297
Sendzimir, Tadeusz, 190, 205
Sert, José Luis, 74, 78, 79, 93
Seville, 116
Sewage, 16, 52, 57, 58, 68, 113, 123, 133, 141–150, 161, 162, 286

Sheffield, 259
Shipping, 24, 26–32, 35, 36, 41, 45, 79
Shopping center, 83, 109, 257, 258, 274
Shopping mall, 8, 83, 109, 110, 120, 158, 159, 160 (fig. 7.5), 217, 263
Sidewalk, 3, 8, 141, 146, 153, 155–158, 257
Silicon Valley, 18, 212, 213, 227, 228
Similarity, 10, 62, 71, 74, 93, 162, 187, 190, 217, 220
Simmel, Georg, 3
Sinclair, Upton, 122, 126, 286
Skyscraper, 1, 3, 18, 87, 88, 235, 263, 282, 298
Slab building, 196–198, 243
Slum, 77, 78, 189, 263
Smog, 288
Social conflict, 1, 136, 139
Social democracy, 261, 262, 273, 274, 294
Social differentiation, 15, 67
Socialism
 municipal, 9, 122, 134, 167
 national (*see* Nazism)
Socialist city, 187–191, 194, 198, 199, 203, 206–208, 233, 234, 236–245, 248, 250–252, 254
Socialization, 110, 122, 134
Social problems, 125, 168, 263, 282, 286, 295
Social structure, 51, 60, 64
Social tension, 119
Sophia-Antipolis, 213, 224, 228
Spa, Belgium, 112
Spencer, Herbert, 272
Sport arenas, 105, 116, 119, 225, 226
Sprawl, urban, 6, 288
Staffanstorp, 278 (fig. 12.4), 295
Stalin, Joseph, 76, 189, 236, 237, 241–244
Stalinism, 187, 203, 233, 234, 238–241, 243, 250
Stam, Mart, 78, 82, 90, 91
Standard
 international, 94, 150, 157
 technical, 10, 45, 64
Standardization, 1, 51, 64, 67, 74, 75, 82, 91, 94, 110, 119, 152, 171, 181, 197, 267

Standard of living, 19, 75, 122, 171, 220, 258
Steamship, 19, 26, 30, 35, 41, 44, 53, 91
Stijl, de, 78, 82, 90, 91
Stinnes, Hugo, 176, 177
Stockholm, 14, 17, 236, 265, 268, 271–274, 275 (fig. 12.2), 276 (fig. 12.3), 277, 289, 291
Strasbourg, 3, 18, 24–29, 32, 34–46, 37 (fig. 2.2)
Streetcar, 6, 8, 15, 52–54, 59, 66, 68, 84, 99, 108, 123, 135, 137, 141–143, 150–153, 157, 161, 168, 173, 175, 185, 272
Street cleaning, 122, 126, 134
Study trip, 11, 16, 72, 85, 88, 126, 136, 151, 188, 268–271, 273
Stuttgart, 9, 91, 107 (fig. 5.2), 240
Suburbanization, 102, 168, 173, 177, 217
 of science, 215–217
Subway, 3, 4, 8, 53, 66, 67, 86, 99, 104, 141, 273
Superblock, 237–239, 251
Sustainable development, 4, 186, 281, 284, 285, 291, 293, 295–297
Symbolic aspects, 15, 50, 52, 58, 59, 67, 68, 99–101, 105, 107, 108, 110, 111, 113, 123, 124, 127, 189, 219, 224, 259, 281, 296, 297
System
 energy, 168–169
 technological, 2 (fig. 1.1), 6, 8, 12, 14, 15, 50, 122, 127, 130, 137, 165, 168, 171, 175–177
Százhalombatta, 233, 245, 247, 250
Sztálinváros, 233, 239–242, 244, 245, 247, 248, 250, 251, 254, 255

Tampere, 294
Tatabánya, 233, 240, 242–246, 246 (fig.11.2), 247, 248, 250, 254
Tati, Jacques, 71
Taut, Bruno, 73, 235
Technocracy, 116, 264–265
Technology, alternative, 290
 communication, 100, 211
 history of, 14–15, 126

household, 171, 182–184
  meaning of, 52, 123–124, 127
  symbolic character of, 101, 123–124, 127
Technology park, 212, 213
Technology transfer, 202–204, 207–208. *See also* Circulation
Technopole, 212–214, 224, 225, 227, 228
Teige, Karel, 77, 89–91, 93, 95 (n. 1)
Telegraph, 14, 50, 68, 104
Terman, Frederick, 227
Thyssen, August, 176, 206
Tiszaújváros, 233, 245, 247, 250, 251, 254
Toronto, 273
Torremolinos, 116, 117, 117 (fig. 5.4)
Tourism, 4, 5, 19, 26, 30, 71, 111–113, 114 (fig. 5.3), 115–117, 119, 120, 141, 270
Trade, 27
Traffic accident, 59
Traffic engineering, 159
Traffic light, 141, 157, 158
Trafford Park, 227
Trans-Atlantic connections, 74–75, 92–93, 123, 135
Transport
  public, 1, 110, 122, 126, 137, 153, 159, 172, 185, 218, 265, 272, 273, 277
  riverine, 24–26, 34, 41
Trieste, 65
Truck, 71, 155–157, 198, 265
Tsukubu, 214
Tuberculosis, 171, 172, 263
Tulla, Johan Gottfried, 32, 36, 39, 40, 44

Uniformity, 10, 15, 17, 72, 120, 152
United States, European views of, 257, 258, 270, 271, 278, 279. *See also* American influence; Americanization; American model
Urban growth, 39, 40, 111, 116, 122
Urbanism, 77, 238, 254, 260
Urbanity, 60, 109, 124, 168
Urbanization, 1–3, 101, 106, 111, 112, 116, 117–119, 131, 188, 207

Utility, 1, 8, 9, 15, 39, 46, 122, 123, 126, 132, 133, 139, 140, 170, 172–179, 181–185, 198, 220, 227
Utopia, 6, 123, 165, 172, 211, 220, 221, 234, 259, 274, 287, 289, 290
Utrecht, 17, 77, 142, 152, 159, 160 (fig. 7.5), 161, 162

Vällingby, 274
Várpalota, 233, 240, 244, 246, 247, 250
Veblen, Thorstein, 287
Venice, 4, 65, 104, 294
Verona, 77
Vienna, 3, 24, 28–30, 66, 85–87, 103, 105, 108, 131, 141, 143, 167, 211, 290, 294

Wagner, Martin, 235
Wagner, Otto, 85, 86
Wall, city, 38, 61, 87, 94, 108, 145
Warsaw, 193, 195, 204, 259
Waste, 4, 10, 12, 14, 23, 71, 132, 133, 135, 139, 144–150, 169, 179, 182, 186, 193, 248, 259, 282, 288, 289, 292, 293, 295
Water closet (WC), 8, 12, 125, 145, 149, 195
Water supply, 10, 26, 50, 52, 53, 56, 62, 64, 66, 67, 68, 113, 121, 122, 133, 139, 141, 143, 144, 146, 198, 286
Weber, Max, 3, 102
Weiner, Tibor, 240, 241, 242 (fig. 11.1), 243
Westernization, 50–52, 56, 59–61, 67–69, 187–190, 234, 243, 254
Williams, Raymond, 286
Women, presence of, 58, 68, 109
Working class, 77, 83, 106, 118, 125, 133, 134, 143, 168, 171, 173, 184, 198, 240, 282, 294
World War I. *See* First World War
World War II. *See* Second World War
Worms, 32, 44
Wright, Frank Lloyd, 85, 88

Zlín, 84, 88
Zoning, 236, 239, 243, 247, 248
Zurich, 74, 89